Methodologies for Improving Antioxidant Properties and Absorption

Methodologies for Improving Antioxidant Properties and Absorption

Guest Editor

Daniela Tofani

MDPI

Basel • Beijing • Wuhan • Barcelona • Belgrade • Novi Sad • Cluj • Manchester

Guest Editor
Daniela Tofani
Department of Science
"Roma Tre" University
Rome
Italy

Editorial Office
MDPI AG
Grosspeteranlage 5
4052 Basel, Switzerland

This is a reprint of the Special Issue, published open access by the journal *Antioxidants* (ISSN 2076-3921), freely accessible at: https://www.mdpi.com/journal/antioxidants/special_issues/W67400DU5Z.

For citation purposes, cite each article independently as indicated on the article page online and as indicated below:

Lastname, A.A.; Lastname, B.B. Article Title. *Journal Name* **Year**, *Volume Number*, Page Range.

ISBN 978-3-7258-7030-1 (Hbk)
ISBN 978-3-7258-7031-8 (PDF)
https://doi.org/10.3390/books978-3-7258-7031-8

Contents

About the Editor

Daniela Tofani

Daniela Tofani is a researcher in organic chemistry at the Department of Science, Roma Tre University, in Rome, Italy.

She conducts research on the synthesis, modification, and functional evaluation of organic molecules with bioactive and antioxidant properties. Her scientific interests span the design and preparation of novel antioxidant derivatives, the investigation of structure–activity relationships, and the development of sustainable strategies for functional compound production, including work on hydroxytyrosol analogues and other phenolic systems with enhanced stability and biological activity. Her work integrates organic synthesis with analytical evaluation to understand how chemical modifications influence antioxidant performance, and she has co-authored studies exploring advanced antioxidant molecules and the applicability of natural and synthetic antioxidants in food, pharmaceutical, and cosmetic contexts.

She is currently involved in research on the water remediation and recovery of antioxidants from natural food industry by-products using filtration through new molecularly imprinted polymers. In addition to her research activities, Daniela Tofani is actively involved in academic teaching, offering courses in foundational and advanced topics in chemistry, including organic chemistry, medicinal natural products chemistry, and polymer/nanopolymer science for undergraduate and doctoral programs at Roma Tre University. She also participates in orientation programs under the European National Recovery and Resilience Plan, supporting secondary school students in their transition to higher education.

Preface

Natural antioxidants have demonstrated significant potential in mitigating oxidative stress and reducing the risk of degenerative diseases, including diabetes, neurodegenerative disorders, and cancer. However, their practical application remains limited due to intrinsic physicochemical constraints such as poor water solubility, low absorption, short biological half-life, and rapid metabolic degradation.

The Special Issue "Methodologies for Improving Antioxidant Properties and Absorption" addresses recent strategies developed to overcome these limitations. Various approaches are highlighted: (i) chemical modification of natural antioxidants, (ii) the use of delivery systems capable of protecting antioxidant molecules and enhancing their absorption until they reach target tissues, and (iii) sustainable strategies for obtaining new antioxidants from waste sources.

Chemical derivatization of antioxidants has led to the creation of new classes of compounds, frequently obtained by combining biologically compatible molecules to ensure safety. Common strategies involve introducing specific biological moieties into the antioxidant scaffold. These modified molecules exhibit improved stability, extended half-life, and promising antioxidant activity.

The second major strategy for improving the efficacy of natural antioxidants focuses on the development of advanced delivery systems to enhance stability, bioavailability, and functional performance. Several delivery platforms have been investigated, including cyclodextrins, liposomes, solid lipid nanoparticles, emulsions, and nanoencapsulation technologies. Their selection depends on the type of antioxidant, the target cells, and the intended application (nutraceutical, cosmetic, or pharmaceutical).

Much recent research also explores sustainable approaches for recovering antioxidants from natural and waste biomass sources. Advanced analytical and biotechnological approaches—such as metabolomic and metagenomic integration—are employed to identify new antioxidant and enzyme-inhibitory molecules and to elucidate their biosynthetic pathways.

Overall, these reprints are intended for researchers and developers in the food, pharmaceutical, and cosmetic fields, showcasing multidisciplinary strategies that combine chemical modification, green extraction, biotechnological production, and advanced carrier systems to optimize antioxidant performance.

Daniela Tofani
Guest Editor

antioxidants

MDPI

Editorial

Methodologies for Improving Antioxidant Properties and Absorption: 1st Edition

Daniela Tofani

Science Department, University of Roma Tre, 00146 Rome, Italy; daniela.tofani@roma3.it; Tel.: +39-0657333371

In recent decades, biological and biochemical research has greatly advanced our understanding of the significant effects of many natural antioxidants on various pathologies and diseases. Antioxidants have shown considerable potential not only in radical scavenging activity but also in reducing—and, in some cases, preventing—the development of degenerative diseases such as diabetes, Alzheimer's disease, Parkinson's disease, and several types of cancer.

Nevertheless, the practical application of natural antioxidants is still in its early stages, mainly due to intrinsic limitations such as low water solubility and absorption, short half-life, and rapid degradation in the gastrointestinal tract [1].

The Special Issue "Methodologies for Improving Antioxidant Properties and Absorption" aims to provide an overview of the most recent research in this field.

Various approaches to overcoming the drawbacks of natural antioxidants have been analyzed, including:

(i) The transformation of natural antioxidants into more stable and/or soluble derivatives;
(ii) The use of different types of delivery systems that can facilitate absorption and protect the molecule until it reaches its target cell or organ.

The first approach has led to the development of a new family of derivatized natural antioxidants, often obtained by combining two biological molecules to ensure the safety of the resulting derivatives. A common strategy involves glycosylation with simple sugars [2] or esterification with fatty acid [3], aimed at enhancing either water solubility or lipophilicity. Accordingly, many fatty acid esters of compounds such as quercetin [4] and hydroxytyrosol [5] have been synthesized and tested for their antioxidant activity and half-life, showing promising results.

In this context, Kim et al. [contribution 1] synthesized the ester of caffeic acid and hydroxytyrosol, which resulted in enhanced ROS scavenging capacity and a significant inhibition of melanin production in B16F10 melanoma cells with lower cytotoxic effect than that of caffeic acid.

More drastic transformations of natural antioxidants into more reactive derivatives can be carried out to enhance specific properties. Ionin et al. [contribute 2] oxidized the Siberian spruce galactoglucomannan–biopolymers, derived from softwood, obtaining an enhancement of heavy metal sorption capacity.

However, in the case of substantial modification of a natural antioxidant intended for pharmaceutical use, such as curcumin for anticancer therapy [6,7], the resulting molecules must be subject to the full range of toxicological evaluations required for the approval of a new drug.

On the other hand, the intense research conducted on delivery systems over the past few decades has found broad application in the field of antioxidants [8,9].

For instance, Mic et al. [contribution 7] used cyclodextrins to incapsulate a catechol hydrazinyl-thiazole derivative in order to reduce degradation and prolong the antioxidant effect over time.

Liposomes [10], solid lipid nanoparticles [11], emulsions, and nanoencapsulation techniques have been employed to enhance the skin absorption of antioxidant compounds in anti-aging cosmetics [contribution 8] as well as to improve the pharmacokinetic profiles of antioxidant drugs administered as adjuvants in various therapies for common age-related disorders, stroke treatment [12], and more severe diseases such as Alzheimer's disease [13] and certain types of cancer [14].

Many studies aim to prolong the shelf life of antioxidants used as food preservatives by employing delivery systems [8,15]. Chen et al. used a Sm-cluster to enhance the performance of cysteine derivatives [contribution 5], while Merlino et el. investigated the antibacterial activity of an emulsion based on caper leaf essential oil [contribution 6].

A different approach to this problem involves the search for new natural derivatives that are more suitable for their intended applications. In this context, research focuses on agricultural and food industry wastes as potential sources of high-value antioxidants [16]. For instance, Poblete et al. analyzed the antioxidant and pro-inflammatory enzyme inhibitory activities of pisco grape pomace waste to identify bioactive molecules with potential nutraceutical applications [Contribution 3]. Similarly, Qamar et al. investigated the enzymatic pretreatment of industrial lignocellulosic waste derived from palm kernel meal to increase the concentration of antioxidant molecules and peptides, making it a valuable ingredient for use as a food supplement in both animal and human diets [Contribution 4].

Whatever the approach to the problem, it is increasingly evident that the full exploitation of antioxidant potential has yet to reach its peak in many areas of our lives—from food, nutraceuticals, and cosmetics to more specific pharmaceutical applications aimed at combating serious and degenerative diseases. Research in these fields is advancing rapidly, yet significant challenges remain before the full therapeutic and preventive potential of antioxidants can be realized. Nevertheless, each contribution adds to a more comprehensive understanding of antioxidant potential in modern science and technology.

Conflicts of Interest: The author declares no conflict of interest.

List of Contributions:

1. Kim, K.; Jeon, J.; Kim, Y.; Lim, K.-M. The Antioxidant and Skin-Brightening Effects of a Novel Caffeic Acid Derivative, Caffeic Acid-3,4-Dihydroxyphenylpropanolester. *Antioxidants* **2025**, *14*, 806. https://doi.org/10.3390/antiox14070806.
2. Ionin, V.A.; Malyar, Y.N.; Borovkova, V.S.; Zimonin, D.V.; Kazachenko, A.S. TEMPO-Oxidized Spruce Galactoglucomannan—Biopolymer with Enhanced Antioxidant Activity and Selective Heavy-Metal Sorption. *Antioxidants* **2025**, *14*, 569. https://doi.org/10.3390/antiox14050569
3. Poblete, J.; Fernández-Martínez, J.; Aranda, M.; Quispe-Fuentes, I. Green Recovery and Identification of Antioxidant and Enzyme Inhibitor Molecules from Pisco Grape Pomace by Targeted Effects Analysis Using Thin-Layer Chromatography, Bioassay, and Mass Spectrometry. *Antioxidants* **2024**, *13*, 1418. https://doi.org/10.3390/antiox13111418
4. Qamar, H.; He, R.; Li, Y.; Song, M.; Deng, D.; Cui, Y.; Yu, M.; Ma, X. Metabolome and Metagenome Integration Unveiled Synthesis Pathways of Novel Antioxidant Peptides in Fermented Lignocellulosic Biomass of Palm Kernel Meal. *Antioxidants* **2024**, *13*, 1253. https://doi.org/10.3390/antiox13101253.
5. Chen, L.; Wang, L.; Ma, L.; Wang, C.; Qin, X.; Wang, M.; Zhang, X.; Yang, R.; Fang, B.; An, J. Synergistic Antioxidant Effects of Cysteine Derivative and Sm-Cluster for Food Applications. *Antioxidants* **2024**, *13*, 910. https://doi.org/10.3390/antiox13080910.
6. Merlino, M.; Condurso, C.; Cincotta, F.; Nalbone, L.; Ziino, G.; Verzera, A. Essential Oil Emulsion from Caper (*Capparis spinosa* L.) Leaves: Exploration of Its Antibacterial and Antioxidant

Properties for Possible Application as a Natural Food Preservative. *Antioxidants* **2024**, *13*, 718. https://doi.org/10.3390/antiox13060718

7. Mic, M.; Pîrňau, A.; Floare, C.G.; Palage, M.D.; Oniga, O.; Marc, G. Inclusion of a Catechol-Derived Hydrazinyl-Thiazole (CHT) in Cyclodextrin Nanocavity and Its Effect on Antioxidant Activity: A Calorimetric, Spectroscopic and Molecular Docking Approach. *Antioxidants* **2023**, *12*, 1367. https://doi.org/10.3390/antiox12071367
8. Ricci, A.; Stefanuto, L.; Gasperi, T.; Bruni, F.; Tofani, D. Lipid Nanovesicles for Antioxidant Delivery in Skin: Liposomes, Ufasomes, Ethosomes, and Niosomes. *Antioxidants* **2024**, *13*, 1516. https://doi.org/10.3390/antiox13121516.

References

1. Gambini, J.; Inglés, M.; Olaso, G.; Lopez-Grueso, R.; Bonet-Costa, V.; Gimeno-Mallench, L.; Mas-Bargues, C.; Abdelaziz, K.M.; Gomez-Cabrera, M.C.; Vina, J.; et al. Properties of Resveratrol: In Vitro and In Vivo Studies about Metabolism, Bioavailability, and Biological Effects in Animal Models and Humans. *Oxidative Med. Cell. Longev.* **2015**, *2015*, 837042. [CrossRef] [PubMed]
2. Zhang, Z.; Bi, X.; Hu, L.; Wu, Y.; Hu, W.; Wang, H.; Zheng, X.; Wu, H.; Mu, D.; Wen, L.; et al. Enhancing Antioxidant Activity of Rice Protein Hydrolysates through Glycosylation Modification. *Front. Nutr.* **2025**, *12*, 1616272. [CrossRef] [PubMed]
3. Hu, J.-N.; Zou, X.-G.; He, Y.; Chen, F.; Deng, Z.-Y. Esterification of Quercetin Increases Its Transport Across Human Caco-2 Cells. *J. Food Sci.* **2016**, *81*, H1825–H1832. [CrossRef] [PubMed]
4. Karimova, E.R.; Baltina, L.A.; Spirikhin, L.V.; Sapozhnikova, T.A.; Gabdrakhmanova, S.F.; Makara, N.S. Methylation of Quercetin by Diazomethane and Hypoglycemic Activity of Its Tetra-O-Methyl Ether. *Chem. Nat. Compd.* **2020**, *56*, 837–841. [CrossRef]
5. Plastina, P. Chapter 45—Hydroxytyrosol and Hydroxytyrosyl Fatty Esters: Occurrence and Anti-Inflammatory Properties. In *Olives and Olive Oil in Health and Disease Prevention*, 2nd ed.; Preedy, V.R., Watson, R.R., Eds.; Academic Press: San Diego, CA, USA, 2021; pp. 547–555. [CrossRef]
6. Kuzminska, J.; Szyk, P.; Mlynarczyk, D.T.; Bakun, P.; Muszalska-Kolos, I.; Dettlaff, K.; Sobczak, A.; Goslinski, T.; Jelinska, A. Curcumin Derivatives in Medicinal Chemistry: Potential Applications in Cancer Treatment. *Molecules* **2024**, *29*, 5321. [CrossRef] [PubMed]
7. Thongsom, S.; Racha, S.; Petsri, K.; Ei, Z.Z.; Visuttijai, K.; Moriue, S.; Yokoya, M.; Chanvorachote, P. Structural Modification of Resveratrol Analogue Exhibits Anticancer Activity against Lung Cancer Stem Cells via Suppression of Akt Signaling Pathway. *BMC Complement. Med. Ther.* **2023**, *23*, 183. [CrossRef] [PubMed]
8. Maqsoudlou, A.; Assadpour, E.; Mohebodini, H.; Jafari, S.M. The Influence of Nanodelivery Systems on the Antioxidant Activity of Natural Bioactive Compounds. *Crit. Rev. Food Sci. Nutr.* **2022**, *62*, 3208–3231. [CrossRef] [PubMed]
9. Molteni, C.; La Motta, C.; Valoppi, F. Improving the Bioaccessibility and Bioavailability of Carotenoids by Means of Nanostructured Delivery Systems: A Comprehensive Review. *Antioxidants* **2022**, *11*, 1931. [CrossRef] [PubMed]
10. Cheng, X.; Yan, H.; Pang, S.; Ya, M.; Qiu, F.; Qin, P.; Zeng, C.; Lu, Y. Liposomes as Multifunctional Nano-Carriers for Medicinal Natural Products. *Front. Chem.* **2022**, *10*, 963004. [CrossRef] [PubMed]
11. Subroto, E.; Andoyo, R.; Indiarto, R. Solid Lipid Nanoparticles: Review of the Current Research on Encapsulation and Delivery Systems for Active and Antioxidant Compounds. *Antioxidants* **2023**, *12*, 633. [CrossRef] [PubMed]
12. Zhang, S.; Peng, B.; Chen, Z.; Yu, J.; Deng, G.; Bao, Y.; Ma, C.; Du, F.; Sheu, W.C.; Kimberly, W.T.; et al. Brain-Targeting, Acid-Responsive Antioxidant Nanoparticles for Stroke Treatment and Drug Delivery. *Bioact. Mater.* **2022**, *16*, 57–65. [CrossRef] [PubMed]
13. Ashok, A.; Andrabi, S.S.; Mansoor, S.; Kuang, Y.; Kwon, B.K.; Labhasetwar, V. Antioxidant Therapy in Oxidative Stress-Induced Neurodegenerative Diseases: Role of Nanoparticle-Based Drug Delivery Systems in Clinical Translation. *Antioxidants* **2022**, *11*, 408. [CrossRef] [PubMed]
14. Sezgin-Bayindir, Z.; Losada-Barreiro, S.; Fernández-Bravo, S.; Bravo-Díaz, C. Innovative Delivery and Release Systems for Antioxidants and Other Active Substances in the Treatment of Cancer. *Pharmaceuticals* **2023**, *16*, 1038. [CrossRef] [PubMed]
15. Ozkan, G.; Franco, P.; De Marco, I.; Xiao, J.; Capanoglu, E. A Review of Microencapsulation Methods for Food Antioxidants: Principles, Advantages, Drawbacks and Applications. *Food Chem.* **2019**, *272*, 494–506. [CrossRef] [PubMed]
16. García-González, A.G.; Rivas-García, P.; Escamilla-Alvarado, C.; Ramírez-Cabrera, M.A.; Paniagua-Vega, D.; Galván-Arzola, U.; Cano-Gómez, J.J.; Escárcega-González, C.E. Fruit and Vegetable Waste as a Raw Material for Obtaining Functional Antioxidants and Their Applications: A Review of a Sustainable Strategy. *Biofuels Bioprod. Biorefin.* **2025**, *19*, 231–249. [CrossRef]

antioxidants

MDPI

Article

The Antioxidant and Skin-Brightening Effects of a Novel Caffeic Acid Derivative, Caffeic Acid-3,4-Dihydroxyphenylpropanolester

Kyu-lim Kim [1,2], Ju-hee Jeon [3], Yeonjoon Kim [4] and Kyung-Min Lim [1,2,*]

1 College of Pharmacy, Graduate School of Pharmaceutical Sciences, Ewha Womans University, Seoul 03760, Republic of Korea; kstarlim@ewha.ac.kr
2 Graduate Program in Innovative Biomaterials Convergence, Ewha Womans University, Seoul 03760, Republic of Korea
3 ACTIVON Co., Ltd., Cheongju 28104, Republic of Korea; jhjeon2110@activon.kr
4 Qmine Co., Ltd., Suwon 16506, Republic of Korea; yeonjoonkim@qmine.co.kr
* Correspondence: kmlim@ewha.ac.kr; Tel.: +82-2-3277-3055; Fax: +82-2-3277-3760

Abstract

Caffeic acid (CA) is a naturally occurring polyphenol antioxidant found in coffee, tea, fruits, and vegetables, known for its strong antioxidant, anti-inflammatory, and anti-aging properties. However, its cosmetic application is limited because of poor dermal absorption due to its high polarity. This study aimed to evaluate the antioxidant and skin-brightening effects of a novel lipophilic CA derivative, CAD (caffeic acid-3,4-dihydroxyphenylpropanolester). CAD was synthesized by conjugating CA with 3,4-DHPEA, a lipophilic antioxidant derived from olive oil. In both DPPH and ABTS assays, CAD exhibited more potent antioxidant activity than CA. In B16F10 melanoma cells, CAD significantly inhibited melanin production without cytotoxicity at concentrations lower than those required for CA. Cellular assays using DCF-DA staining demonstrated that CAD effectively reduced intracellular ROS levels. Mechanistic studies revealed that CAD inhibited tyrosinase activity and down-regulated the expression of TYR, TRP-1, and TRP-2. Additionally, CAD suppressed MITF phosphorylation, along with reduced phosphorylation of ERK and JNK, elucidating its anti-melanogenic mechanism. Importantly, CAD showed dose-dependent skin-brightening effects in the 3D human skin model Melanoderm™, as evidenced by increased lightness and histological evaluation. In conclusion, CAD demonstrates strong potential as a safe and effective antioxidant and skin-brightening agent for cosmetic applications.

Keywords: caffeic acid; 3,4-dihydroxyphenylpropanolester (3,4-DHPEA); antioxidant activity; skin brightening

1. Introduction

ROS, reactive oxygen species, are a diverse family of oxidants derived from molecular oxygen. Excessive ROS production can cause adverse effects and cytotoxicity through non-specific reactions [1]. Antioxidants can eliminate ROS, and their effects are contingent upon the cell penetration, stability, kinetics, and mechanism of ROS scavenging [2]. Antioxidants are available in both synthetic and natural forms. Natural antioxidants used in cosmetics are derived from a variety of sources, including spices, plants, grains, and fruits [3]. Natural antioxidants can be classified into categories such as phenolic compounds, vitamins, and carotenoids. Phenolic compounds range from simple molecules like caffeic acid to complex polyphenols such as tannins and flavonoids [4]. Vitamins, including vitamin E

https://doi.org/10.3390/antiox14070806

and vitamin C, are widely recognized for their antioxidant properties, while carotenoids like beta-carotene and lycopene also contribute to antioxidant activity [5].

Antioxidants are widely used in industries such as food, cosmetics, pharmaceuticals, and nutritional supplements. Especially, antioxidants such as vitamins C and E are widely used to mitigate the detrimental effects of ROS on the skin, and a variety of antioxidants are being used in cosmetics [6]. These antioxidants are also used for skin-brightening cosmetics, that target the activation of melanocytes. Melanocytes reside in the basal layer of the epidermis [7], and melanin synthesis can be initiated via several stimuli, including UV radiation or α-melanocyte-stimulating hormone (α-MSH). Melanogenesis occurs within the melanosome, a cellular organelle responsible for the synthesis, storage, and transport of melanin. Melanogenic enzymes such as tyrosinase, TRP-1, and TRP-2 located in the membrane of the melanosome are responsible for the synthesis of melanin [8–10]. Several natural antioxidant compounds are being utilized as skin-brightening agents. Notable examples include the polyphenolic compounds resveratrol and ferulic acid [11], niacinamide [12], and the carotenoid compound astaxanthin [13]. However, these natural antioxidants generally exhibit weak brightening activity [14] that is attributable to poor cell penetration.

Caffeic acid, a phenolic acid within the polyphenol antioxidant family, exhibits strong antioxidant properties. When used in cosmetics, it helps stimulate collagen synthesis and prevent premature skin aging [15]. Another phenolic compound, 3,4-dihydroxyphenylethanol (3,4-DHPEA), a compound found in virgin olive oil, exhibits strong antioxidant potency and enhances the stability of refined olive oil [16]. However, caffeic acid and 3,4-DHPEA have low lipophilicity (XLogP3, 1.2 and −0.7, respectively), limiting cellular uptake and skin absorption. Indeed, caffeic acid tends to remain in the stratum corneum, with only a minimal fraction reaching the dermis (less than 5%) [17,18]. 3,4-DHPEA may also have limited efficacy due to low skin absorption [19]. Therefore, it is necessary to improve cellular uptake and skin absorption of these water-soluble antioxidants to expect better efficacy.

In this study, we synthesized a conjugated form of caffeic acid and 3,4-DHPEA, caffeic acid-3,4-dihydroxyphenylpropanolester, CAD (first figure in Section 3) with increased lipophilicity (XLogP3, 3.34) in an effort to improve antioxidant activity and skin absorption. We examined the antioxidant and anti-melanogenic activity of CAD and compared with CA. Finally, we explored its applicability to skin-brightening cosmetics using a 3D human-pigmented epidermis skin model, Melanoderm™.

2. Materials and Methods

2.1. Materials

Caffeic acid+3,4-DHPEA was synthesized in Qmine Co. (Suwon, Republic of Korea). A stock solution of the chemical was made in dimethyl sulfoxide (Sigma-Aldrich, St Louis, MO, USA) at a concentration of 1 mg/mL.

2.1.1. Synthesis of (E)-3-(3,4-Diacetoxyphenyl)Acrylic Acid

(E)-3-(3,4-Dihydroxyphenyl)acrylic acid(4.5 g, 24.9 mmol) was dissolved in pyridine (6.0 mL, 3.0 eq.) and cooled to 0 °C, treated with acetic anhydride (23.6 mL, 10.0 eq.), which was stirred at room temperature overnight. After the reaction is complete, the reaction mixture was concentrated via rotary evaporation, and the residue was diluted with ethyl acetate and washed with saturated sodium bicarbonate and brine. The organic layer was dried over anhydrous magnesium sulfate and concentrated under reduced pressure. Crude product was purified via silica gel column chromatography to yield (E)-3-(3,4-diacetoxyphenyl)acrylic acid (6.3 g, 95%), (Scheme 1).

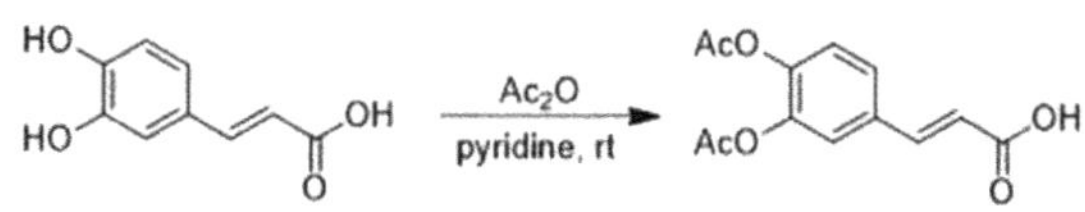

(*E*)-3-(3,4-Dihydroxyphenyl)acrylic acid **(*E*)-3-(3,4-diacetoxyphenyl)acrylic acid**

Scheme 1. The first step in the synthesis of the target compound: synthesis of (E)-3-(3,4-Diacetoxyphenyl)acrylic acid.

2.1.2. Synthesis of (E)-4-(3-(3,4-Dihydroxyphenethoxy)-3-oxoprop-1-en-1-yl)-1,2-Phenylene Diacetate

To a solution of (E)-3-(3,4-diacetoxyphenyl)acrylic acid (13.2 g, 50.0 mmol) and 4-(2-hydroxyethyl)benzene-1,2-diol (8.1 g, 52.5 mmol, 1.05 eq.) in anhydrous dichloromethane (100 mL) under argon atmosphere, N,N-dicyclohexylcarbodiimide (12.4 g, 59.9 mmol, 1.2 eq.) and 4-dimethylaminopyridine (3.1 g, 25.0 mmol, 0.5 eq.) were added. After undergoing stirring at room temperature overnight, the precipitate was filtered off and washed with dichloromethane. The filtrate was collected and washed with 0.1 N hydrochloric acid, saturated sodium bicarbonate and brine. The organic layer was dried over anhydrous magnesium sulfate and concentrated under reduced pressure. A crude product was purified by silica gel column chromatography to give (E)-4-(3-(3,4-dihydroxyphenethoxy)-3-oxoprop-1-en-1-yl)-1,2-phenylene diacetate (5.0 g, 25%) as a white solid, (Scheme 2).

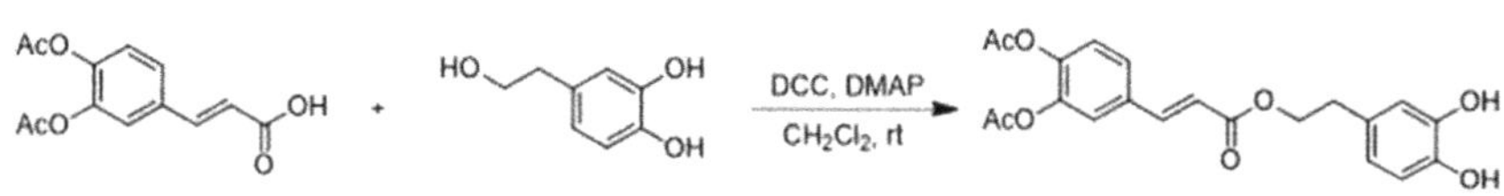

(*E*)-3-(3,4-diacetoxyphenyl)acrylic acid **4-(2-hydroxyethyl)benzene-1,2-diol** **(*E*)-4-(3-(3,4-dihydroxyphenethoxy)-3-oxoprop-1-en-1-yl)-1,2-phenylene diacetate**

Scheme 2. The second step in the synthetic pathway: synthesis of (E)-4-(3-(3,4-Dihydroxyphenethoxy)-3-oxoprop-1-en-1-yl)-1,2-phenylene diacetate.

2.1.3. Synthesis of 3,4-Dihydroxyphenethyl (E)-3-(3,4-dihydroxyphenyl)acrylate

To a solution of (E)-4-(3-(3,4-dihydroxyphenethoxy)-3-oxoprop-1-en-1-yl)-1,2-phenylene diacetate (1.0 g, 2.5 mmol) in 1,4-dioxane (10 mL), aqueous 2 N NaOH (3.7 mL, 3.0 eq.) was added and the solution was stirred at 50 °C for 3 h. After cooling to room temperature, 10% HCl (10 mL) and ethyl acetate (15 mL) were added to the reaction mixture and further stirred for 30 min. The organic layer was separated, dried over anhydrous magnesium sulfate, and concentrated under reduced pressure. The residue was purified via silica gel column chromatography with heptane/ethyl acetate as an eluent to provide 3,4-dihydroxyphenethyl (E)-3-(3,4-dihydroxyphenyl)acrylate (0.5 g, 68%). 1H NMR (400 MHz, DMSO-d6): δ 7.46 (d, 1H), 7.06 (s, 1H), 6.81 (d, 1H), 6.65–6.50 (m, 4H), 6.29 (d, 1H), 4.48 (t, 2H), 2.96 (t, 2H). Mass (ESI+): m/z 317.12 (M + H)+, (Scheme 3).

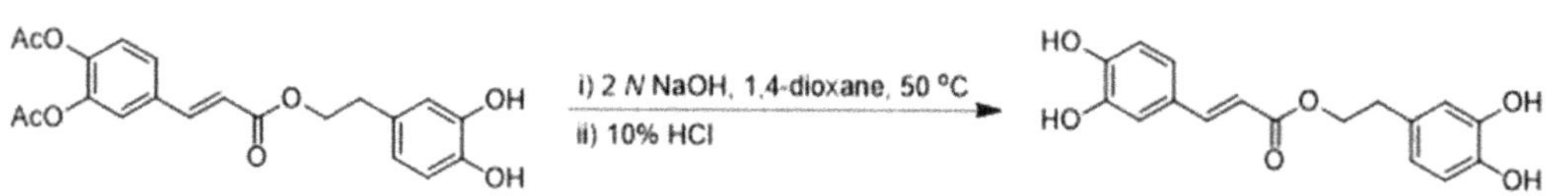

(*E*)-4-(3-(3,4-dihydroxyphenethoxy)-3-oxoprop-1-en-1-yl)-1,2-phenylene diacetate **3,4-dihydroxyphenethyl (*E*)-3-(3,4-dihydroxyphenyl)acrylate**

Scheme 3. Final step in the synthesis of 3,4-Dihydroxyphenethyl (E)-3-(3,4-Dihydroxyphenyl)acrylate.

2.2. Cell Culture

Urine melanoma B16F10 cells were purchased from ATCC (Manassas, VA, USA) and cultured in Dulbecco's Modified Eagle's Medium (DMEM; ATCC) containing 10% fetal bovine serum (FBS; ATCC) and 1% penicillin-streptomycin (Hyclone, South Logan, UT, USA). The cells were maintained every other day and sub-cultured by detaching them using a 0.05% trypsin-EDTA solution (Gibco, Thermo Fisher Scientific, Waltham, MA, USA).

2.3. DPPH Assay and ABTS Assay

Additionally,. 100 μM 2,2-diphenyl-1-picrylhydrazyl (DPPH) solution 247.5 μL dissolved in ethanol was mixed with 2.5 μL of the test substance. Ascorbic acid (AA), 10 μg/mL, was used as a positive control. The final treatment concentrations of CA and CAD are 1, 10, 100, and 1000 μg/mL. Experiments were performed in 96-well plates, as described earlier [20]. The samples were added and allowed to react for 0, 15, 30, and 60 min at room temperature in the dark. Then, the absorbance was measured at 517 nm using Infinite M200 Pro microplate reader (Tecan Group Ltd., Mannedorf, Switzerland) equipped at Ewha Drug Development Research Core Center to determine the color change to yellow due to scavenging of free radicals [21]. Ethanol, 250 μL, was used as the blank for the experiment. The data were analyzed using the following equation:

$$\text{Radical Scavenging } (\%) = \left\{1 - \frac{(Abs\ experiment - Abs\ blank)}{Abs\ Control}\right\} \times 100$$

A 7 mM 2,2′-azino-bis(3-ethylbenzothiazoline-6-sulfonic acid) (ABTS) solution and a 2.45 mM potassium persulfate solution, both dissolved in PBS, were mixed at a 1:1 ratio in 50 mL conical tube and incubated at room temperature in the dark for 16 h. The following day, the mixture was diluted with PBS to reach an absorbance of 0.700 (±0.002) at 734 nm. Then, 200 μL of the substance was combined with 800 μL of the absorbance-adjusted mixture and allowed to react for 5 min at room temperature. The final treatment concentrations of CA and CAD were 1, 10, 100, and 1000 μg/mL. Experiments were performed in 96-well plates, as described previously [22,23]. The absorbance was measured at 734 nm using a microplate spectrophotometer. PBS 200 μL was used as the blank for the experiment. The data were analyzed using the following equation:

$$\text{ABTS activity } (\%) = \left\{1 - \left(\frac{Abs\ experiment}{Abs\ Control}\right)\right\} \times 100$$

2.4. Intracellular ROS Level

Intracellular ROS was evaluated using the 2′,7′-dichlorofluoresceindiacetate (DCF-DA, Invitrogen, Eugene, OR, USA) fluorescence staining. DCF-DA was first dissolved in anhydrous DMSO to a concentration of 500 μM and subsequently diluted with PBS to a final concentration of 5 μM. B16F10 cells were seeded at a density of 1×10^5 cells per 2 mL in 35 mm culture dishes. B16F10 cells were treated with 0.2 μM α-melanocyte stimulating hormone (α-MSH) or UVA (1 J/cm^2), followed by treatment with CA or CAD for 24 h. After incubation, cells were stained with DCF-DA solution for 5 min to assess intracellular ROS levels. UVA irradiation (1 J/cm^2) was performed using a Biosun UV irradiation system (Vilber Lourmat, Marne-la-Vallée, France). Additionally, 10 μg/mL of AA was used as a positive control. Fluorescently illuminated cell images were taken using the Softmax 5.2 program and an Axiovert 200 M microscope (Carl Zeiss, Land Baden Württemberg, Germany).

2.5. Mitochondrial Superoxide Assay

Intracellular mitochondrial superoxide was evaluated via double staining with MitoSOX red (Invitrogen, Eugene, OR, USA) and Mitotracker green (Invitrogen, Eugene, OR, USA). MitoSOX was diluted in anhydrous DMSO to 5 mM and then diluted in Hank's Buffered Salt Solution (HBSS, Gibco, Thermo Fisher Scientific, Waltham, MA, USA) to a final concentration of 1 μM. Mitotracker green was diluted in anhydrous DMSO to 1 mM and diluted with HBSS to a final concentration of 200 nM. B16F10 cells were seeded at a density of 1×10^4 cells per 400 μL in 4-well culture plates and treated with 0.2 μM α-MSH or UVA (1 J/cm^2), followed by treatment with CA or CAD for 24 h. Moreover, 10 μg/mL of AA was used as a positive control. After incubation, cells were stained for 30 min and gently washed once with HBSS. The fluorescently stained cells were mounted with fluoromount-G and imaged as a confocal microscope (Zeiss LSM 880 with Airyscan, Carl Zeiss, Oberkochen, Germany) equipped at the Ewha Fluorescence Core Imaging Center.

2.6. Melanin Assay and Cell Viability Assay

For melanin content and cell viability measurements, B16F10 cells were seeded at a density of 2×10^4 cells per 600 μL in 48-well plates. After 24 h of seeding, cells were treated for 48 h with CA and CAD in the media containing 0.2 μM α-MSH. Those treated with α-MSH alone were used as the negative control, while 50 μg/mL arbutin-treated cells served as the positive control. To determine melanin content, cells were treated with 40 mg/mL of NaOH solution diluted with 3DW for 1 h and absorbance was measured at 405 nm in a microplate spectrophotometer. Additionally, 40 mg/mL of NaOH solution 100 μL was used as the blank for the experiment. For cell viability determination, cells were treated with 0.2 mg/mL MTT solution diluted in phenol red-free medium for 3 h. After incubation, the medium was removed, and 300 μL of DMSO was added to each well to dissolve the resulting formazan crystals by shaking at 300 rpm for 30 min at 25 °C. The supernatant was then collected, and absorbance was measured at 540 nm using a microplate spectrophotometer. DMSO 150 μL was used as the blank for the experiment.

2.7. Mushroom Tyrosinase Inhibitory Assay

To evaluate the inhibition of tyrosinase activity of the substances, the mushroom tyrosinase assay was performed. Experiments were performed in 96-well plates. Moreover, 50 μg/mL of kojic acid (KA) was used as a positive control. The reaction mixture (total volume: 300 μL) consisted of 220 μL of 0.1 M phosphate buffer, 20 μL of test substance (CA or CAD), 20 μL of mushroom tyrosinase (1500 U/mL), and 40 μL of 1.5 mM L-tyrosine. For the blank, 20 μL of DMSO (used as the solvent for the test substances) was added instead of the test substance, and 20 μL of 0.1 M phosphate buffer was used in place of the tyrosinase enzyme solution. The final treatment concentrations of CA and CAD are 1, 10, 100, and 1000 μg/mL. After 15 min of reaction in a dark 37 °C incubator, the absorbance was measured at 490 nm using a microplate spectrophotometer.

2.8. mRNA Extraction, cDNA Synthesis, and Real-Time PCR

To analyze the gene expression following CAD treatment, B16F10 cells were seeded at a density of 2×10^5 cells per 2 mL in 6-well plates and treated with CAD for 24 h. After each well was washed with phosphate-buffered saline, Trizol (Invitrogen, Waltham, MA, USA) was added to lyse the cells. Chloroform was added, and the supernatant was collected after centrifugation (13,000 rpm, 15 min, 4 °C). To the supernatant, isopropanol was added, mixed by inverting, and centrifuged (13,000 rpm, 15 min, 4 °C). The isopropanol was carefully removed, and 70% ethanol was added, followed by centrifugation again (13,000 rpm, 5 min, 4 °C). After careful removal of 70% ethanol, the RNA pellet was dissolved in RNase-free diethylpyrocarbonate (DEPC)-treated water. RNA concentration and purity were measured with a nanodrop spectrophotometer (NanoDrop ND1000, Thermo Fisher Scientific, Wilmington, DE, USA). Complementary DNA (cDNA) was synthesized using 1250 ng of RNA and Reverse Transcription Master Premix (Elpis Biotech, Daejeon, Republic of Korea). Relative mRNA expression levels were assessed through real-time PCR, performed with Power SYBR Green PCR Master Mix (Applied Biosystems, Foster City, CA, USA) on a StepOnePlus™ Real-Time PCR System (Applied Biosystems, Foster City, CA, USA). The expression levels of target genes were normalized to β-actin, used as a housekeeping gene. The sequences of the primers were as follows:

Forward β-actin 5′-AGG GAA ATC GTG CGT GAC AT-3′
Reverse β-actin 5′-GGA AAA GAG CCT CAG GGC AT-3′
Forward Tyrosinase 5′-GGG CCC AAA TTG TAC AGA GA-3′
Reverse Tyrosinase 5′-ATG GGT GTT GAC CCA TTG TT-3′
Forward TRP-1 5′-GTT CAA TGG CCA GGT CAG GA-3′
Reverse TRP-1 5′-CAG ACA AGA AGC AAC CCC GA-3′
Forward TRP-2 5′-TTA TAT CCT TCG AAA CCA GGA-3′
Reverse TRP-2 5′-GGG AAT GGA TAT TCC GTC TTA-3′

The following mouse gene accession numbers were used for real-time PCR analysis: β-actin (NM_007393.5), Tyrosinase (NM_001317397.2), TRP-1 (NM_001282014.1), and TRP-2 (NM_010024.3).

2.9. Western Blot Analysis

To examine protein expression following CAD treatment, B16F10 cells were seeded at a density of 2×10^5 cells per 2 mL (24 h), 4×10^5 cells per 2 mL (1 h or 4 h), in 6-well plates and treated with CAD for 1, 4, and 24 h. After treatment, a lysis buffer containing RIPA buffer, 1 tablet of protease inhibitor cocktail, and 1 tablet of phosphatase inhibitor cocktail was added to each well. Cells were scraped, and lysates were centrifuged at 12,000 rpm for 10 min at 4 °C. The supernatant was collected, and protein concentrations were determined using the BCA assay. Samples were prepared by mixing a lysis buffer, 4× dye, a 10× reducing agent, and 14 ng of protein, followed by heating at 95 °C. Proteins were separated with 10% sodium dodecyl sulfate-polyacrylamide gel electrophoresis (SDS-PAGE) and transferred to membranes using the Trans-Blot Turbo Transfer System and Trans-Blot Turbo Transfer Pack (Bio-Rad, Hercules, CA, USA). Membranes were blocked with 5% BSA for 2 h at room temperature. The primary antibody against the target protein was bound in 5% BSA overnight at 4 °C. After washing three times with 1× Tris-buffered saline containing 0.1% Tween 20 (TBST), membranes were bound with the secondary antibody in 5% BSA for 1 h at room temperature. After additional washes with TBST, protein detection was performed using the ECL Western Blotting Detection Reagent (Cytiva, Marlborough, MA, USA). Protein bands were visualized using an Amersham Imager 600 (GE Healthcare Life Sciences, Stevenage, UK). BLUEstain™ protein ladder (GoldBio, St. Louis, MO, USA; 11–245 kDa) was used as a molecular size marker. The β-actin antibody (ab8226; Abcam, Cambridge, UK) was used as a housekeeping gene, TYR antibody tyrosinase (ab170905; Abcam, Cambridge, UK), TRP-1 antibody tyrosinase (ab235447; Abcam, Cambridge, UK), TRP-2 antibody tyrosinase (ab221144; Abcam, Cambridge, UK), MITF antibody (ab20663; Abcam, Cambridge, UK), p-MITF antibody (LS-C199259; LSBio, Seattle, WA, USA), p-ERK antibody (4370S; Cell Signaling Technology, Danvers, MA, USA), ERK antibody (4695S; Cell Signaling Technology, Danvers, MA, USA), p-JNK antibody (9251S; Cell Signaling Technology, Danvers, MA, USA), JNK antibody (9252S; Cell Signaling Technology, Danvers, MA, USA), p-p38 antibody (4511S; Cell Signaling Technology, Danvers, MA, USA), and p38 antibody (9212S; Cell Signaling Technology, Danvers, MA, USA) were used.

2.10. 3D Human-Pigmented Epidermis Model, Melanoderm™

Melanoderm™ and the EPI-100-NMM-113 maintenance medium were obtained from MatTek Corporation (Ashland, MA, USA). On the day of arrival, tissues were pre-incubated with 5 mL of NMM-113 maintenance medium in a humidified incubator at 37 °C and 5% CO_2 for 18 h. Following pre-incubation, Melanoderm™ was treated with CA, CAD, and 1% kojic acid (KA) every other day for 14 days, with 1% KA serving as the positive control. Photos were taken before each treatment to assess tissue pigmentation, and ΔL values were measured. After 14 days, tissues were fixed in 10% neutral-buffered formalin and analyzed through Hematoxylin and Eosin (H&E) staining and Fontana-Masson staining.

2.11. Statistical Analysis

The data are expressed as means ± standard errors of the means (SEM) obtained from three or four independent experiments. Statistical significance was assessed using Student's t-test, and a p-value of less than 0.05 was considered indicative of a significant difference.

3. Results

3.1. Antioxidant Effects of CA and CAD in Cell-Free Systems

The antioxidant activities of caffeic acid (CA, Figure 1a) and caffeic acid + 3,4-DHPEA (CAD, Figure 1b) were compared via a DPPH assay and an ABTS assay. The DPPH assay employs a free radical scavenging method and is primarily suitable for evaluating lipophilic antioxidants, whereas the ABTS assay utilizes an anion radical scavenging method and can assess both hydrophilic and lipophilic antioxidants. Conducting both assays together provides complementary results, enabling a more comprehensive and accurate evaluation of antioxidant capacity. As a positive control, 10 μg/mL ascorbic acid (AA) was used. To compare the antioxidant activities of CA and CAD, their radical scavenging activity (RSA%) was evaluated relative to that of the positive control (AA) at the same concentration of 10 μg/mL. In the DPPH assay, the RSA% of CAD was higher than that of CA at 60 min after the reaction at 10 μg/mL (Figure 1c). Similarly, in the ABTS assay, CAD showed higher RSA% than CA at 30 min at 10 μg/mL (Figure 1d). These findings suggest that, in a cell-free system, CAD exhibits more potent antioxidant capacity than CA.

3.2. Comparison of Anti-Melanogenesis Effects of CA and CAD in α-MSH Induced B16F10 and Cell-Free Tyrosinase Assay

To evaluate the anti-melanogenesis activity of CA and CAD, melanin contents were measured in B16F10 cells stimulated with α-MSH after treatment with CA or CAD. Arbutin 50 μg/mL was used as a positive control. Following initial experiments conducted across three concentration ranges, two non-cytotoxic concentrations were selected for further analysis, excluding the concentration that exhibited cytotoxicity. As a result, CAD exhibited anti-melanogenesis activity more potent than CA while maintaining cell viability above 80% at the two selected concentrations (Figure 2a–d). The mushroom tyrosinase assay was performed to assess the ability of CA and CAD to directly inhibit tyrosinase activity. While CA failed to inhibit tyrosinase activity up to 10 μg/mL, CAD suppressed it significantly from 1 μg/mL (KA, kojic acid 50 μg/mL as a positive control, Figure 2e).

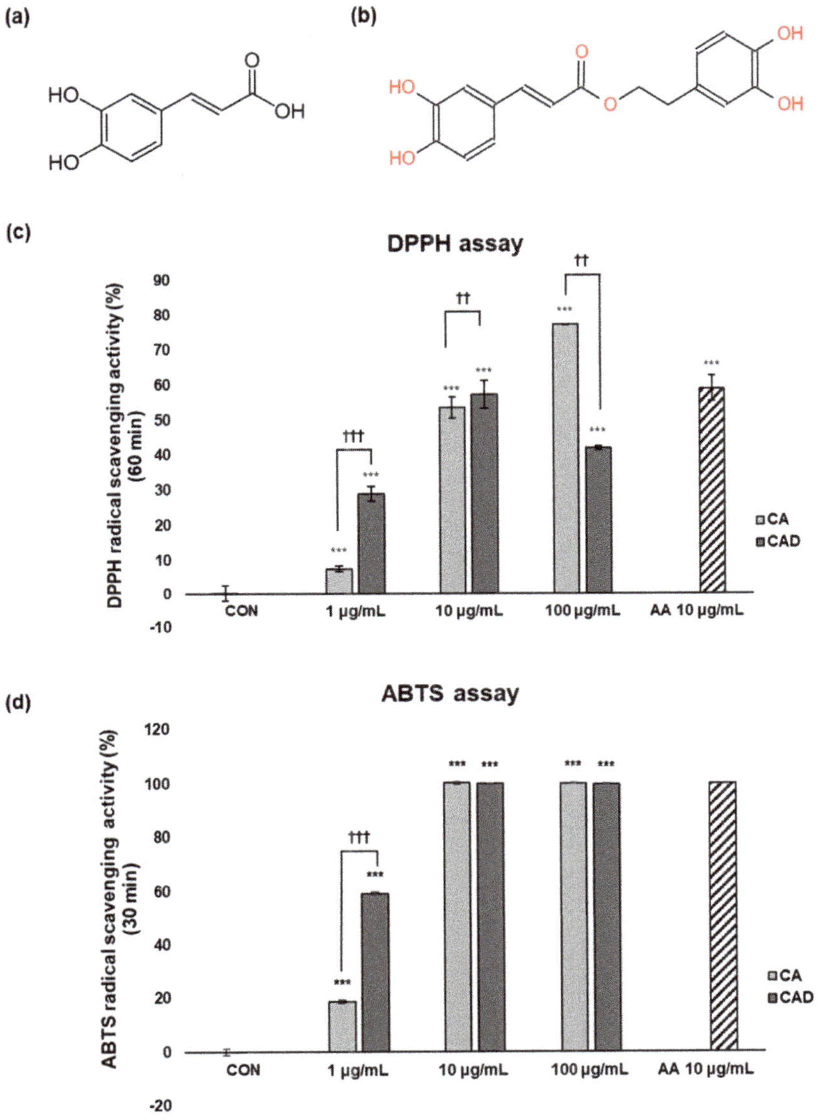

Figure 1. Structures of CA and CAD and their antioxidant activity assessed with DPPH and ABTS assays. Chemical structure of (**a**) caffeic acid (CA) and (**b**) caffeic acid + 3,4-DHPEA (CAD). Antioxidant activity of CA and CAD was measured with (**c**) a DPPH radical scavenging activity assay and (**d**) an ABTS radical scavenging activity assay. In both assays, the percentage of radical scavenging activity was measured at 517 nm. 10 μg/mL of ascorbic acid (AA) was used as a positive control. Data are presented as the means ± SDs ($n = 4$, *** $p < 0.001$ compared with the control, †† $p < 0.01$, ††† $p < 0.001$ comparing CA with CAD).

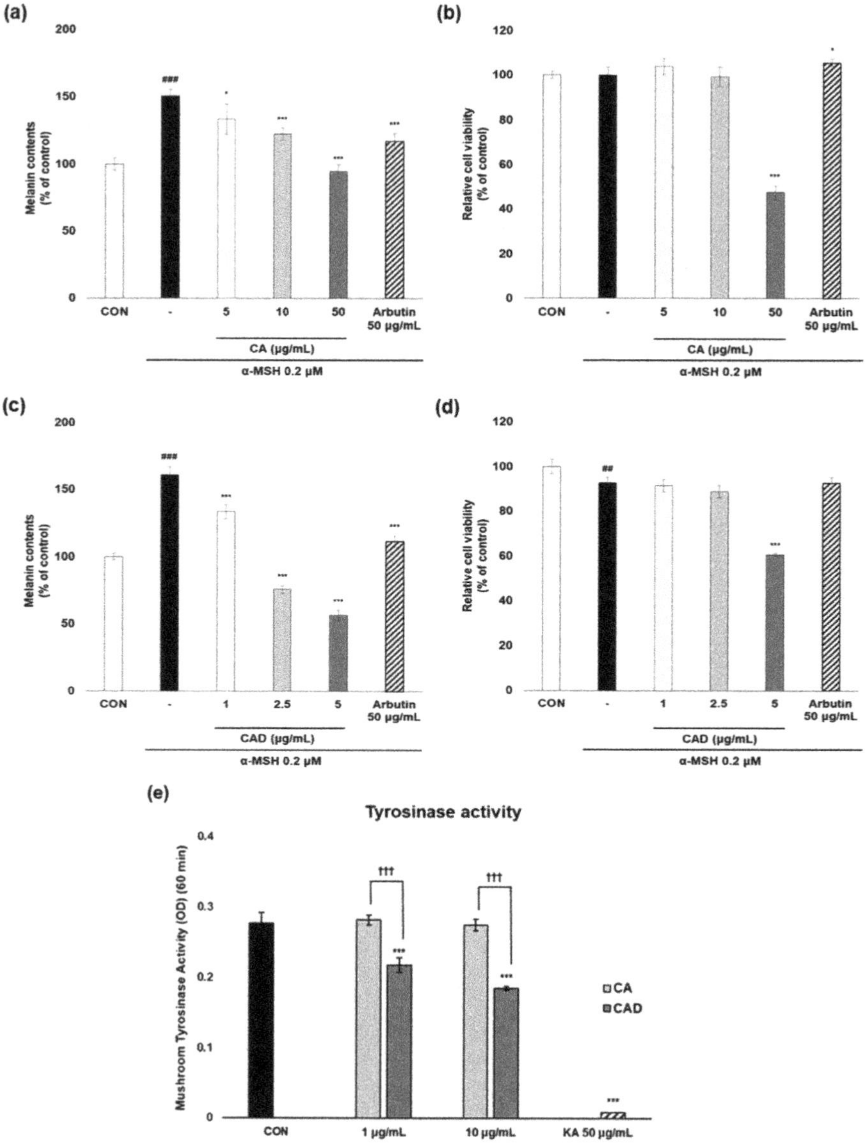

Figure 2. Anti-melanogenesis activities of CA and CAD in cell-based and cell-free system. B16F10 cells stimulated with 0.2 µM α-MSH were treated with CA and CAD for 48 h. (**a**,**c**) Melanin contents were measured using a melanin assay, and (**b**,**d**) cell viability was investigated using an MTT assay. Tyrosinase enzymatic activity in (**e**) CA and CAD was measured using mushroom tyrosinase assay with L-tyrosine as a substrate. Fifty µg/mL of kojic acid (KA) was used as a positive control. Data are presented as the means ± SDs of OD ($n = 4$, * $p < 0.05$, *** $p < 0.001$ compared with the negative control, ## $p < 0.01$, ### $p < 0.001$ compared with the control, ††† $p < 0.001$ comparing CA with CAD).

3.3. CAD Reduces Intracellular but Not Mitochondrial ROS Production in B16F10 Stimulated via UVA and α-MSH

To evaluate whether CA and CAD can exert antioxidant activities in melanocytes, B16F10 cells, and murine melanoma cells, stimulated with UVA or α-MSH after treatment with CA or CAD and DCF-DA/MitoSOX fluorescence staining was performed. Negative control cells were treated with UVA and α-MSH alone, while positive control cells were treated with 10 μg/mL AA. In all experiments, UVA stimulation induced higher ROS levels than α-MSH stimulation. Fluorescence intensity was quantified using Image J software (https://imagej.net/ij/, accessed on 1 September 2023). To facilitate a quantitative comparison between CA and CAD, ROS inhibition levels were calculated and expressed as a percentage using the following formula: % ROS inhibition = [(UVA or α-MSH-treated (negative control) − CA or CAD-treated)/(UVA or α-MSH-treated (negative control) − untreated control)] × 100, as presented in Figure 3a,c. The DCF-DA fluorescent staining revealed that CA inhibited intracellular ROS generation more effectively than CAD in cells stimulated with either UVA or α-MSH but CAD exerted antioxidant effects at a lower concentration (10 vs. 2 μg/mL) (Figure 3a–d). To further investigate the site of ROS inhibition via CAD, double staining with DCF-DA and Mitotracker red, a mitochondrial dye or MitoSOX red, a mitochondrial ROS-specific dye, and Mitotracker green was conducted. The results indicated that CAD can reduce the total intracellular ROS in the cells stimulated with either UVA or α-MSH, as shown in the merged fluorescence images (Figure 3e,g,h,j). In contrast, mitochondrial ROS levels were not significantly increased via a UVA or α-MSH treatment, and merged fluorescence images confirmed that CAD did not significantly suppress mitochondrial ROS production (Figure 3f,g,i,j). These findings indicate that CAD reduces intracellular ROS levels in cells induced via UVA or α-MSH but could not affect mitochondrial ROS. This localized ROS inhibition of CAD suggests that CAD may modulate oxidative stress through cytosolic pathways without significantly altering mitochondrial ROS production.

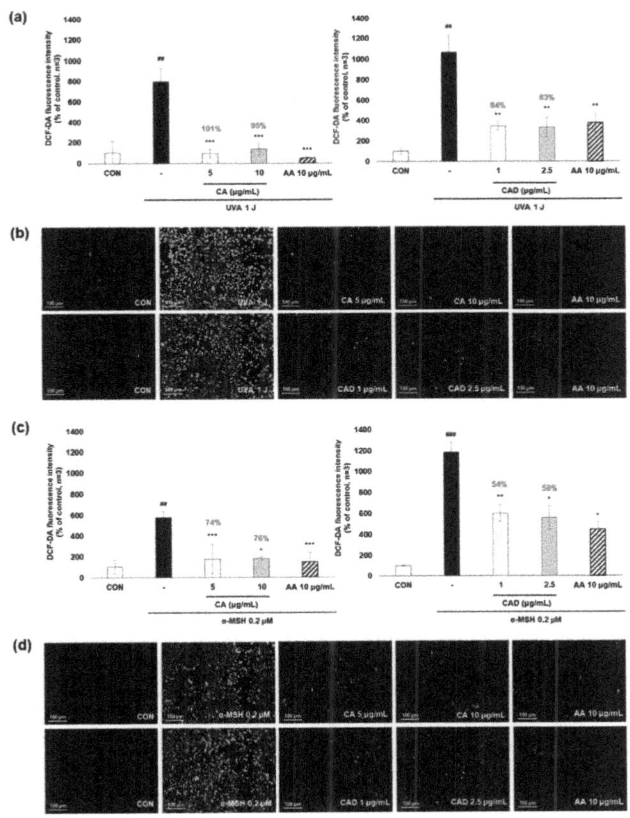

Figure 3. *Cont.*

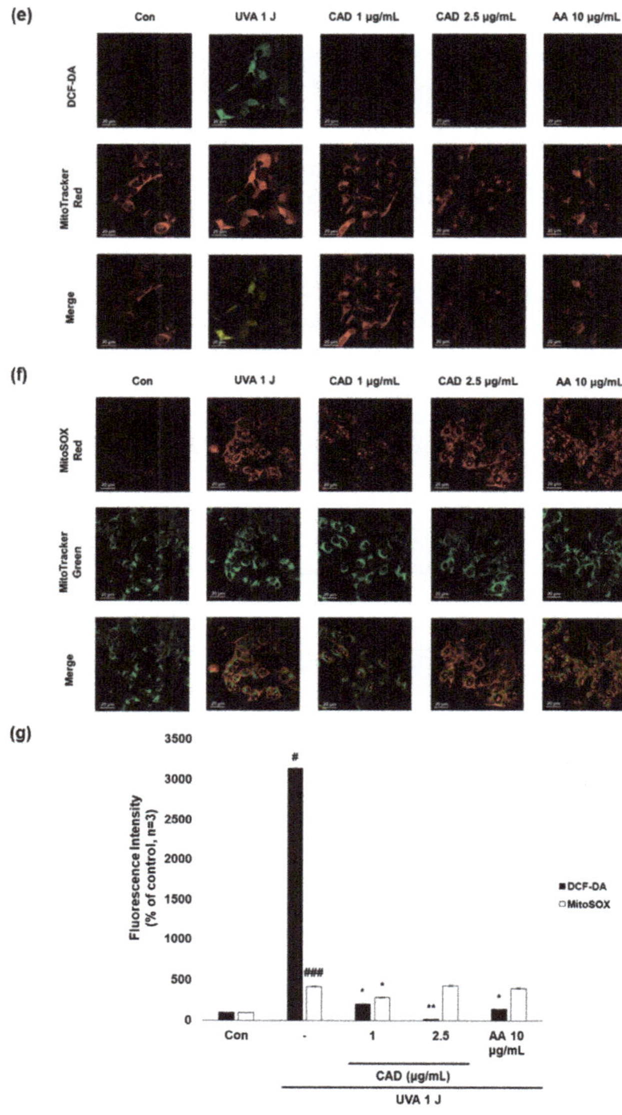

Figure 3. *Cont.*

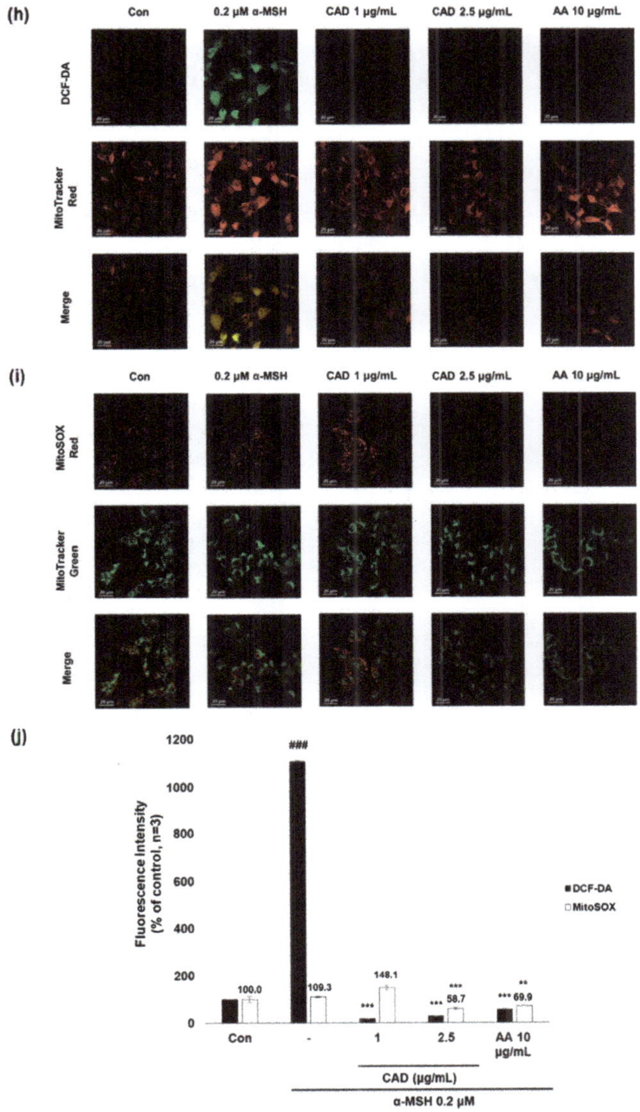

Figure 3. The effects of CA and CAD on ROS production in B16F10 cells induced via α-MSH and UVA. The cells treated with α-MSH and UVA served as the negative control, while those treated with 10 μg/mL AA were used as the positive control. B16F10 cells were irradiated with 1 J UVA or α-MSH at 0.2 μM. CA and CAD were treated for 24 h. (**b**) B16F10 cells stimulated with UVA or (**d**) α-MSH were stained with DCF-DA. (**e**,**f**) B16F10 cells stimulated with UVA or (**h**,**i**) α-MSH were double-stained with DCF-DA green-Mitotracker red and MitoSOX red-Mitotracker green. Representative images were captured via confocal microscopy (400×). (**a**,**c**,**g**,**j**) Fluorescence intensity was analyzed using the Image J software (https://imagej.net/ij/, accessed on 1 September 2023). Data are presented as means ± SDs (n = 3, * $p < 0.05$, ** $p < 0.01$, *** $p < 0.001$ compared with the negative control, # $p < 0.05$, ## $p < 0.01$, ### $p < 0.001$ compared with the control).

3.4. CAD Downregulates the Gene and Protein Expression of Melanogenic Enzymes

The effects of CAD on the gene and protein expression of tyrosinase (TYR), tyrosinase-related protein 1 (TRP-1), and tyrosinase-related protein 2 (TRP-2), key melanogenic enzymes were assessed using real-time PCR and Western blot analysis. Treatment with α-MSH alone significantly upregulated the gene expression of TYR, TRP-1, and TRP-2. In contrast, treatment with CAD significantly downregulated TRP-1 mRNA expression (Figure 4a). A protein expression analysis revealed a significant decrease in TYR and TRP-1 levels and an insignificant but marked decrease in TRP-2 (Figure 4b,c). These findings indicate that CAD inhibits the expression of key melanogenic enzymes.

(a)

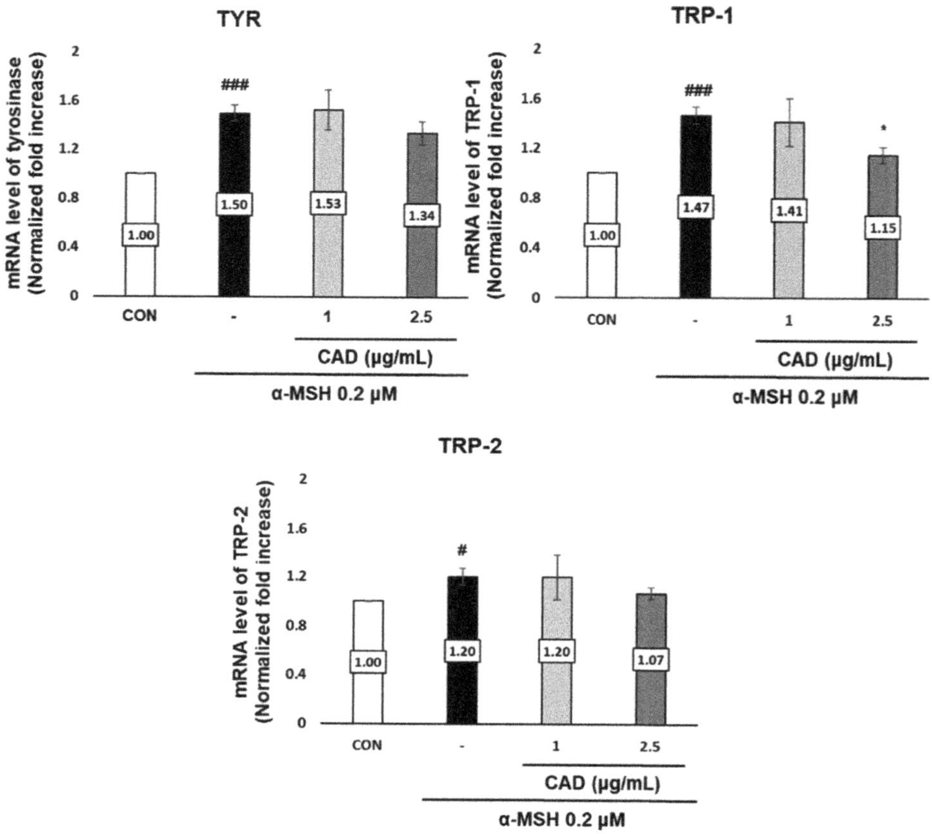

Figure 4. *Cont.*

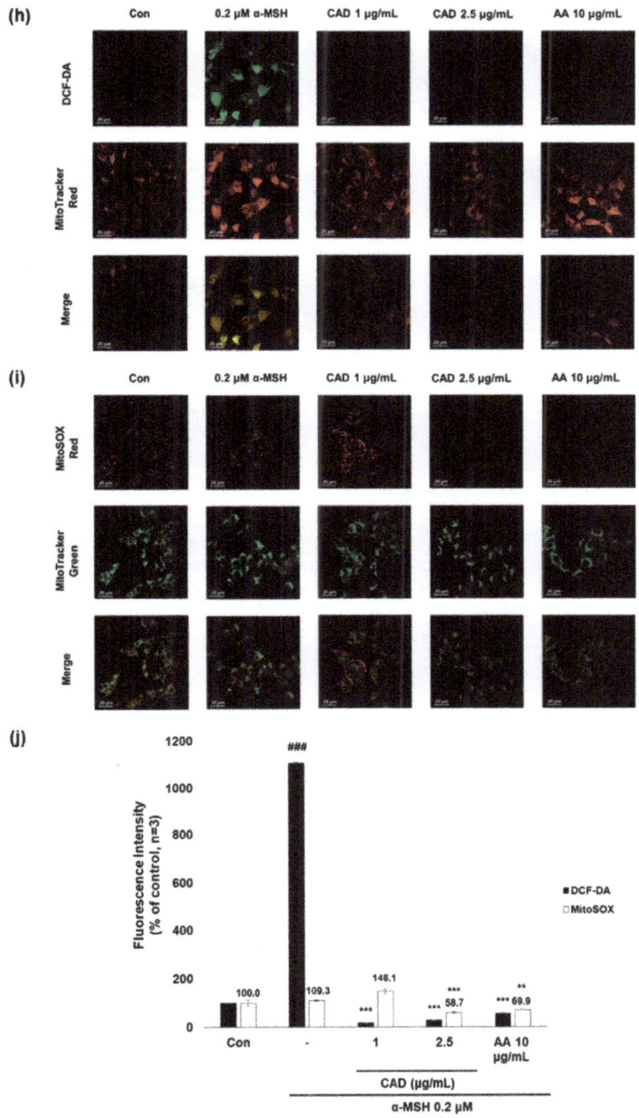

Figure 3. The effects of CA and CAD on ROS production in B16F10 cells induced via α-MSH and UVA. The cells treated with α-MSH and UVA served as the negative control, while those treated with 10 μg/mL AA were used as the positive control. B16F10 cells were irradiated with 1 J UVA or α-MSH at 0.2 μM. CA and CAD were treated for 24 h. (**b**) B16F10 cells stimulated with UVA or (**d**) α-MSH were stained with DCF-DA. (**e**,**f**) B16F10 cells stimulated with UVA or (**h**,**i**) α-MSH were double-stained with DCF-DA green-Mitotracker red and MitoSOX red-Mitotracker green Representative images were captured via confocal microscopy (400×). (**a**,**c**,**g**,**j**) Fluorescence intensity was analyzed using the Image J software (https://imagej.net/ij/, accessed on 1 September 2023). Data are presented as means ± SDs ($n = 3$, * $p < 0.05$, ** $p < 0.01$, *** $p < 0.001$ compared with the negative control, # $p < 0.05$, ## $p < 0.01$, ### $p < 0.001$ compared with the control).

3.4. CAD Downregulates the Gene and Protein Expression of Melanogenic Enzymes

The effects of CAD on the gene and protein expression of tyrosinase (TYR), tyrosinase-related protein 1 (TRP-1), and tyrosinase-related protein 2 (TRP-2), key melanogenic enzymes were assessed using real-time PCR and Western blot analysis. Treatment with α-MSH alone significantly upregulated the gene expression of TYR, TRP-1, and TRP-2. In contrast, treatment with CAD significantly downregulated TRP-1 mRNA expression (Figure 4a). A protein expression analysis revealed a significant decrease in TYR and TRP-1 levels and an insignificant but marked decrease in TRP-2 (Figure 4b,c). These findings indicate that CAD inhibits the expression of key melanogenic enzymes.

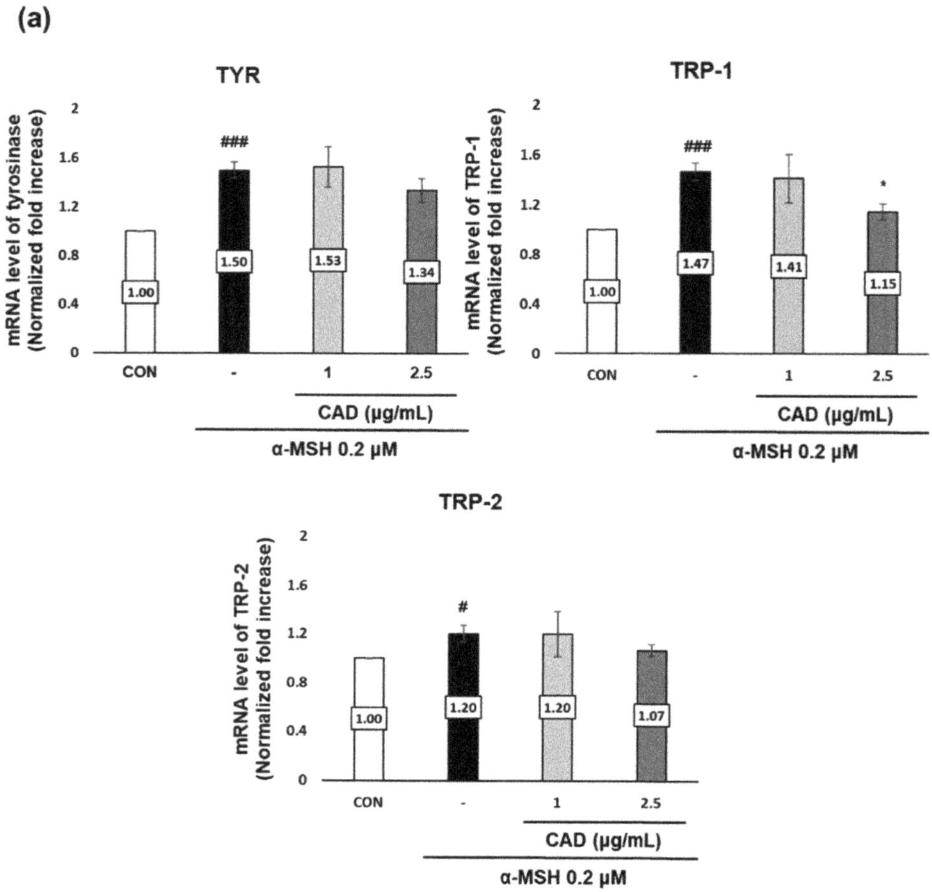

Figure 4. *Cont.*

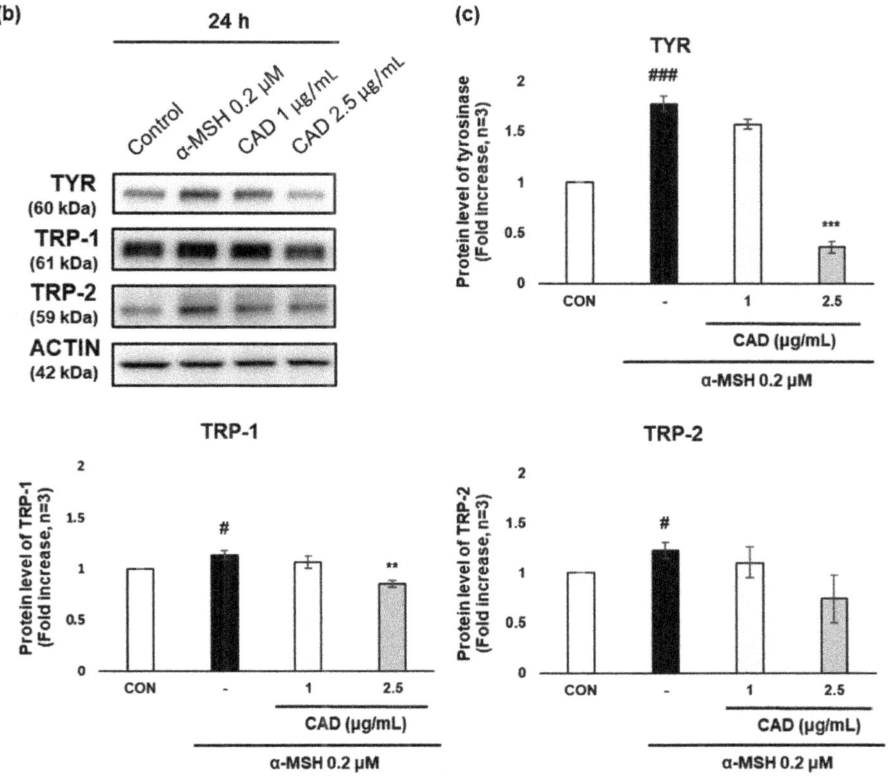

Figure 4. Effects of CAD on mRNA and protein levels of melanogenesis-related molecular targets in B16F10 cells. B16F10 cells stimulated with 0.2 μM of α-MSH were treated with 1 μg/mL and 2.5 μg/mL of CAD for 24 h. Treatment with α-MSH alone served as the negative control. (**a**) mRNA expression levels and (**b**) protein levels of TYR, TRP-1, and TRP-2 in B16F10 cells were analyzed via (**a**) real-time PCR and (**b**) Western blot. (**c**) Quantitation of TYR, TRP-1, and TRP-2 protein levels from the Western blot was performed using the Image J software. For real-time PCR, β-actin was used as the housekeeping gene, and ACTIN was used for Western blot analysis. All data were normalized to the expression of the respective housekeeping genes. Data are presented as the mean ± SD ($n = 3$, * $p < 0.05$, ** $p < 0.01$, *** $p < 0.001$ compared with negative control, # $p < 0.05$, ### $p < 0.001$ compared with control).

3.5. CAD Inhibits the Phosphorylation of MITF and MAPK Signaling Molecules

The effects of CAD on MITF, an upstream regulator of TYR, TRP-1, and TRP-2 that are involved in melanocyte stimulation, and on ERK, JNK, and p38, key components of the MAPK signaling pathway, were analyzed. Phospho-MITF (p-MITF) expression increased after 4 h of α-MSH treatment alone but was significantly reduced by CAD. In contrast, the expression of total MITF (MITF) remained unchanged (Figure 5a). As a result, the protein level of p-MITF relative to total MITF was significantly reduced in a concentration-dependent manner via CAD (Figure 5b). Phosphorylation of ERK (p-ERK), JNK (p-JNK), and p38 (p-p38) increased after 1 hr of α-MSH treatment alone. In contrast, CAD significantly reduced p-ERK and p-JNK levels, while it did not significantly affect p-p38, a key factor in the MAPK signaling pathway. The expression of total JNK (JNK) and total p38 (p38) remained stable, whereas the total ERK (ERK) expression followed a similar trend to p-ERK (Figure 5c,d). These findings indicate that CAD effectively inhibits the phosphorylation of MITF and significantly modulates ERK and JNK phosphorylation, key components of the MAPK signaling pathway, contributing to its regulation of melanogenesis.

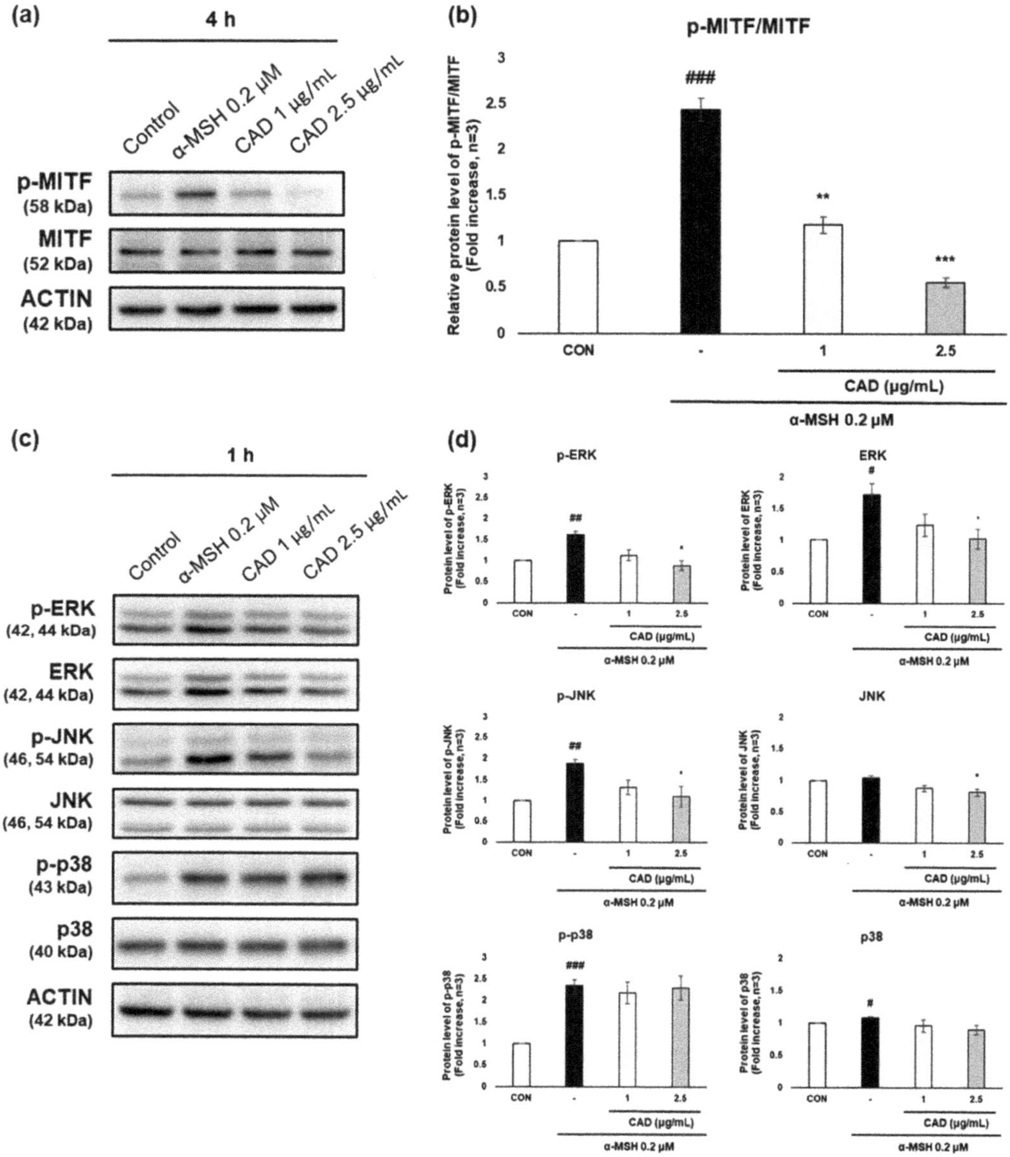

Figure 5. Effects of CAD on phosphorylation of MITF through the MAPK signaling pathway in B16F10 cells. B16F10 cells stimulated with 0.2 μM of α-MSH were treated with 1 μg/mL and 2.5 μg/mL of CAD for 1 h and 4 h. Treatment with α-MSH alone served as the negative control. The protein levels of (**a**) p-MITF and MITF, (**c**) ERK, JNK, p38, and their phosphorylated form were analyzed by Western blot. (**b**,**d**) Quantitation of protein levels from the Western blot was performed using Image J software. (**b**) The data are presented as the comparison of phosphorylated forms to their respective total forms, and (**d**) the data are normalized to the expression of the respective housekeeping genes. Data are expressed as means ± SDs ($n = 3$, * $p < 0.05$, ** $p < 0.01$, *** $p < 0.001$ compared with the negative control, # $p < 0.05$, ## $p < 0.01$, ### $p < 0.001$ compared with the control).

3.6. Brightening Effects of CA and CAD on Melanoderm™, an Artificial 3D Pigmented Human Epidermis Model

The anti-melanogenic effects of CAD were evaluated in comparison to CA using the 3D pigmented human epidermis model, Melanoderm™. Both CA (150, 300 mg/mL) and CAD (30, 75 mg/mL) were applied every other day, and the tissue brightening was observed on day 14 compared to day 0 (Figure 6a,d). The L-value, representing skin brightness, indicated a reduction in melanin content in both low and high concentrations of CA and CAD (Figure 6b,e). On day 14, skin tissues were fixed and analyzed using Hematoxylin and Eosin (H&E) and Fontana–Masson (FM) staining. The staining results revealed that melanin particles in both CA- and CAD-treated tissues were smaller and more finely distributed compared to the control group (Figure 6c,f). These findings suggest that CAD can manifest skin-brightening effects at concentration ranges than CA.

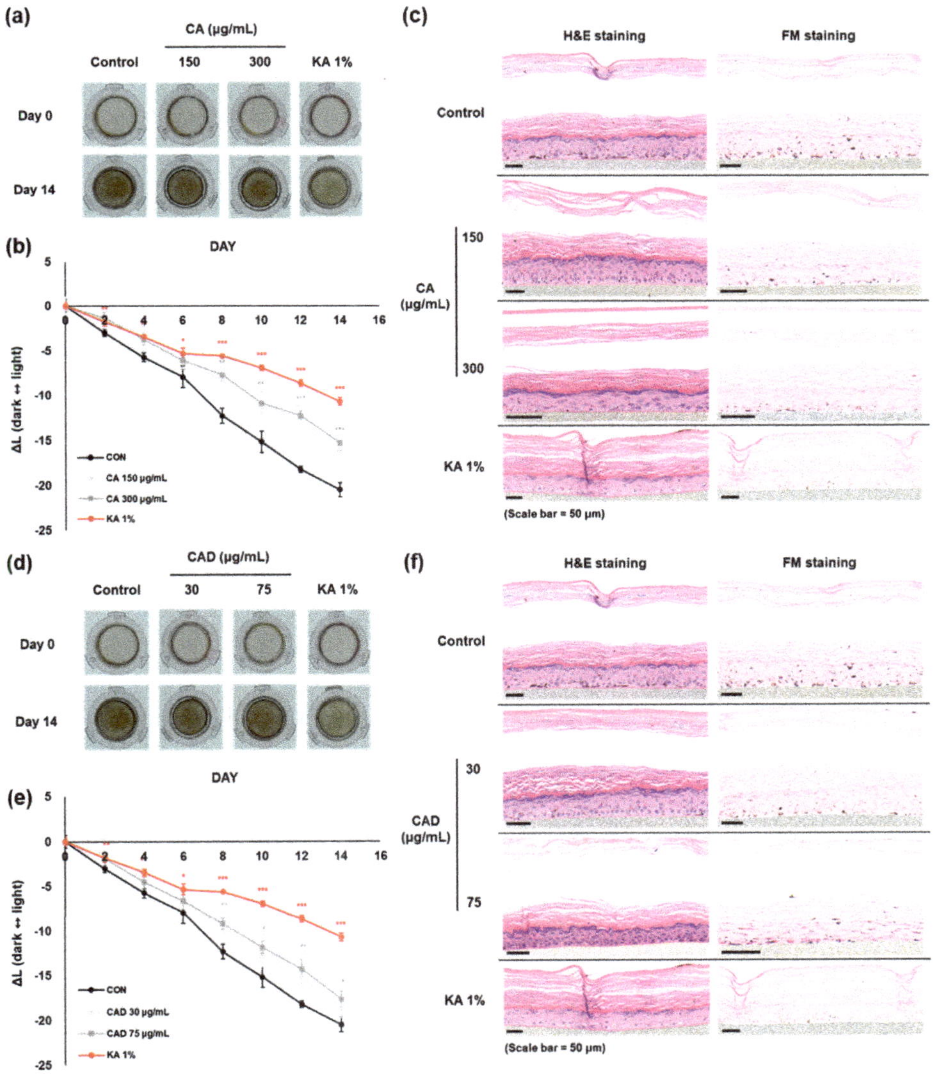

Figure 6. Effects of CAD on Melanoderm™, an artificial 3D pigmented human epidermis model. Melanoderm™, 3D human-pigmented epidermis model was treated with CA 150 μg/mL, 300 μg/mL

and CAD 30 μg/mL, 75 μg/mL and KA 1% every other day for 14 days. KA 1% was used as positive control. (**a**,**d**) Observed changes in tissue color on Day 0 and Day 14. (**b**,**e**) The degree of brightness of the tissues was measured every other day as the ΔL-value from Day 0 to Day 14. (**c**,**f**) Tissues on Day 14 were stained with H&E and FM. Data are presented as means ± SDs ($n = 3$, * $p < 0.05$, ** $p < 0.01$, *** $p < 0.001$, compared with the control).

4. Discussion

In this study, we synthesized a novel antioxidant compound, caffeic acid-3,4-dihydroxyphenylpropanol ester (CAD), and investigated its antioxidant and skin-brightening effects. CAD is a derivative of caffeic acid (CA) conjugated with 3,4-dihydroxyphenylethanol (3,4-DHPEA). In silico analysis revealed that CAD had a higher XLogP3 value compared to CA (3.34 vs. 1.2), suggesting greater lipophilicity and potentially improved dermal absorption [24]. Consistent with this prediction, CAD demonstrated significantly stronger antioxidant and anti-melanogenic activities than CA.

Conventional skin-brightening agents such as kojic acid, arbutin, and hydroquinone act primarily through direct inhibition of tyrosinase, a key enzyme in melanogenesis [25]. While CA failed to inhibit tyrosinase activity, CAD significantly inhibited it at 1 μg/mL, indicating that conjugation with 3,4-DHPEA enhances tyrosinase inhibitory potency. In addition, CAD downregulated both mRNA and protein expression of TYR, TRP-1, and TRP-2. Notably, the mRNA level of TRP-1 and the protein levels of TYR and TRP-1 were significantly reduced. The observed discrepancies between mRNA and protein expression may be attributed to differences in the timing of gene transcription versus protein translation and the limited number of sampling time points.

To further elucidate the mechanism underlying CAD's anti-melanogenic effect, we examined its influence on MITF, the master transcription factor regulating melanocyte function, including differentiation, pigmentation, and survival [26,27], as well as its upstream regulator, the MAPK signaling pathway. CAD significantly suppressed the phosphorylation of MITF, along with that of ERK and JNK—two key components of the MAPK cascade implicated in melanogenic regulation.

Reactive oxygen species (ROS) serve as intracellular signaling molecules, but excessive ROS accumulation induces oxidative stress [28]. ROS can activate the MAPK pathway both directly, via kinases such as apoptosis signal-regulating kinase 1 (ASK1) [29,30], and indirectly, by inactivating phosphatases that dephosphorylate JNK [31,32]. As an antioxidant, CAD effectively reduced intracellular ROS levels, which may contribute to the suppression of ERK and JNK phosphorylation and, consequently, the inhibition of MITF activity and melanogenesis.

The MAPK pathway plays a pivotal role in melanogenesis by regulating MITF. Specifically, phosphorylation of MITF at serine 73 by ERK and JNK enhances its ubiquitination and proteasomal degradation, thereby downregulating the transcription of melanogenic enzymes such as TYR, TRP-1, and TRP-2 [33,34]. Conversely, transient or low-level activation of ERK and JNK may stabilize MITF and promote melanin synthesis. In this study, CAD markedly suppressed ERK and JNK phosphorylation, thereby reducing MITF phosphorylation. This suppression likely underlies the observed downregulation of melanogenic genes and supports the conclusion that CAD mediates its skin-brightening effects through inhibition of the ERK/JNK-MITF axis within the MAPK signaling cascade.

To evaluate CAD's efficacy in a physiologically relevant system, we utilized a 3D pigmented human epidermis model, Melanoderm™. CAD exhibited skin-brightening effects at both low and high concentrations. An increase in L-value compared to the control indicated reduced melanin content. Histological analysis further confirmed that melanin granules in CAD-treated tissues were smaller and more fragmented. Importantly, CAD demonstrated a brightening effect even at lower concentrations than CA, suggesting its potential effectiveness in human skin applications.

In summary, CAD exhibits strong antioxidant and anti-melanogenic activities, supported by in vitro, in silico, and 3D tissue model data. These findings highlight CAD as a promising candidate for cosmetic applications aimed at skin brightening and antioxidant protection.

Author Contributions: Conceptualization, K.-l.K., J.-h.J. and K.-M.L.; methodology, K.-l.K.; formal analysis, K.-l.K. and K.-M.L.; investigation, K.-l.K. and Y.K.; resources, J.-h.J. and Y.K.; writing—original draft preparation, K.-l.K.; writing—review and editing, K.-M.L.; supervision, K.-M.L.; project administration, K.-M.L.; funding acquisition, K.-M.L. All authors have read and agreed to the published version of the manuscript.

Funding: This research was funded by the National Research Foundation of Korea (Grant No. 2021R1A2C2013347 and 2021R1A6C101A442) and the Korea Health Technology R&D Project through the Korea Health Industry Development Institute (KHIDI), funded by the Ministry of Health & Welfare, Republic of Korea (grant number: RS-2024-00344565).

Institutional Review Board Statement: Not applicable.

Informed Consent Statement: Not applicable.

Data Availability Statement: The data presented in this study are available upon request from the corresponding author. The data are not publicly available since they are raw data.

Conflicts of Interest: The author J.-h.J. is affiliated with ACTIVON Co., and the author Y.K. is affiliated with Qmine Co. These companies played no role in the design of the study, in the collection, analysis, or interpretation of data, in the writing of the manuscript, or in the decision to publish the results. The other authors declare no conflicts of interest.

References

1. Abbas, M.; Saeed, F.; Anjum, F.M.; Afzaal, M.; Tufail, T.; Bashir, M.S.; Ishtiaq, A.; Hussain, S.; Suleria, H.A.R. Natural polyphenols: An overview. *Int. J. Food Prop.* **2017**, *20*, 1689–1699. [CrossRef]
2. Sies, H.; Belousov, V.V.; Chandel, N.S.; Davies, M.J.; Jones, D.P.; Mann, G.E.; Murphy, M.P.; Yamamoto, M.; Winterbourn, C. Defining roles of specific reactive oxygen species (ROS) in cell biology and physiology. *Nat. Rev. Mol. Cell Biol.* **2022**, *23*, 499–515. [CrossRef] [PubMed]
3. Murphy, M.P.; Bayir, H.; Belousov, V.; Chang, C.J.; Davies, K.J.A.; Davies, M.J.; Dick, T.P.; Finkel, T.; Forman, H.J.; Janssen-Heininger, Y.; et al. Guidelines for measuring reactive oxygen species and oxidative damage in cells and in vivo. *Nat. Metab.* **2022**, *4*, 651–662. [CrossRef] [PubMed]
4. Hoang, H.T.; Moon, J.-Y.; Lee, Y.-C. Natural antioxidants from plant extracts in skincare cosmetics: Recent applications, challenges and perspectives. *Cosmetics* **2021**, *8*, 106. [CrossRef]
5. Stahl, W.; Sies, H. Antioxidant activity of carotenoids. *Mol. Asp. Med.* **2003**, *24*, 345–351. [CrossRef] [PubMed]
6. Gălbău, C.-Ș.; Irimie, M.; Neculau, A.E.; Dima, L.; da Silva, L.P.; Vârciu, M.; Badea, M. The Potential of Plant Extracts Used in Cosmetic Product Applications—Antioxidants Delivery and Mechanism of Actions. *Antioxidants* **2024**, *13*, 1425. [CrossRef]
7. Kang, S.-H.; Jeon, Y.-D.; Cha, J.-Y.; Hwang, S.-W.; Lee, H.-Y.; Park, M.; Lee, B.-R.; Shin, M.-K.; Kim, S.-J.; Shin, S.-M.; et al. Antioxidant and skin-whitening effects of aerial part of Euphorbia supina Raf. extract. *BMC Complement. Altern. Med.* **2018**, *18*, 256. [CrossRef]
8. Gillbro, J.; Olsson, M. The melanogenesis and mechanisms of skin-lightening agents–existing and new approaches. *Int. J. Cosmet. Sci.* **2011**, *33*, 210–221. [CrossRef]
9. Hwang, K.-S.; Yang, J.Y.; Lee, J.; Lee, Y.-R.; Kim, S.S.; Kim, G.R.; Chae, J.S.; Ahn, J.H.; Shin, D.-S.; Choi, T.-Y.; et al. Anovel anti-melanogenic agent, KDZ-001, inhibits tyrosinase enzymatic activity. *J. Dermatol. Sci.* **2018**, *89*, 165–171. [CrossRef]
10. Kim, K.; Yoon, J.; Lim, K.-M. Syringaresinol Attenuates α-Melanocyte-Stimulating Hormone-Induced Reactive Oxygen Species Generation and Melanogenesis. *Antioxidants* **2024**, *13*, 876. [CrossRef]
11. Boo, Y.C. Human skin lightening efficacy of resveratrol and its analogs: From in vitro studies to cosmetic applications. *Antioxidants* **2019**, *8*, 332. [CrossRef] [PubMed]

12. Marques, C.; Hadjab, F.; Porcello, A.; Lourenço, K.; Scaletta, C.; Abdel-Sayed, P.; Hirt-Burri, N.; Applegate, L.A.; Laurent, A. Mechanistic insights into the multiple functions of niacinamide: Therapeutic implications and cosmeceutical applications in functional skincare products. *Antioxidants* **2024**, *13*, 425. [CrossRef] [PubMed]
13. Wang, Y.; Zhao, H.; Lin, C.; Ren, J.; Zhang, S. Potential Application of Natural Bioactive Compounds as Skin-Whitening Agents: A Review. *Molecules* **2022**, *27*(19), 6577.
14. Lourenço, S.C.; Moldão-Martins, M.; Alves, V.D. Antioxidants of natural plant origins: From sources to food industry applications. *Molecules* **2019**, *24*, 4132. [CrossRef]
15. Spagnol, C.M.; Di Filippo, L.D.; Isaac, V.L.; Correa, M.A.; Salgado, H.R. Caffeic acid in dermatological formulations: In vitro release profile and skin absorption. *Comb. Chem. High Throughput Screen.* **2017**, *20*, 675–681. [CrossRef]
16. Baldioli, M.; Servili, M.; Perretti, G.; Montedoro, G.F. Antioxidant activity of tocopherols phenolic compounds of virgin olive oil. *J. Am. Oil Chem. Soc.* **1996**, *73*, 1589–1593. [CrossRef]
17. Spagnol, C.M.; Zaera, A.M.; Isaac, V.L.B.; Corrêa, M.A.; Salgado, H.R.N. Release permeation profiles of spray-dried chitosan microparticles containing caffeic acid. *Saudi Pharm. J.* **2018**, *26*, 410–415. [CrossRef]
18. Katuwavila, N.P.; Perera, A.D.L.C.; Karunaratne, V.; Amaratunga, G.A.J.; Karunaratne, D.N. Improved Delivery of Caffeic Acid through Liposomal Encapsulation. *J. Nanomater.* **2016**, *2016*, 9701870. [CrossRef]
19. Ray, N.B.; Lam, N.T.; Luc, R.; Bonvino, N.P.; Karagiannis, T.C. Cellular and molecular effects of bioactive phenolic compounds in olives and olive oil. In *Olive and Olive Oil Bioactive Constituents*; Elsevier: Amsterdam, The Netherlands, 2015; pp. 53–91.
20. da Silva, S.P.; da Costa, C.B.L.; de Freitas, A.F.S.; da Silva, J.D.F.; Costa, W.K.; da Silva, W.S.F.L.; Machado, J.C.B.; da Silva, S.M.S.; Ferreira, M.R.A.; Soares, L.A.L.; et al. Saline extract of Portulaca elatior leaves with photoprotective and antioxidant activities does not show acute oral and dermal toxicity in mice. *Toxicol. Res.* **2023**, *39*, 179–190. [CrossRef]
21. Gulcin, I.; Alwasel, S.H. DPPH radical scavenging assay. *Processes* **2023**, *11*, 2248. [CrossRef]
22. Shah, P.; Modi, H.A. Comparative study of DPPH, ABTS and FRAP assays for determination of antioxidant activity. *Int. J. Res. Appl. Sci. Eng. Technol.* **2015**, *3*, 636–641.
23. Ağan, K.; Kaya, S.T.; Ağan, A.F.; Ağyar-Yoldaş, P.; Yoldaş, T.; İkinci-Keleş, A.; Çaprazlı, T.; Arıca, E.; Kekeçoglu, M. Alleviating doxorubicin-induced reproductive toxicity: Protective and androgenic effects of drone larvae on sperm morphology and hormonal balance. *Toxicol. Res.* **2025**, *41*, 149–165. [CrossRef] [PubMed]
24. Kim, B.; Cho, H.-E.; Moon, S.H.; Ahn, H.-J.; Bae, S.; Cho, H.-D.; An, S. Transdermal delivery systems in cosmetics. *Biomed. Dermatol.* **2020**, *4*, 10. [CrossRef]
25. Philipp-Dormston, W.G. Melasma: A Step-by-Step Approach Towards a Multimodal Combination Therapy. *Clin. Cosmet. Investig. Dermatol.* **2024**, *ume17*, 1203–1216. [CrossRef] [PubMed]
26. Levy, C.; Khaled, M.; Fisher, D.E. MITF: Master regulator of melanocyte development and melanoma oncogene. *Trends Mol. Med.* **2006**, *12*, 406–414. [CrossRef]
27. Niu, C.; Aisa, H.A. Upregulation of melanogenesis and tyrosinase activity: Potential agents for vitiligo. *Molecules* **2017**, *22*, 1303. [CrossRef]
28. Liu, Z.; Ren, Z.; Zhang, J.; Chuang, C.-C.; Kandaswamy, E.; Zhou, T.; Zuo, L. Role of ROS and nutritional antioxidants in human diseases. *Front. Physiol.* **2018**, *9*, 360203. [CrossRef]
29. McCubrey, J.A.; LaHair, M.M.; Franklin, R.A. Reactive oxygen species-induced activation of the MAP kinase signaling pathways. *Antioxid. Redox Signal.* **2006**, *8*, 1775–1789. [CrossRef]
30. Matsuzawa, A.; Ichijo, H. Redox control of cell fate by MAP kinase: Physiological roles of ASK1-MAP kinase pathway in stress signaling. *Biochim. Biophys. Acta Gen. Subj.* **2008**, *1780*, 1325–1336. [CrossRef]
31. Kamata, H.; Honda, S.-I.; Maeda, S.; Chang, L.; Hirata, H.; Karin, M. Reactive oxygen species promote TNFα-induced death and sustained JNK activation by inhibiting MAP kinase phosphatases. *Cell* **2005**, *120*, 649–661. [CrossRef]
32. Hong, Y.; Boiti, A.; Vallone, D.; Foulkes, N.S. Reactive oxygen species signaling and oxidative stress: Transcriptional regulation and evolution. *Antioxidants* **2024**, *13*, 312. [CrossRef] [PubMed]
33. Lee, J.H.; Lee, J.; Dej-Adisai, S.; Hwang, J.S. Piperine Regulates Melanogenesis through ERKActivation Proteasomal Degradation of, MITF. *Biomol. Ther.* **2025**, *33*, 408–414. [CrossRef] [PubMed]
34. Ko, G.-A.; Cho, S.K. Phytol suppresses melanogenesis through proteasomal degradation of MITF via the ROS-ERK signaling pathway. *Chem. Biol. Interact.* **2018**, *286*, 132–140. [CrossRef] [PubMed]

antioxidants

MDPI

Article

TEMPO-Oxidized Spruce Galactoglucomannan–Biopolymer with Enhanced Antioxidant Activity and Selective Heavy-Metal Sorption

Vladislav A. Ionin [1,2], Yuriy N. Malyar [1,2,*], Valentina S. Borovkova [1,2], Dmitriy V. Zimonin [1,2] and Aleksandr S. Kazachenko [1,2,3]

1 Institute of Chemistry and Chemical Technology, Krasnoyarsk Science Center, Siberian Branch Russian Academy of Sciences, Akademgorodok 50/24, Krasnoyarsk 660036, Russia; ionin.va@icct.krasn.ru (V.A.I.); bing0015@mail.ru (V.S.B.); zimonind89@mail.ru (D.V.Z.); askazachenko@sfu-kras.ru (A.S.K.)

2 School of Non-Ferrous Metals, Siberian Federal University, pr. Svobodny 79, Krasnoyarsk 660041, Russia

3 Institute of Chemical Technologies, Reshetnev Siberian State University of Science and Technology, Mira St 82, Krasnoyarsk 660049, Russia

* Correspondence: yumalyar@gmail.com; Tel.: +7-9082065517

Abstract: This study examines galactoglucomannan, a well-studied biopolymer isolated from Siberian spruce (*Picea obovata* Ledeb). Due to its structure, abundant with hydroxyl groups, galactoglucomannan's properties, such as heavy-metal ion affinity, are considered to be mediocre. Nevertheless, there are various ways to enhance its functionality via oxidative TEMPO/NaBr/NaOCl processing. This work is concerned with the determination of the oxidation effect on the structure and performance properties, such as thermal decomposition behavior, antioxidant activity, and selective heavy-metal sorption. In the results, TEMPO-oxidized galactoglucomannan yields vary in the range of 78.3 ± 6.4 wt.%. The carboxylate group in the oxidized derivative represents up to 0.084 g/1 g of the sample. According to antioxidant activity tests, the oxidized galactoglucomannan exceeds the initial sample in terms of hydroxyl radical scavenging ability. The spectral characteristics of the initial and oxidized galactoglucomannan samples reveal the differences in absorption units (1725, 1610, and 1371 cm^{-1}). The preservation of the polymeric structure was confirmed by the gel permeation chromatography analysis results. The heavy-metal ion capacity of galactoglucomannan is higher for the oxidized derivative, which demonstrated Cd^{2+}, Fe^{2+}, Cu^{2+}, and Pb^{2+} adsorption values of 166.8 mg/g, 142.8 mg/g, 150.0 mg/g, and 199.2 mg/g, accordingly. The obtained result of the competitive heavy-metal ion adsorption of oxidized galactoglucomannan also exceeds its initial form, as characterized by its summary 143.1 mg/g capacity.

Keywords: galactoglucomannan; oxidation; 2,2,6,6-tetramethylpiperidin-1-oxyl; heavy-metal adsorption; 1,1-diphenyl-2-picrylhydrazyl; hydroxyl radicals

1. Introduction

In nature, heavy metals (HMs) such as cadmium, lead, copper, or iron occur at low concentrations depending on the source of the matter, or due to the consequences of weathering, volcanic activities, etc. [1–3]. Nevertheless, with the increasing role and scale of anthropogenic activities around the world, which produce approximately 70% of global waste [3], a redistribution of resources takes place, leading to local increases in pollutant concentrations, which result in the contamination of soils used by humanity to raise crops [4]. Even low concentrations of heavy metals easily absorbed by plant cells

https://doi.org/10.3390/antiox14050569

make an impact on plant growth and development [5]. Therefore, plants, in the course of evolution, developed a defensive mechanism, aiding to avoid the negative effects of heavy metals, by the immobilization or accumulation of heavy metals in cells by forming chelating complexes or by binding with polysaccharides and other biologically active compounds [6].

The recent achievements in the chemical modification of polysaccharides are dedicated to improving functional properties via the mild, non-aggressive intrinsic tuning structure of macromolecules with a varied degree of complexity [7]. Previous successes in the synthesis of sulfated polysaccharide derivatives with a 12–13% sulfur content could have been promising due to their pronounced high antioxidant abilities [8,9]. The preservation of inherited anticoagulant, antimicrobial, and antitumor effects on the physiological functions influenced by the initial polysaccharides' biological properties has been described earlier [10–13]. Despite sulfated polysaccharides having some favorable properties due to their biocompatibility, antioxidant activity, and reduced toxicity [14], the sulfation commonly employs harsh conditions for the reaction to take place, affecting the polysaccharides' degradation due to the decay of vulnerable linkages [15,16]. Upon such modifications, the sulfated polysaccharides' physical and biological properties are difficult to manage because they undergo alterations [17].

On the other hand, one of the most promising directions for modification is the regioselective oxidation of the primary hydroxyls of polysaccharides using 2,2,6,6-tetramethylpiperidin-1-oxyl (TEMPO) [18–20], which could enhance the suitability of the properties of polysaccharides for various applications, such as water solubility and emulsification abilities in synthesis [21,22]. In addition to the abovementioned aspects, as a result of TEMPO-catalyzed oxidation the neutral basis of the original polysaccharide is altered to a highly negatively charged state, due to the conversion of the primary hydroxyls into carboxylates, impacting the affinity with HM ions, which could be subsequently used in the development of novel adsorptive functional materials, such as hydrogels and composite films [23,24].

Here, we focus on one of the common types of hemicelluloses obtained from softwoods—galactoglucomannan (GGM). This neutral polysaccharide's structure is represented mainly by β-D-mannopyranoses (Man*p*), commonly substituted by α-D-glucopyranosic (Glc*p*) and β-D-galactopyranosic (Gal*p*) units, linked through β-(1→4) glycosidic linkages [25]. GGM's solubility and chemical activity depend on different ratios of substitution by these units, affecting its accessibility and high reactivity.

Thus, this study focused on the oxidation of galactoglucomannan obtained from Siberian spruce (*Picea obovata* Ledeb) wood via a TEMPO-catalyzed medium. The impact of the oxidation on the GGM's molecular weight and process dynamics was evaluated by the gel permeation chromatography (GPC) method. The dynamics of oxidation were also followed by spectrophotometric and titration methods of determining the formation of uronic acids. The structural changes in the obtained samples were registered by Fourier transform infrared spectroscopy (FTIR) and nuclear magnetic resonance (NMR) spectroscopy. Important biological properties such as antioxidant activity (AOA) were determined using a model of free and hydroxyl radical compounds. Finally, the oxidized GGM's derivative adsorption capacity of heavy-metal ions (Fe^{2+}, Cu^{2+}, Cd^{2+}, or Pb^{2+}) was examined to propose further applications.

2. Materials and Methods

2.1. Materials

In course of this study, the galactoglucomannan was isolated from Siberian spuce (*Picea obovata* Ledeb) wood. TEMPO, 1,1-diphenyl-2-picrylhydrazyl (DPPH), and carbazole (analytical grade) were purchased from Sigma-Aldrich (St. Louis, MO, USA). The other chemical reagents (acetic acid, hydrogen peroxide, solvents, and metal salts) used in this study were purchased from Khimreaktivsnab (AO "Khimreaktivsnab", Ufa, Russian Federation).

2.2. Isolation of Spruce Galactomannan and Its Monosaccharide Structure

In this study, as an initial material, isolated spruce galactoglucomannan was used for the following oxidation. Isolation, in a similar manner to that carried out in an earlier study [26], was performed as follows: 10 g of air-dried spruce wood sawdust was loaded into a three-necked glass flask equipped with a reflux condenser and an overhead stirrer. Then, a mixture containing acetic acid (30 wt.%), hydrogen peroxide (6 wt.%), and distilled water (liquid/solid ratio—15) was added, heated up to 100 °C and stirred at 300 rpm for 3 h. The obtained hot solution was separated from the wood residue via filtration using a Buchner funnel. After the cooling, the solution was concentrated on a rotary evaporator to a 1/5 volume. The galactoglucomannan was precipitated with a fourfold greater volume of ice-cold ethanol (96%) under slow stirring. After the precipitation, the obtained mixture was kept in ice-cold conditions for 12 h. The obtained precipitate further separated by filtration and were then dried in an "Iney-6" (Iney, Moscow, Russia) freeze-dryer. After the drying, the precipitate was mechanically ground in a mortar. The resulting medium yields of the GGM consist of up to 10 wt.% from the raw biomass material.

The monosaccharide composition of the initial GGM was determined by following a commonly used method [27,28], conducting the sample hydrolysis with the following modification to obtain trimethylsilyl derivatives. These derivatives were further analyzed by gas chromatography (GC). A VARIAN-450 GC-system equipped with a flame ionization detector and a VF-624ms (30 × 0.32 mm) capillary column were used. The analysis of the derivatives was performed in helium as a carrier gas, through an injector heated up to 250 °C, with an initial column temperature of 50 °C (5 min), rising up to 180 °C at a rate of 10 °C/min with a following retention at 180 °C for 37 min. The peaks on the chromatogram were identified based on an internal standard sample—the sorbitol, in accordance with the retention times of the monosaccharide's tautomeric forms.

2.3. Galactoglucomannan TEMPO-Catalyzed Oxidation

A TEMPO-catalyzed oxidation of the isolated galactoglucomannan was conducted in a stirred three-necked flask using techniques described in [18,29], with some improvements. In a typical procedure, 0.8 g of NaBr and 0.04 g of TEMPO were added to 2 g of the air-dried GGM dissolved in 520 mL of distilled water under constant stirring (400 rpm). The reaction started simultaneously with the gradual addition of 80 mL of a freshly prepared 15 wt.% NaOCl solution, followed by a mixed pH adjustment to 9.5 via a 1 M HCl solution. The reaction mixture temperature was constantly cooled at 5 °C by an ice bath and pH was maintained at 9.5 via the gradual addition of a 0.1 M NaOH solution every 15 min. Consumption of the NaOH solution was permanently monitored as a function of time. Once the consumption remains unchanged for more than 3 periods of time in a row, 0.3 g of $NaBH_4$ was added into the reaction mixture to reduce the carbonyl byproducts and stirred for another 30 min. After that, to quench the oxidation, the reaction medium pH was adjusted to 3 by a 1 M HCl solution and 60 mL of ethanol, with the addition of 0.06 g of NaCl. An obtained aqueous solution was evaporated approximately to a volume of 80 mL under vacuum conditions and further dialyzed using a 46 mm width cut-off dialysis bag, MF-5030-46 (MFPI, Seguin, TX, USA), with a 3.5 kg/mol pore size for 24 h to transfer the low-molecular-weight byproducts and salt residues into deionized water. Finally, the obtained solution was then poured into a Petri dish and dried at 60 °C in an oven to a constant weight.

The GGM's oxidation was controlled by portions of the reaction medium analysis using spectrophotometric and titration methods. To evaluate the impact of oxidation, air-dried samples of the initial and oxidized GGM derivative (GGM-T) were studied via FTIR, NMR spectroscopy, and GPC. AOA assays were performed using compounds modeling free DPPH and hydroxyl radicals. The ability of the GGM samples to adsorb heavy-metal ions was determined using ICP-MS.

2.4. Galactoglucomannan Hydrolysis and Spectrophotometric Determination of the Formed Uronic Acids

The evaluation of the uronic acids formed by the strong acid hydrolysis of an oxidizing GGM's reaction media was performed based on the modified carbazole method [30] using an Ecoview UV 6900 spectrophotometer (Shanghai Mapada Instruments Co., Ltd., Shanghai, China) with a created calibration curve similar to that in a previous study [18]. The analysis procedure was conducted as follows:

A 1 mL sample of the oxidizing GGM reaction medium was placed in a test tube, after which 6 mL of a borate-sulfuric acid agent (0.0125 M solution of sodium tetraborate in concentrated sulfuric acid) was slowly added under constantly ice-cold conditions. After that, the sample was vortexed, cooled, and 0.5 mL of a 0.015 wt.% solution of the carbazole-ethanol reagent was added, after which it was thoroughly mixed again and incubated at 80 °C for 5–10 min until a stable color change occurred. Then, the tube was cooled to room temperature, followed by the spectrophotometric measurement of the sample (λ = 530 nm), zeroed on distilled water as a reference solution.

2.5. Determination of the Initial and Oxidized Galactoglucomannan Carboxylic Content with a Potentiometric Titration

The determination of the initial and TEMPO-oxidized GGM carboxyl groups was performed based on a potentiometric titration method [31], as follows: Firstly, 5 mL of 0.1 M NaCl and 1 mL of 0.01 M HCl aqueous solutions were added to 0.3 g of the initial and oxidized GGM samples, dissolved in 100 mL of deionized water. The obtained mixtures were continuously stirred at 300 rpm for 6 h at room temperature until a well-dispersed solution was obtained. Furthermore, it was titrated with a 0.01 M NaOH solution at a 0.1 mL/min rate to adjust the pH to 10, performed with a Radelkis pH-stat (Radelkis, Budapest, Hungary). To evaluate the carboxylate content in the studied samples, the following Dalton's law equation was used:

$$C_1V_1 = C_2V_2, \quad (1)$$

where V_1 and V_2—volumes of the titrated substance and titrant, respectively, mL; C_1—concentration of the carboxylate groups, mM/L; and C_2—the alkali concentration, mM/L.

2.6. Fourier Transform Infrared Spectroscopy

The IR spectra were recorded by using a Shimadzu IRTracer-100 FTIR spectrometer (Shimadzu, Kyoto, Japan). The recorded spectrum of pure ZnSe crystal absorption is hereinafter referred to as a background spectrum. The initial and oxidized GGM samples were placed on a crystal, completely covering its surface, and the sample's IR spectra were recorded in the range of 4000–630 cm^{-1} with a 4 cm^{-1} shooting resolution and 32 repeating scans. The resulting spectrum is the difference between the sample and a background spectra, which were processed further by using the OPUS 7.5 software package (Bruker BioSpin, Rheinstetten, Germany).

2.7. Nuclear Magnetic Resonance Measurements

The GGM sample's chemical structure was estimated by recording the ^1H and ^{13}C nuclear magnetic resonance spectra by using a Bruker Avance III 600 spectrometer (Bruker BioSpin, Rheinstetten, Germany). Before the analysis, both the initial and oxidized samples were air-dried, ground, and completely dissolved in the D_2O, and placed into a 5 mm NMR tube at room temperature.

2.8. Gel Permeation Chromatography

To determine the initial and oxidized GGM samples' molecular weight characteristics (average molecular weights and polydispersity index (PD)), a multi-detector GPC/SEC Agilent 1260 Infinity II system (Agilent Technologies, Santa Clara, CA, USA) was used. The system was equipped with Agilent PL Aquagel-OH Mixed-M and Agilent PL Aquagel-OH-30 columns, using a 0.1 M $NaNO_3$ solution combined with 250 ppm NaN_3 dissolved in deionized water as a mobile phase. Before the analysis, the samples were dissolved in a mobile phase (5 mg/mL) and poured through a 0.45 μm Agilent PES membrane filter (Millipore, Burlington, MA, USA). Then, 100 μL of the studied sample was analyzed in an eluent flow with a 1 mL/min rate.

The calibration was performed using an Agilent EasiVial PEG/PEO polyethylene glycol standard (Agilent, Santa Clara, CA, USA). All of the collected data were processed by Agilent GPC/SEC MDS software version 2.2.

2.9. Thermogravimetric Analysis

A thermogravimetric analysis (TGA) of the initial and oxidized GGM samples was performed using a NETZSCH STA 449 F1 Jupiter simultaneous thermal analysis instrument (NETZSCH STA 449 F1 Jupiter instrument, Netzsch, Selb, Germany) in an argon medium. The sample was placed in an Al_2O_3 cylindrical crucible with a perforated cover and analysis started with a heating rate of 10 °C/min in a 30–700 °C temperature range. The protective and blowout gas flow rates were 20 and 50 mL per minute, respectively. An empty corundum crucible with a cover was used as a reference. The calibration of the instrument was conducted according to the specifications, with the use of reference substances supplied by the instrument data. The sample weight for the analysis was determined on a Sartorius BP121S analytical lab-scale digital balance. The measurement data were processed using NETZSCH Proteus Thermal Analysis 5.1.0 software supplied by the instrument.

2.10. Antioxidant Activity

2.10.1. DPPH Radical Scavenging Assay

The ability of the initial and oxidized GGM samples to scavenge DPPH radicals was determined according to the method described in previous studies [18,26], with some improvements. The absorbance of the samples was measured at 517 nm against a blank, using an Ecoview UV 6900 spectrophotometer (Shanghai Mapada Instruments Co., Ltd., Shanghai, China).

The DPPH radical scavenging ability of the initial and oxidized GGM samples was determined using the following equation:

$$\text{DPPH Radical Scavenging Ability } (\%) = \left(1 - \frac{A_S - A_B}{A_C}\right) * 100\%, \quad (2)$$

where A_C represents the absorbance of the DPPH solution without a sample, A_S is the absorbance of the test sample mixed with the DPPH solution, and A_B is the absorbance of the sample without the DPPH solution.

2.10.2. Hydroxyl Radical Scavenging Assay

The ability of the initial and oxidized GGM samples to scavenge the hydroxyl radicals was determined as previously reported [18,26]. The absorbance of the mixture was measured at 510 nm against the blank.

The capability to scavenge the hydroxyl radical was calculated as follows:

$$\text{Hydroxyl Radical Scavenging Ability } (\%) = \left(1 - \frac{A_S - A_B}{A_C}\right) * 100\% \quad (3)$$

where A_C is the absorbance of the mixture without the sample, A_S is the absorbance of the test sample mixed with a reaction solution, and A_B is the absorbance of the sample without a salicylic acid solution.

2.11. *Heavy-Metal Affinity Tests*

In a typical procedure, 20 mg of the dried initial or oxidized GGM sample (25 mg) was added to 10 mL of the aqueous solutions containing heavy-metal ions (Fe^{2+}, Cu^{2+}, Cd^{2+}, or Pb^{2+}) alone with a 500 mg/L concentration. The adsorption process was carried out in an oscillator at room temperature and 300 rpm for 1 h. Then, 20 mL of ice-cold ethanol was added to the aqueous solution. This was followed by the mixing and removal of the solid precipitate via centrifugation and decantation. The obtained solution was poured through a 0.45 μm PTFE membrane. Finally, ethanol was evaporated, and the remaining heavy-metal ion concentration in the solution was determined by using an Agilent 7900 ICP-MS spectrometer (Agilent Technologies, Santa Clara, CA, USA) with a He collision cell to eliminate atomic interference. The solutions without an additional dilution were injected into argon plasma as a dry aerosol. The analysis was performed using a SemiQuant mode with atomic-scale calibration with a tuning solution. Sensitivity was up to 5 ng/L.

The adsorption degree (AD) of the initial and oxidized GGM samples was calculated as follows:

$$AD\ (\%) = (C_{in} - C_f)/C_{in} \times 100, \quad (4)$$

where C_{in} (mg/L) and C_f (mg/L) are the initial and final concentrations of the investigated heavy-metal ions.

The heavy-metal ion adsorption capacity (AC) of both the initial and oxidized GGM samples (mg/g) were calculated using the following equation:

$$AC = (C_{in} - C_f)V/m, \quad (5)$$

where C_{in} (mg/L) and C_f (mg/L) are the initial and final concentrations of the investigated heavy-metal ions in the solutions; V is the volume of the heavy-metal ion solution (L); and m is the mass of the tested GGM sample.

The competitive adsorption experiments were conducted using the same amount of the initial or oxidized GGM sample (25 mg), which was added to 8 mL of an aqueous solution containing a mixture of Fe^{2+}, Cu^{2+}, Cd^{2+}, or Pb^{2+} metal ions. This mixture contained 2 mL of each HM stock solution with a concentration of 500 mg/L, resulting in their initial concentration in the competitive experiment of up to 125 mg/L. The other conditions of the tests remained unchanged and the AC was determined to be similar to that in the abovementioned equation.

The specific metal selectivity (%), defined as the ratio of the specific metal concentration that was adsorbed during the competitive experiment to the concentration of all metals adsorbed [24] by the initial and oxidized GGM samples. For example, the specific metal selectivity for one of the investigated heavy-metal ions was calculated as follows:

$$\text{Selectivity (\%)} = (C_{in}\,HM_1 - C_f\,HM_1)/((C_{in}\,HM_1 - C_f\,HM_1) + (C_{in}\,HM_2 - C_f\,HM_2) + (C_{in}\,HM_3 - C_f\,HM_3) + (C_{in}\,HM_4 - C_f\,HM_4)) \quad (6)$$

where C_{in} (mg/L) and C_f (mg/L) are the initial and final concentrations of the investigated heavy-metal ions.

2.12. Statistical Processing of Data

According to the standard practices and analytical methods, all of the measurements of the potentiometric titration, spectrophotometric evaluation of uronic acids, determination of the antioxidant activity, and heavy-metal affinity tests were conducted in triplicates. The collected data were statistically processed by using Statgraphics Centurion XVI, DOE block (Design of Experiment) software (v. 16.2.04, Statpoint Technologies Inc., Warrenton, VG, USA).

3. Results

3.1. Galactoglucomannan Oxidation Dynamics

The role of the heterogeneous nature of an isolated GGM was investigated from TEMPO-catalyzed oxidation mechanism perspectives. Analysis of the GGM monosaccharide composition, according to a proven method [27,28], revealed the presence of large amounts of mannose, galactose, and glucose, with some degree of arabinose and xylose in a ratio of 5:3:2:1:1. Trace amounts of the xylose and arabinose could be caused by the presence of an arabinogalactan fraction, meaning that not all of the Gal*p* connected with the Man*p* [32]. An important structural feature was the mannose and glucose residues partially substituted by the acetyl groups—one per three–four hexose units, on average. In summary, according to the registered GC analysis data, a proposed structure of the GGM elementary units is represented in Figure 1.

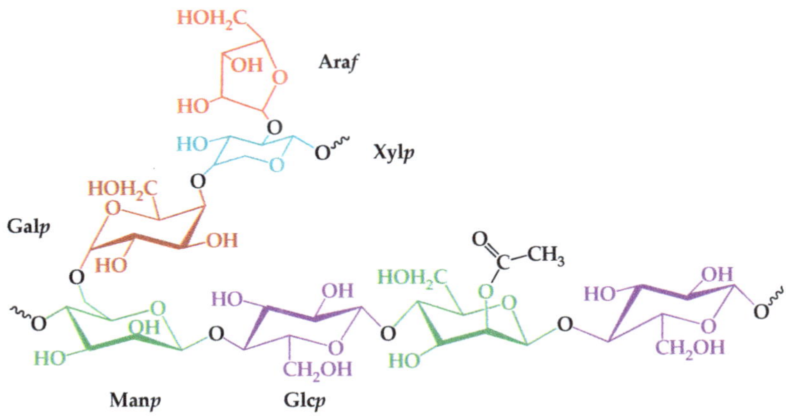

Figure 1. The proposed structure of isolated galactoglucomannan elementary units.

Considering the moment of the addition of the total volume of the NaClO solution as a starting point for the reaction, a pH-declining rate of oxidation, catalyzed by NaOH consumption, reflects its kinetics. It is generally accepted that the expense of the NaOH could be partially considered an indirect and effective method for comparing the TEMPO-oxidation rates [33]. In addition, the physicochemical properties of GGM-T, as a polysaccharide derivative, depend on the amount of the charged groups caused by carbohydrates with carboxylic acids (uronic acids) [34], which could be evaluated by hydrolysis.

Figure 2 depicts the kinetics of 0.1 M NaOH consumption and the uronic acid content, formed by the strong acid hydrolysis of the reaction media in the course of the TEMPO-catalyzed oxidation.

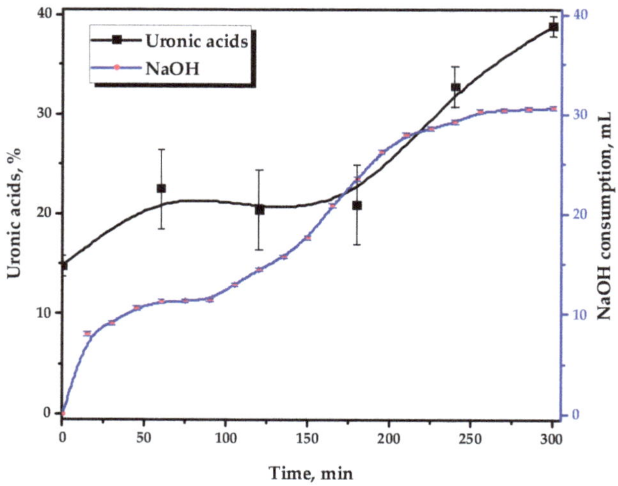

Figure 2. The 0.1 M NaOH consumption by the reaction media and its uronic acid content during the TEMPO-catalyzed oxidation.

It has been demonstrated that GGM oxidation occurs in stages, with variable dynamics. This could also be indirectly traced by the performed uronic acid content determination in the reaction medium hydrolysates using the modified "carbazole" method [30]. It has been revealed that during the oxidation of GGM that the uronic acid content in its reaction medium hydrolysates increases from 14.8 to 38.9%, which is achieved as a result of a 5 h reaction.

However, while assessing the uronic acid content in the reaction media of the oxidizing GGM samples using the modified "carbazole" method, atypical changes in the hydrolysates' color is observed. During the starting periods until 3 h of reaction, the hydrolysates are characterized by a dark-green color instead of violet/purple (Figure 3), typical for carbazole–uronic acid complexes. This fact probably could be a reason of a significant error in the spectrophotometric determination of the uronic acid content in the GGM oxidizing reaction media hydrolysates.

It should be mentioned that this fact is only important for understanding the underlying mechanisms of the TEMPO-catalyzed GGM oxidation. Once the third hour of the reaction is reached, this anomaly is not observed and has no visible effect on the resulting product, so the investigated hydrolysates acquire a stable violet color. Obviously, the nature of this anomaly is most likely related to the specific structural features of the initial GGM: NaOH consumption increases at first and then reaches a plateau, observed in a range of 60–90 min.

Primarily, this could be explained by the oxidative cleavage during the TEMPO-catalyzed oxidation of the arabinoxylose (lacking primary hydroxyl groups) side chains. Its presence adversely affects carboxylate group formation by limiting chemical accessibility [33,35,36]. After 90 min of the oxidation, NaOH consumption by the reaction media continues and gradually reaches a novel plateau after 3–4 h of the reaction, which remains practically unchanged past this point.

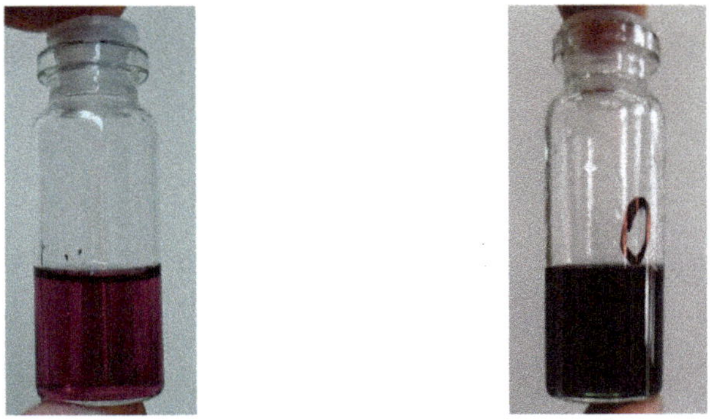

1 2

Figure 3. The reaction media hydrolysates, mixed with a carbazole solution: (**1**) "normal" type of the carbazole–uronic acid complex; (**2**) an atypical color change in the carbazole–uronic acid complex.

Thus, the proposed route of the TEMPO-catalyzed oxidation of the isolated galactoglucomannan is shown in Figure 4.

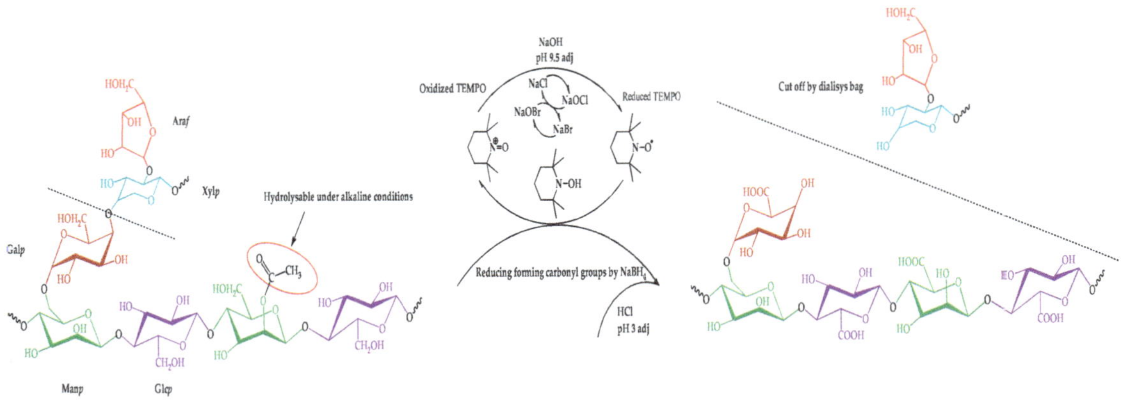

Figure 4. Scheme of the TEMPO-catalyzed oxidation of galactoglucomannan.

The resulting mass yields of the product after the separation of the byproducts and inorganic salts by a cut-off dialysis bag vary in a range of 78.3 ± 6.4 wt.%. The obtained yields remain challenging due to the heterogeneous nature of the initially isolated galactoglucomannan.

3.2. Determination of Carboxylic Acid Group Content

As a result of the TEMPO-catalyzed GGM oxidation, the formation of carboxylic acid groups on the polysaccharide is expected, which could be estimated by potentiometric titration, expressed by the curve (Figure 5).

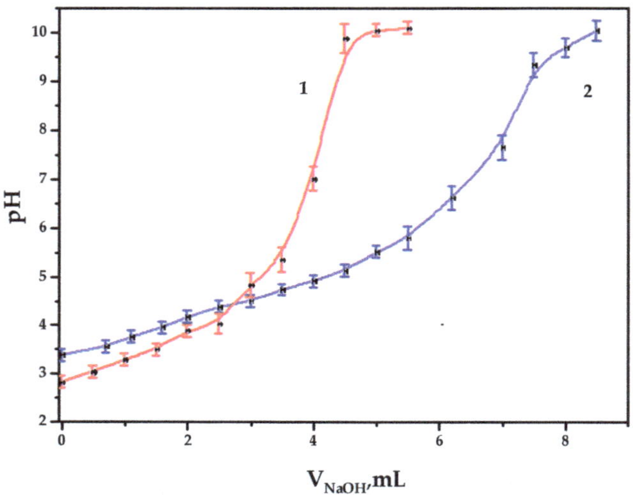

Figure 5. The potentiometric titration curves of the galactoglucomannan samples: (**1**) GGM initial; (**2**) GGM-T.

As it could be observed, 4.0 and 6.7 mL of NaOH were spent on the titration of GGM and GGM-T samples, respectively. Since the alkali could be spent not only on reaction with carboxyl groups but also on the neutralization of pre-added hydrochloric acid, we therefore subtract 1 mL from V_2, which results in 3.0 and 5.7 mL of alkali consumption by the sample. Consequently, the concentration of the acid groups in the GGM and GGM-T solutions in terms of the monobasic acid was 0.30 ± 0.01 and 0.57 ± 0.02 mmol/L, respectively. In terms of the carboxylate group content in 1 g of the studied GGM or GGM-T sample polysaccharide, COO- represents up to 0.044 ± 0.001 and 0.084 ± 0.002 g, respectively (4.4 ± 0.1 and 8.4 ± 0.2 wt.%).

3.3. FTIR Spectra of Initial and Oxidized GGM Samples

The FTIR spectra of the GGM samples represented in Figure 6 contain all of the bands characterizing polysaccharides, in which the following specific features should be distinguished. A detailed assignment of the characteristic absorption bands is given in Table S1.

As was expected, the IR spectra depict characteristic broad absorption bands in the region of 3400 cm^{-1}, commonly associated with the stretching vibrations of the O-H bonds. The 2900 cm^{-1} region absorption bands could be attributed to the stretching vibrations of the aliphatic -CH_2 groups [37] connected with the primary alcohols. Its intensity changes insignificantly in the course of the TEMPO-catalyzed GGM oxidation.

The observed intensity perseverance of bands in the ~1725 cm^{-1} region corresponds to vibrations of the alkyl esterified carboxyl groups [38], typical both for the initial and oxidized GGM. In fact, this region could possibly be assigned to the C=O stretching vibration and the C- due to acetyl partial substitution of the mannose and glucose residues in the initially isolated sample. It has already been reported that spruce wood galactoglucomannan is characterized by the high degree of acetylation [39,40]. Nevertheless, the presence of C=O acetyl stretching in the oxidized GGM-T sample was completely excluded by the $NaBH_4$ specific reduction, affecting only the carbonyl, not the carboxyl groups.

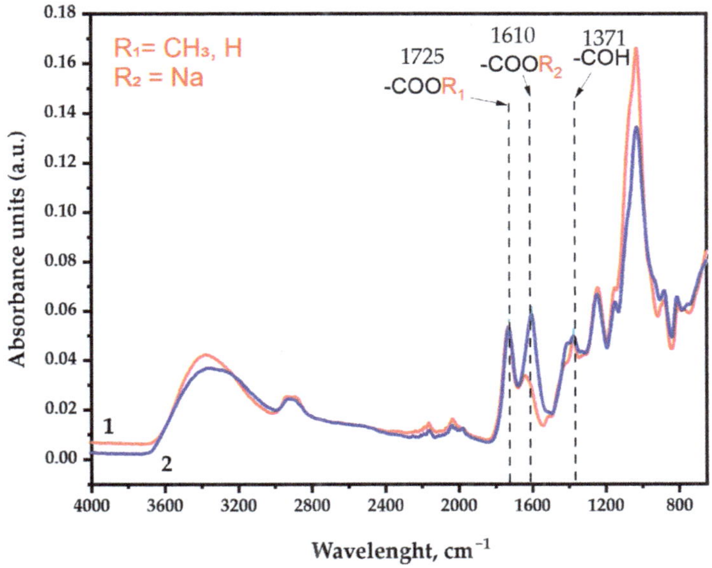

Figure 6. The FTIR spectra of the galactoglucomannan sample absorbance units: (**1**) GGM initial; (**2**) GGM-T.

The most important difference between the initial and oxidized GGM was the significant increase in the band intensity in the ~1610 cm^{-1} region, which was attributed to the asymmetric vibration of the formed -COONa groups as a result of the suggested selective oxidation [37,41] of primary hydroxyl groups of Man*p*/Glc*p*/Gal*p* residues in the initial GGM. Moreover, the a.u. at this region could also be connected with the formation of intra- and intermolecular hydrogen bonds with water [37,42], which also indicates the high presence of carboxylic acid salts in the resulting product.

The ~1370 cm^{-1} region is the "fingerprint" zone for all plant-type polysaccharides. The mutual presence of the a.u. in this region is related to the C–OH deformation vibrations with the O–C–O bonds symmetrical vibrations caused by the high-uronic-acid-carboxylate group's contribution [42]. The uronic acids are commonly present in the isolated spruce GGM structure in trace amounts, and, as could be observed, visually transform in the case of oxidized GGM.

The absorption band in the 1250 cm^{-1} region could be associated with a variety of modeling compounds containing the C–H bonds. However, the most probable are the C–H planar deformation vibrations of the alkyl functional groups [37,42], obtained as a result of precipitation at the end of the reaction via quenching with an ethanol.

Wide absorption bands in the 1150–1040 cm^{-1} region indicate the C-O-C stretching vibration presence characterizing the glycosidic bonds, inherent for GGM units [43]. As was already mentioned above, these functional groups could undergo rearrangement due to the partial arabinoxylose cleavage in a polar medium during the TEMPO-catalyzed oxidation. Obviously, such decay lowers the intensity of the a.u. in this region for the oxidized GGM. At the same time, the a.u. in the regions of 900 cm^{-1} and 780 cm^{-1} corresponds to the signals of stretching vibrations characterizing the nature of β-galactopyranose and β-glucopyranose bonds [37,44,45]. These regions are common for both the initial and oxidized GGM, witnessing mild oxidation take over with the preservation of the C-O-C bonds [37] crosslinking the "skeleton" of the polysaccharide.

3.4. NMR Spectra of Initial and Oxidized GGM Samples

Still, as a result of GGM isolation from spruce sawdust via oxidative delignification with further ethanol precipitation, side processes transform the functional groups of its terminal units. The subsequent TEMPO-catalyzed oxidation of GGM mainly affects the hexose C-6 groups, converting the primary hydroxyl functional groups into carboxyl groups (Figure 4). Moreover, the quenching of the TEMPO-catalyzed oxidation of the GGM using a mineral acid could also provoke changes in the resulting product. Thus, to evaluate the changes in the structure of the initial and oxidized GGM functional groups, NMR spectroscopy was performed (Figure 7).

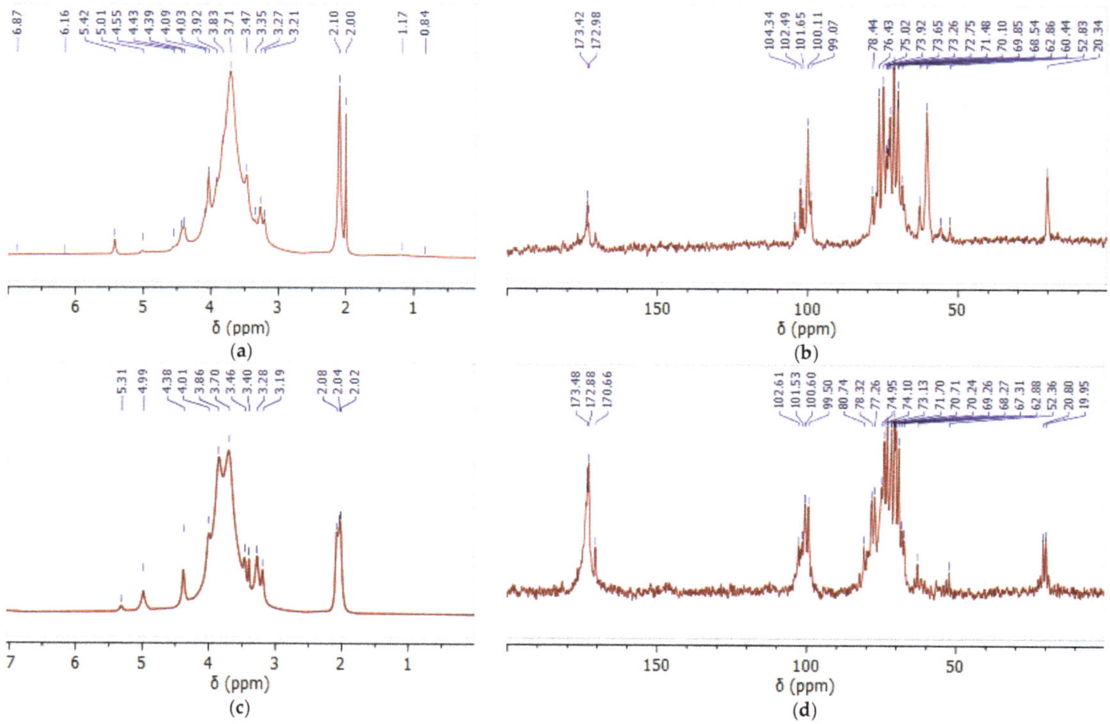

Figure 7. The ^1H (**a**,**c**) and ^{13}C (**b**,**d**) NMR spectra of initial (**a**,**b**) and oxidized (**c**,**d**) GGM recorded in D_2O.

In the ^1H NMR spectrum (Figure 7a) of the initial GGM, the main proton signals characteristic for polysaccharides of this type were determined (H-2, 3.7 ppm; H-3, 3.9 ppm; H-4, 4.1 ppm; and H-5, 4.4 ppm) [46]. The hydrogen atoms, H-1, of the galactosyl and mannosyl residues are separated in the spectrum, and correspond to the signals at 5.3 and 5.0 ppm, respectively [47].

Similar results are observed in the ^{13}C spectra, where the main carbon atoms of the polysaccharide were identified (Figure 7b). The significant differences in the proton spectra of the samples are observed in the 20 ppm region (Figure 7a,c). In the initial GGM the two bands are observed, probably corresponding to the methoxyl and ethoxyl protons, indicating a carboxyl group's esterification. After the TEMPO-catalyzed oxidation, these peaks are preserved, but their intensity lowers, obviously due to the hydrolysis of the ester bonds.

Moreover, this was also confirmed by the signal intensity changes at 60.44 ppm and 62.86 ppm, characterizing the methoxyl and ethoxyl groups' presence, respectively. It is obvious that after the TEMPO-catalyzed oxidation its intensity significantly decreases, confirming the hypothesis of partial hydrolysis of the bonds (Figure 7b,d). At the same time, intense new signals appear in the 170–174 ppm region, corresponding to the C6 units of the uronic acids (mannuronic, galacturonic, and glucuronic) in esterified and protonated form [48]. The overlapping signals at δ 99–104 ppm in the anomeric region were attributed to the C-1 units of hexoses in the initial GGM, and after the TEMPO-catalyzed oxidation this region changes due to the increase in uronic acid signals.

3.5. Molecular Weight Distribution Characteristics of Initial and Oxidized GGM

Polysaccharides' molecular weight characteristics are crucial, determining many properties of such complexed systems. The observation of the GGM TEMPO-catalyzed oxidation (Figure 8) kinetics from the point of view of molecular weight changes on a polymer basis allows for controlling the reaction course. Moreover, its evaluation aids in selecting the experimental conditions that preserve the initial polymer structure into the future.

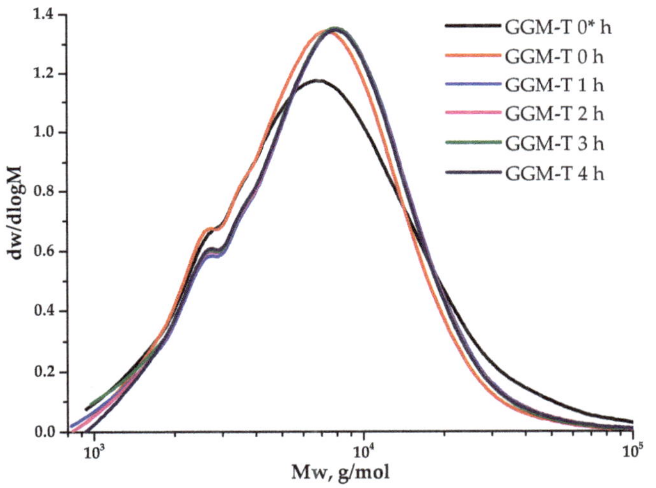

Figure 8. The dynamic molecular weight distribution of the reaction medium of the TEMPO-catalyzed GGM. *—before NaOCl addition.

As could be observed, the initial GGM is characterized by a monomodal MWD curve profile with Mw of 11,203 g/mol and polydispersity of 2.43. The analysis of the reaction medium, containing TEMPO and NaBr, immediately after adding NaOCl to the mixture (GGM-T 0 h), reveals the transformation of polymer MWD. It is mainly connected with the formation of an intermediate complex compound. At this stage, an increase in the proportion of the low-molecular-weight component is observed, resulting in an overall Mw decrease to 8744 g/mol and polydispersity to 1.59 (Table 1). After 1 h of the TEMPO-catalyzed oxidation, the product MWD profile remains virtually unchanged, but shifts toward the higher-molecular-weight region. Such a change may indicate the polymer structure preservation, inherited from the initial GGM. At the same time, an increase in mass subsequently occurs due to the selective hydroxyl groups' oxidation. Nevertheless, the oxidation duration from 1 to 4 h does not affect the product MWD, possibly connected with the oxidation already completed in the first hour, which partially corresponds with the data, described in previous sections.

Table 1. Molecular weight characteristics of the GGM samples.

Scheme	M_p (g/mol)	M_n (g/mol)	M_w (g/mol)	PD
GGM initial	7376	4611	11,203	2.43
GGM-T pH = 3	8945	6716	11,407	1.70
GGM-T 0 * h	6692	5711	10,595	1.86
GGM-T 0 h	7451	5510	8744	1.59
GGM-T 1 h	8047	5913	9639	1.63
GGM-T 2 h	8047	5863	9588	1.63
GGM-T 3 h	7864	5830	9457	1.62
GGM-T 4 h	8047	5855	9694	1.66

*—before NaOCl addition.

Despite a slight decrease in the mass yields of the target substances, the removal of the low-molecular-weight reaction byproducts from the reaction medium is an important stage for oxidized GGM isolation, affecting its quality and performance characteristics, which can be described further. The purge was carried out using a dialysis cut-off bag with a pore diameter of 3.5 kDa (Section 2.2.). Figure 9 depicts the MWD profiles for both the initial and oxidized GGM samples.

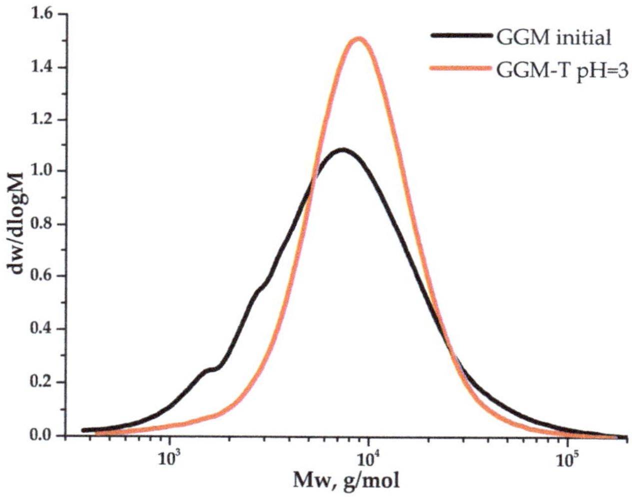

Figure 9. The molecular weight distribution of the galactoglucomannan samples.

3.6. Thermal Decomposition Properties of the Initial and Oxidized GGM Samples

As is observable, the thermal decomposition behavior of polysaccharides could reveal its heterogeneous nature by characterizing the underlying chemical reactions that proceed. Simultaneous bond degradation could encourage changes in the polysaccharide conformation and phase conversion, which is crucial to determine the possible area of application for the biopolymer-based materials [49]. The TGA/DTG curves, represented by Figure 10, are registered for the GGM and GGM-T in the 30–700 °C range, where the following common and different features of the observed thermal events should be mentioned.

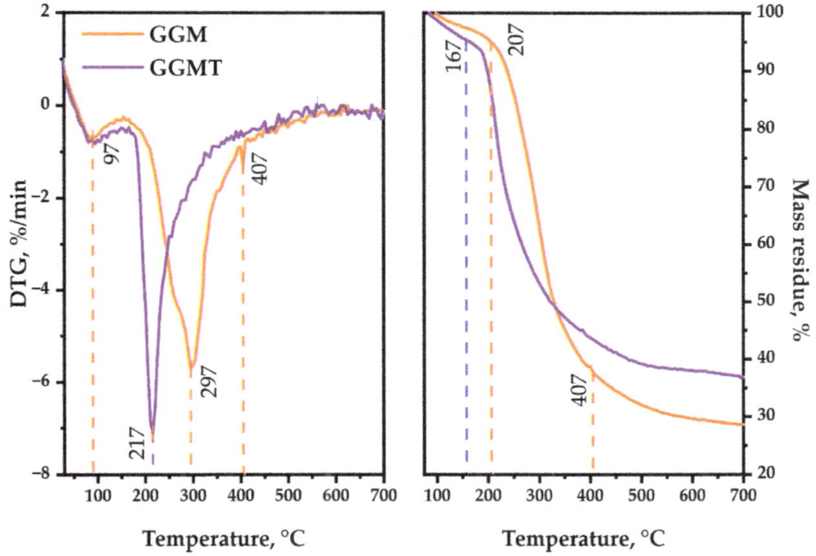

Figure 10. TGA/DTG thermal degradation profiles of the initial and oxidized GGM samples.

The first stage (approximately at 92 °C) of thermolysis is connected with the evaporation of sorbed and crystallized moisture by the GGM samples [50], which typically leads to an insignificant change in weight loss. The temperature for 1 wt.% loss of the oxidized sample is 97 °C, whereas for the initial sample it was 117 °C. A further increase in the temperature, approximately up to 207 °C, has almost no effect on the initial GGM structure. It reaches 5 wt.% loss at that point, which could be explained by the hydrogen bond decay of the polar functional groups of polysaccharides. The oxidized GGM-T reaches that value earlier, at 167 °C.

As could be observed, the second stage of initial GGM thermolysis (217–407 °C) reaches 10 wt.% loss at 237 °C, with the maximum at 297 °C followed by intensive mass loss of up to 37.8 wt.% at the end of the stage. Meanwhile, the GGM-T decomposition starts earlier, resulting in 10 wt.% loss at 202 °C, reaching the maximum decomposition rate at 217 °C. Nevertheless, as with its initial form, the GGM-T thermolysis slows down at 407 °C with a 43.5 wt. % mass loss of the residue.

The third stage of the initial GGM decomposition (407–700 °C) is characterized by less weight loss, which corresponds to its carbonization processes, leading to the formation of coke residue [51], resulting in 28.6 wt.% mass residue at the end of the thermolysis. The oxidized GGM-T sample degradation slightly differs—the mass of its residue contains up to 36.9 wt.% at the end of the decomposition.

Based on the obtained data, the main-stage thermal decomposition activation energy of the GGM and GGM-T samples was calculated. The initial GGM thermal decomposition range lies in 217–407 °C, with an activation energy value of 37.4 kJ/mol. At the same time, the GGM-T decomposition activation energy in the main stage (197–400 °C) was characterized by a lower values, 18.1 kJ/mol.

Such GGM-T thermal decomposition behavior may be a consequence of the presence of disordered regions, caused by ester group formation [52]. As a result, the degradation process of the GGM-T initiates earlier at lower temperatures compared to the initial GGM, but the ordered regions of the oxidized polysaccharide resist the decomposition, which leads to higher thermal stability with the perseverance of sample mass.

3.7. Antioxidant Activity of Initial and Oxidized Galactoglucomannan

Due to their non-toxic nature, polysaccharides, derived from plant sources, are useful in terms of their antioxidant activity. DPPH and hydroxyl radical scavenging assays are among the most commonly used methods for assessing antioxidant activity. Early studies have shown that the antioxidant properties of plant polysaccharides depend on their structural features, including the composition and ratio of the monosaccharide's structure, molecular weight characteristics, uronic acid content, and other factors [53].

Antioxidant activity is commonly expressed in terms of IC_{50} values (mg/mL)—the concentration of half-maximal radical inhibition, i.e., the concentration of an antioxidant required to reduce the concentration of radicals by 50% [54]. The IC_{50} values for the GGM and GGM-T samples with respect to DPPH and hydroxyl radical scavenging are represented in Table 2.

Table 2. The GGM and GGM-T samples' IC_{50} values in relation to DPPH and hydroxyl radical scavenging.

Sample	IC_{50} Values (mg/mL)	
	DPPH	$•OH^-$
GGM	1.28	4.71
GGM-T	3.26	3.54

Based on the data, it is evident that the IC_{50} values obtained for the GGM and GGM-T samples differ significantly. The IC_{50} of the DPPH for GGM-T is twice as high as that for GGM (Table 2). However, GGM-T is characterized by lower scavenging activity, expressed by a lower IC_{50} value, which is undoubtedly affected by the GGM structure transformation during the TEMPO-catalyzed oxidation. Figure 11 clearly demonstrates that, regardless of the method used for assessing the antioxidant capacity, the scavenging of the DPPH and hydroxyl radicals is dose-dependent. Nevertheless, the presence of the reactive hydroxyl groups in the initial GGM structure, as well as a lower molecular weight (11,203 g/mol), are not the only sources for the high AOA. A more branched structure of the initial GGM, in comparison to the oxidized derivative, is obviously the reason for the greater scavenging capacity for the DPPH free radicals, reaching a maximum of ~74% (Figure 11a). However, such features introduce some limitations into the scavenging of hydroxyl radicals (Figure 11b).

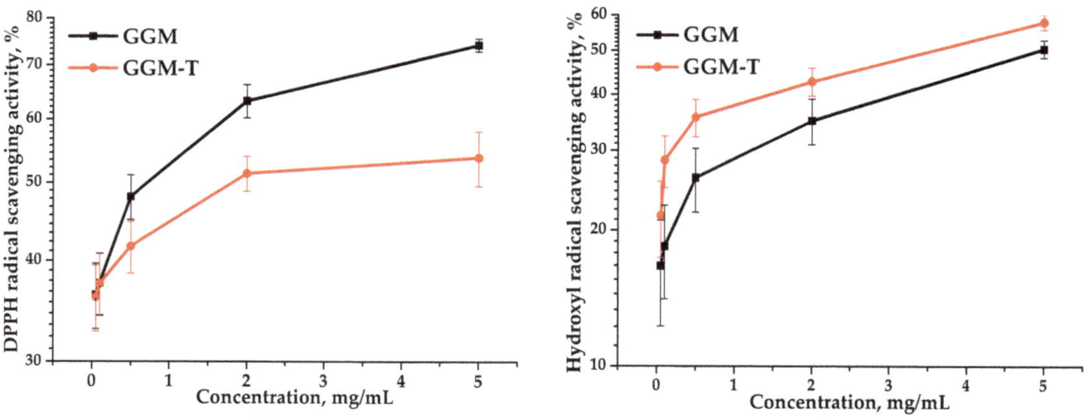

(a) (b)

Figure 11. Ability of the GGM and GGM-T samples to scavenge (**a**) DPPH and (**b**) hydroxyl radicals.

Despite the fact that the oxidized GGM has a higher molecular weight (11,407 g/mol) compared to the initial GGM, the transformation of the polymer structure significantly affected the ability to scavenge free radicals, reaching a maximum of 53% (Figure 11a). It is important to note that the higher carboxyl group content of GGM-T formed by oxidation significantly increases its ability to scavenge hydroxyl radicals from 50 to 58% (Figure 11b).

Thus, it was found that the efficiency of radical scavenging by the oxidized GGM was directly connected with the presence of certain functional groups and inherent structural features. Additionally, the obtained results of AOA assays undoubtedly indicate the prospects for the application of plant polysaccharides and their derivatives in medicine and pharmacology.

3.8. Impact of TEMPO-Catalyzed Oxidation on GGM Heavy-Metal Adsorption Capability

Heterogeneous structure polysaccharides, such as GGM, abundant with a number of functional groups on their backbones, could provide numerous binding sites to adsorb the heavy-metal ions in aqueous solutions [23,24]. The most common functional groups that could be involved in adsorption processes are the primary and side hydroxyl groups of the GGM elementary units. It was expected that the presence of insignificant amounts of the acetylated groups in the initial GGM would not impact the adsorption processes. On the other hand, the acetylated groups are subjected to alkaline hydrolysis [55], which could occur in the course of TEMPO-mediated oxidation. It was expected that the following possible acetylated group conversion into hydroxyl groups could enhance the adsorption capacities of its oxidized derivative.

It is important to mention that solution pH is among the most crucial factors affecting a substance's adsorption properties. A pH = 4 of the tested GGM sample solution was chosen, according to data that report better HM adsorption capacities in the pH = 2–5 range. First of all, it is directly connected with HM ion hydrolysis and its conversion into forms such as $Pb_2(OH)^{3+}$, $Pb_3(OH)^{4+}$, and $Cd(OH)^+$ [56,57], which negatively affect the absorption capacity.

Nevertheless, the cation sorption to the different polysaccharides occurs preferentially via the acidic groups on the latter [24], because of the low steric effect for interaction with metal ions [58]. In this course, consideration of bivalent, specifically HM ion, complexation mechanisms with functional groups of the TEMPO-oxidized GGM was investigated preferably, due to its high affinity (Figure 12).

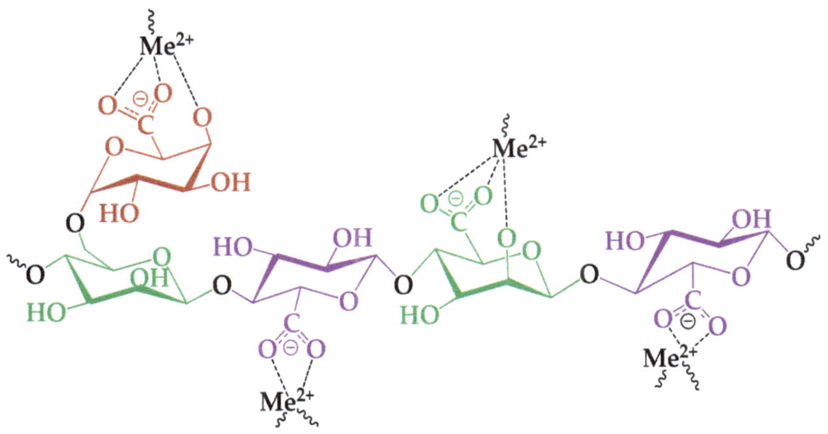

Figure 12. The proposed mechanism of the HM ion adsorption by the GGM-T.

Thus, to investigate the effect of the GGM TEMPO-catalyzed oxidation performed earlier, a series of adsorption experiments were conducted: with single bivalent HM ions and in terms of the competitive tests. The obtained results of the affinity tests are represented in Figures 13 and 14. The numerical values of the sorption properties are represented in Table S2.

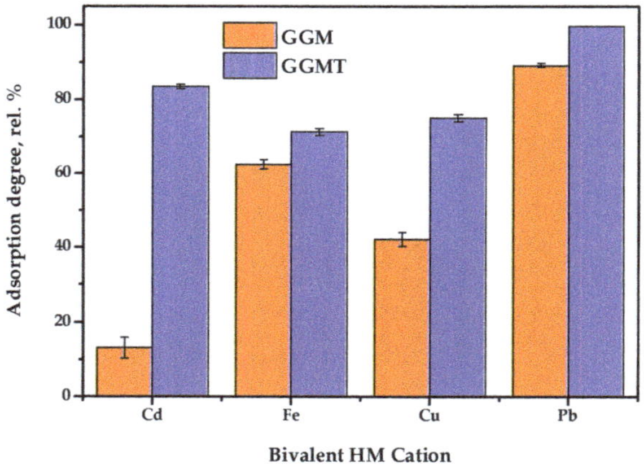

Figure 13. Single heavy-metal ion adsorption degrees of initial and oxidized GGMs.

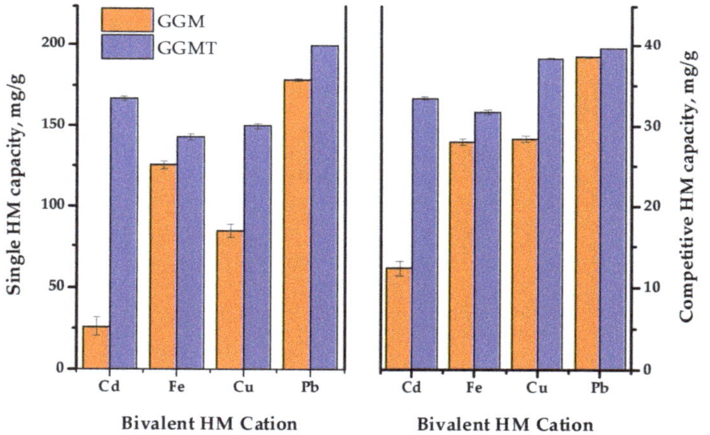

Figure 14. The single and competitive heavy-metal ion adsorption capacities of the initial and oxidized GGMs.

The results of adsorption tests in conditions of the single metal ion type used allow us to state that the conducted TEMPO-catalyzed oxidation strongly affects the degree of HM adsorption.

As is shown in Figure 13, the initial GGM was characterized by a low Cd^{2+} affinity with 13.0 rel.% AD, compared to the oxidized derivative, GGM-T, with 83.4 rel.% AD. Expressed in terms of quantity, the AC of Cd^{2+} was valued at 26.0 and 166.8 mg/g for the initial and oxidized samples, respectively. The smaller but significant difference between the initial and oxidized GGM sample AD rates is also observed for Cu^{2+} ions—42.3 and 75.0 rel.%, respectively. The smallest difference in the AD value rates of the initial and oxidized GGM samples is observed for Fe^{2+} and Pb^{2+}; nevertheless, the Pb^{2+} adsorption values are approaching the possible maximum, while the specifics of Fe^{2+} adsorption remain unclear, which will be described below.

In order to reveal the adsorption performance of the initial and oxidized GGM samples in a multi-component system, a mixture of the selected HM ions was tested in terms of the competitive experiment. The obtained results are depicted in Figure 14.

As was expected, the summary adsorption capacities for both the initial and oxidized GGM samples in most cases were lower to some extent, compared to the single HM ion AC. Nevertheless, since the same amount of the total HM ion concentrations was used in both single and competitive tests, it is important to note that the summary capacity of the initial GGM in conditions of a competitive experiment was higher than the capacity of single Cd^{2+} and Cu^{2+}—107.3 ± 1.7 mg/g compared to 26.0 ± 5.7 and 84.6 ± 3.8 mg/g, respectively. As for the oxidized GGM derivative, its summary capacity in terms of the competitive tests was higher, in comparison to the initial GGM—143.1 ± 0.6 mg/g. However, due to the overall GGM-T greater single HM adsorption capacity, its summary AC values in a competitive experiment exceeds the AC values registered for Fe^{2+}. This fact obviously demonstrates a competitive relationship between the investigated bivalent HM ion adsorption.

Noticeably, the relative difference ratio of Fe^{2+} adsorption by the GGM samples in conditions of single HM and competitive experiments has an equaling value. This could be due to the specifics of Fe^{2+} ions, which could interact with water in solutions, or caused by the substitution of hydrogen from the carboxyl groups formed by the TEMPO-catalyzed GGM oxidation. Moreover, it was expected that the Fe^{2+} ion adsorption selectivity would increase at the cost of less active ions such as Cu^{2+}, which should be easily substituted, but on the contrary, that has no effect.

As is shown in Figure 14, Pb^{2+} adsorption by both the initial and oxidized GGM samples was predominant in terms of the single HM ion and competitive experiments. It is obvious that GGM was initially inclined to adsorb Pb^{2+} with a high degree of selectivity and good adsorption capacities, even in a mixture with a relatively strong affinity for Pb^{2+}, which could be connected with its lower hydration radius [59].

Thus, as has been shown by the obtained data, the oxidized GGM-T adsorbent exhibited superior adsorption properties and affinity for multiple metal ions in terms of the competitive experiments compared to its original, the initial GGM. The relative selectivity of HM adsorption in terms of the competitive tests undergo rearrangement, which was obviously caused by the modification of the initial GGM structure. As can be seen from Table S2, the HM ion adsorption selectivity values of the initial GGM varies in the order of $Cd^{2+} < Fe^{2+} \approx Cu^{2+} < Pb^{2+}$. Meanwhile, the TEMPO-oxidized derivative GGM-T selectivity values increase in the order of $Fe^{2+} < Cd^{2+} < Cu^{2+} < Pb^{2+}$.

According to the existing data, the cellulose, starch, chitosan, sodium alginate, and their derivatives are traditionally observed as promising adsorbents. As for the polysaccharides, they are most commonly represented by hydrogels, synthesized on a guar gum, agar, or hyaluronic acid basis. It was found that the oxidized GGM-T derivative adsorption capacity of single Cu^{2+} (150.0 mg/g) exceeds alginate/carboxymethyl cellulose (105.93 mg/g) [60] and alginate/polyvinyl/ZIF-9 (98.98 mg/g) [61] materials' adsorption capacities. As for Pb^{2+}, the GGM-T adsorption capacity (199.2 mg/g) also exceeds alginate/carboxymethyl cellulose [60], polyvinyl alcohol/alginate [62], and *E. prolifera*-based double network hydrogel (166.7 mg/g) [63] materials.

Nevertheless, in comparison to synthetic polymer adsorption capacity, GGM-T was characterized by a lower efficiency. One possible path for improving existing adsorption capabilities lies in the application of GGM-T to the synthesis of composite materials. For example, materials based on biodegradable compounds like chitosan/sodium alginate/calcium ion double network hydrogels exhibit greater Pb^{2+}, Cu^{2+}, and Cd^{2+} affinities in terms of the competitive tests—176.50, 70.83, and 81.25 mg/g, respectively [64].

The applications of such biopolymers could vary significantly, based on the obtained results. It is obvious that oxidized GGM-T has great potential in the synthesis of composite materials with multiple functional groups or polyelectrolyte complexes, also used as drug carriers [65]. Such affinity to the cations, combined with high AOA and excellent water solubility, allows for proposing oxidized GGM in trials of HM uptake from living organisms. Due to the low toxicity of the polymer basis, metal complexes represent potential for the development of products used as preparation for the protection of skin and hair from the harmful action of UV radiation [66]. In addition, the complexation of polysaccharides with cations is applicable in analytical methods of HM determination, with its easy removal achievable via centrifugation and filtration. Moreover, such properties could find their application in the field of electrodialysis and reverse osmosis processes for the removal of heavy metals from aqueous solutions.

4. Conclusions

A novel bio-based polymer was synthesized from isolated spruce galactoglucomannan via TEMPO-catalyzed oxidation. The dynamics of the process were controlled using the well-known titration, gel permeation chromatography, and spectrophotometric methods. It was observed that GGM oxidation proceeds in a mild way; nevertheless, the partial cleavage of glycoside bonds linking arabinoxylose's units, the presence of which is due to the techniques used for GGM isolation, was also confirmed by FTIR. Despite a slight decrease in mass yields of the oxidized GGM (78.3 $\pm$ 6.4 wt.%), the removal of the low-molecular-weight reaction byproducts is an important stage for target substance isolation from the reaction medium, affecting both quality and performance characteristics.

It has been revealed that during the oxidation of GGM, the uronic acid content multiplies by more than twice (from 14.8 to 38.9%), which indirectly points to oxidation dynamics, corresponding with the NaOH consumption rate during the process. Expressed by a quantity, the carboxylate group content in the polysaccharide contains up to 8.4 wt.%, indicating the efficiency of the conducted TEMPO-catalyzed oxidation.

As was expected, the thermal degradation of the oxidized GGM-T begins earlier at lower temperatures compared to the initially isolated sample. Nevertheless, we assume that the ordered regions of the oxidized polysaccharide resist decomposition, which leads to a higher thermal stability, in general, with a greater mass perseverance.

In this study, the initial and oxidized GGMs were successfully tested to adsorb various bivalent HMs, singled or in terms of the competitive tests (Fe^{2+}, Cu^{2+}, Cd^{2+}, or Pb^{2+}). The adsorption capabilities of the initial and oxidized GGM were determined by the physical crosslinking of polysaccharide chains with HM ions. The obtained complexes were further precipitated by ethanol and easily separated from the solution as a hydrogel via centrifugation and filtration through a PTFE membrane. It was found that the HM ions' adsorption capability of the initial GGM was mainly influenced by the hydroxyl-HM ions' migrating activity and varies in the order of $Cd^{2+} < Fe^{2+} \approx Cu^{2+} < Pb^{2+}$. The presence of the acetylated groups in the isolated initial GGM insignificantly affects adsorption properties. Additionally, the TEMPO-oxidized derivative GGM-T adsorption capability is mostly influenced by the presence of both the hydroxyl and formed carboxyl groups. Their presence modifies the HM ions' migrating activity, which increases in the order of $Fe^{2+} < Cd^{2+} < Cu^{2+} < Pb^{2+}$. In accordance with the existing data, the GGM-T derivative adsorption capacity exceeds a number of traditionally studied HM adsorbent materials. Nevertheless, the obtained bio-based polymer is still inferior in comparison to the synthetically obtained adsorbents in terms of the competitive adsorption tests, which constitutes a new challenge. The synthesis of composite materials based on GGM-T is a possible route with which to improve the biopolymer properties, which should be considered for the further studies.

Supplementary Materials: The following supporting information can be downloaded at: https://www.mdpi.com/article/10.3390/antiox14050569/s1, Table S1: Absorbance units of initial and oxidized GGM registered by FTIR; Table S2: Sorption properties of initial and oxidized GGM with various bivalent heavy metals.

Author Contributions: Conceptualization, V.A.I. and Y.N.M.; methodology, V.A.I., Y.N.M., V.S.B. and A.S.K.; software, Y.N.M., V.A.I. and V.S.B.; validation, Y.N.M., V.A.I. and V.S.B.; formal analysis, V.A.I., V.S.B., D.V.Z. and A.S.K.; investigation, V.A.I. and V.S.B.; resources, Y.N.M.; data curation, V.A.I., Y.N.M., V.S.B. and D.V.Z.; writing—original draft preparation, V.A.I., V.S.B. and Y.N.M.; writing—review and editing, V.A.I., V.S.B. and Y.N.M.; visualization, V.A.I., V.S.B. and Y.N.M.; supervision, Y.N.M.; project administration, Y.N.M.; funding acquisition, Y.N.M. All authors have read and agreed to the published version of the manuscript.

Funding: This study was supported by the Russian Science Foundation, project no. 22-73-10212, https://rscf.ru/en/project/22-73-10212/ (accessed on 21 March 2025).

Institutional Review Board Statement: Not applicable.

Informed Consent Statement: Not applicable.

Data Availability Statement: All data generated during this study are included in the article.

Acknowledgments: This study was carried out using the equipment of the Krasnoyarsk Regional Centre for Collective Use, Krasnoyarsk Scientific Center, Siberian Branch of the Russian Academy of Sciences. Furthermore, we would like to thank O.Yu. Fetisova, researcher of ICCT SB RAS, for valuable contributions to the thermogravimetric analysis of studied samples.

Conflicts of Interest: The authors declare no conflicts of interest.

References

1. Hayat, M.T.; Nauman, M.; Nazir, N.; Ali, S.; Bangash, N. Chapter 7—Environmental Hazards of Cadmium: Past, Present, and Future. In *Cadmium Toxicity and Tolerance in Plants*; Hasanuzzaman, M., Prasad, M.N.V., Fujita, M., Eds.; Academic Press: Cambridge, MA, USA, 2019; pp. 163–183.
2. Wu, Z.Q.; Li, H.Y.; Lü, L.Y.; Liang, G.J.; Wu, T.T.; Zhu, J.X. Distributions and risk assessment of heavy metals in solid waste in lead-zinc mining areas and across the soil, water body, sediment and agricultural product ecosystem in their surrounding areas. *China Geol.* **2025**, *8*, 92–106. [CrossRef]
3. Cai, J.; Zheng, W.; Luo, M.; Kuang, C.; Tang, X. Characterization of copper (II) chemical forms and heavy metal distribution in chemical looping gasification of municipal solid waste. *J. Energy Inst.* **2021**, *96*, 140–147. [CrossRef]
4. Jaishankar, M.; Tseten, T.; Anbalagan, N.; Mathew, B.B.; Beeregowda, K.N. Toxicity, mechanism and health effects of some heavy metals. *Interdiscip. Toxicol.* **2014**, *7*, 60–72. [CrossRef] [PubMed]
5. Song, L.-Y.; Liu, X.; Zhang, L.-D.; Hu, W.-J.; Xu, C.-Q.; Li, J.; Song, S.-W.; Guo, Z.-J.; Sun, C.-Y.; Tang, H.-C.; et al. Proteomic analysis reveals differential responsive mechanisms in Solanum nigrum exposed to low and high dose of cadmium. *J. Hazard. Mater.* **2023**, *448*, 130880. [CrossRef]
6. Kaur, R.; Kaur, H.; Sharma, A. Uptake and toxicity of heavy metals: The protective frontiers of metal binding proteins. *J. Geochem. Explor.* **2025**, *271*, 107673. [CrossRef]
7. dos Santos Lima, A.; de Oliveira Pedreira, F.R.; Bento, N.A.; Novaes, R.D.; dos Santos, E.G.; de Almeida Lima, G.D.; de Almeida, L.A.; Belo, T.C.A.; Vieira, F.V.; Mohammadi, N.; et al. Digested galactoglucomannan mitigates oxidative stress in human cells, restores gut bacterial diversity, and provides chemopreventive protection against colon cancer in rats. *Int. J. Biol. Macromol.* **2024**, *277*, 133986. [CrossRef]
8. Malyar, Y.N.; Vasilyeva, N.Y.; Kazachenko, A.S.; Borovkova, V.S.; Skripnikov, A.M.; Miroshnikova, A.V.; Zimonin, D.V.; Ionin, V.A.; Kazachenko, A.S.; Issaoui, N. Modification of Arabinogalactan Isolated from Larix sibirica Ledeb. into Sulfated Derivatives with the Controlled Molecular Weights. *Molecules* **2021**, *26*, 5364. [CrossRef]
9. Kazachenko, A.S.; Vasilieva, N.Y.; Borovkova, V.S.; Fetisova, O.Y.; Issaoui, N.; Malyar, Y.N.; Elsuf'ev, E.V.; Karacharov, A.A.; Skripnikov, A.M.; Miroshnikova, A.V.; et al. Food Xanthan Polysaccharide Sulfation Process with Sulfamic Acid. *Foods* **2021**, *10*, 2571. [CrossRef]

10. Tang, S.; Wang, T.; Huang, C.; Lai, C.; Fan, Y.; Yong, Q. Sulfated modification of arabinogalactans from Larix principis-rupprechtii and their antitumor activities. *Carbohydr. Polym.* **2019**, *215*, 207–212. [CrossRef]
11. Mourão, P.A.S. Perspective on the Use of Sulfated Polysaccharides from Marine Organisms as a Source of New Antithrombotic Drugs. *Mar. Drugs* **2015**, *13*, 2770–2784. [CrossRef]
12. Zhang, H.; Wang, Z.-Y.; Yang, L.; Yang, X.; Wang, X.; Zhang, Z. In Vitro Antioxidant Activities of Sulfated Derivatives of Polysaccharides Extracted from Auricularia auricular. *Int. J. Mol. Sci.* **2011**, *12*, 3288–3302. [CrossRef]
13. Ganenko, T.; Tantsyrev, A.; Sapozhnikov, A.; Khutsishvili, S.; Vakulskaya, T.; Fadeeva, T.; Sukhov, B.; Trofimov, B. Nanocomposites of silver with arabinogalactan sulfate: Preparation, structure, and antimicrobial activity. *Russ. J. Gen. Chem.* **2015**, *85*, 477–484. [CrossRef]
14. Alfinaikh, R.S.; Alamry, K.A.; Hussein, M.A. Sustainable and biocompatible hybrid materials-based sulfated polysaccharides for biomedical applications: A review. *RSC Adv.* **2025**, *15*, 4708–4767. [CrossRef] [PubMed]
15. Arlov, Ø.; Rütsche, D.; Korayem, M.; Öztürk, E.; Zenobi, M. Engineered Sulfated Polysaccharides for Biomedical Applications. *Adv. Funct. Mater.* **2021**, *31*, 2010732. [CrossRef]
16. Gurpilhares, D.d.B.; Moreira, T.R.; Bueno, J.d.L.; Cinelli, L.P.; Mazzola, P.G.; Pessoa, A.; Sette, L.D. Algae's sulfated polysaccharides modifications: Potential use of microbial enzyme. *Process Biochem.* **2016**, *51*, 989–998. [CrossRef]
17. Li, Z.-W.; Du, Z.-M.; Wang, Y.-W.; Feng, Y.-X.; Zhang, R.; Yan, X.-B. Chemical Modification, Characterization, and Activity Changes of Land Plant Polysaccharides: A Review. *Polymers* **2022**, *14*, 4161. [CrossRef]
18. Ionin, V.A.; Malyar, Y.N.; Borovkova, V.S.; Zimonin, D.V.; Gulieva, R.M.; Fetisova, O.Y. Inherited Structure Properties of Larch Arabinogalactan Affected via the TEMPO/NaBr/NaOCl Oxidative System. *Polymers* **2024**, *16*, 1458. [CrossRef]
19. de Nooy, A.E.J.; Besemer, A.C.; van Bekkum, H. Selective oxidation of primary alcohols mediated by nitroxyl radical in aqueous solution. Kinetics and mechanism. *Tetrahedron* **1995**, *51*, 8023–8032. [CrossRef]
20. Bragd, P.L.; van Bekkum, H.; Besemer, A.C. TEMPO-Mediated Oxidation of Polysaccharides: Survey of Methods and Applications. *Top. Catal.* **2004**, *27*, 49–66. [CrossRef]
21. Yong, Q.; Xu, J.; Wang, L.; Tirri, T.; Gao, H.; Liao, Y.; Toivakka, M.; Xu, C. Synthesis of galactoglucomannan-based latex via emulsion polymerization. *Carbohydr. Polym.* **2022**, *291*, 119565. [CrossRef]
22. Halahlah, A.; Abik, F.; Lahtinen, M.H.; Kemppinen, A.; Kaipanen, K.; Kilpeläinen, P.O.; Granato, D.; Ho, T.M.; Mikkonen, K.S. Effects of pH and temperature of ultrafiltration on the composition and physicochemical properties of hot-water-extracted softwood galactoglucomannans. *Ind. Crops Prod.* **2023**, *198*, 116656. [CrossRef]
23. Liu, S.; Cai, Y.; Du, M.; Liu, J.; Zhang, X.; Yuan, H.; Xu, Z.; Chen, X.; Kan, X.; Zhang, S.; et al. Research on Adsorption of Heavy Metal Ions by Polysaccharide Hydrogels. *ChemistrySelect* **2024**, *9*, e202401752. [CrossRef]
24. Levy-Ontman, O.; Yanay, C.; Alfi, Y.; Paz-Tal, O.; Wolfson, A. Selective Sorption of Heavy Metals by Renewable Polysaccharides. *Polymers* **2023**, *15*, 4457. [CrossRef]
25. Konkol, Y.; Vuorikoski, H.; Tuomela, J.; Holmbom, B.; Bernoulli, J. Galactoglucomannan-rich hemicellulose extract from Norway spruce (*Picea abies*) exerts beneficial effects on chronic prostatic inflammation and lower urinary tract symptoms in vivo. *Int. J. Biol. Macromol.* **2017**, *101*, 222–229. [CrossRef] [PubMed]
26. Borovkova, V.S.; Malyar, Y.N.; Sudakova, I.G.; Chudina, A.I.; Skripnikov, A.M.; Fetisova, O.Y.; Kazachenko, A.S.; Miroshnikova, A.V.; Zimonin, D.V.; Ionin, V.A.; et al. Molecular Characteristics and Antioxidant Activity of Spruce (*Picea abies*) Hemicelluloses Isolated by Catalytic Oxidative Delignification. *Molecules* **2022**, *27*, 266. [CrossRef]
27. Ruiz-Matute, A.I.; Hernández-Hernández, O.; Rodríguez-Sánchez, S.; Sanz, M.L.; Martínez-Castro, I. Derivatization of carbohydrates for GC and GC–MS analyses. *J. Chromatogr. B* **2011**, *879*, 1226–1240. [CrossRef] [PubMed]
28. Sychev, V.; Miroshnikova, A.; Kazachenko, A.; Skripnikov, A.; Taran, O.; Tarabanko, V. Reductive catalytic fractionation of flax shives by using nickel formate for in situ catalyst generation and the shives as a support for the metal. *Biomass Convers. Biorefinery* **2024**, 1–10. [CrossRef]
29. Wu, F.; Yan, N.; Guo, Y.; Yu, X.; Yi, L.; Ouyang, Y.; Wang, X.; Zhang, Z. Pattern of Specific Oxidation of Konjac Glucomannan with TEMPO/NaBr/NaClO system. *Carbohydr. Res.* **2022**, *515*, 108558. [CrossRef]
30. Kumar, V.; Nagar, S. Studies on Tinospora cordifolia monosugars and correlation analysis of uronic acids by spectrophotometric methods and GLC. *Carbohydr. Polym.* **2014**, *99*, 291–296. [CrossRef]
31. Saito, T.; Isogai, A. TEMPO-Mediated Oxidation of Native Cellulose. The Effect of Oxidation Conditions on Chemical and Crystal Structures of the Water-Insoluble Fractions. *Biomacromolecules* **2004**, *5*, 1983–1989. [CrossRef]
32. Berglund, J.; Azhar, S.; Lawoko, M.; Lindström, M.; Vilaplana, F.; Wohlert, J.; Henriksson, G. The structure of galactoglucomannan impacts the degradation under alkaline conditions. *Cellulose* **2019**, *26*, 2155–2175. [CrossRef]
33. Meng, Q.; Fu, S.; Lucia, L.A. The role of heteropolysaccharides in developing oxidized cellulose nanofibrils. *Carbohydr. Polym.* **2016**, *144*, 187–195. [CrossRef] [PubMed]

34. Batsoulis, A.; Nacos, M.; Pappas, C.; Tarantilis, P.; Mavromoustakos, T.; Polissiou, M. Determination of Uronic Acids in Isolated Hemicelluloses from Kenaf Using Diffuse Reflectance Infrared Fourier Transform Spectroscopy (DRIFTS) and the Curve-Fitting Deconvolution Method. *Appl. Spectrosc.* **2004**, *58*, 199–202. [CrossRef]
35. Fontes, A.; Santos, A.; Aguado, R.; Tarrés, Q.; Fiol, N.; Pèlach, M.; Delgado-Aguilar, M. Kinetic study and real-time monitoring strategy for TEMPO-mediated oxidation of bleached eucalyptus fibers. *Cellulose* **2022**, *30*, 1421–1436. [CrossRef]
36. Syverud, K.; Chinga-Carrasco, G.; Toledo, J.; Toledo, P.G. A comparative study of Eucalyptus and Pinus radiata pulp fibres as raw materials for production of cellulose nanofibrils. *Carbohydr. Polym.* **2011**, *84*, 1033–1038. [CrossRef]
37. Li, Z.; Zhang, L.; Mao, C.; Song, Z.; Li, X.; Liu, C. Preparation and characterization of konjac glucomannan and gum arabic composite gel. *Int. J. Biol. Macromol.* **2021**, *183*, 2121–2130. [CrossRef] [PubMed]
38. Chen, X.; Qi, S.; Chuanhe, Z.; Wang, Q. Effect of ultrasound on the properties and antioxidant activity of hawthorn pectin. *Int. J. Biol. Macromol.* **2019**, *131*, 273–281. [CrossRef]
39. Lundqvist, J.; Teleman, A.; Junel, L.; Zacchi, G.; Dahlman, O.; Tjerneld, F.; Stålbrand, H. Isolation and characterization of galactoglucomannan from spruce (*Picea abies*). *Carbohydr. Polym.* **2002**, *48*, 29–39. [CrossRef]
40. Willför, S.; Sjöholm, R.; Laine, C.; Roslund, M.; Hemming, J.; Holmbom, B. Characterisation of water-soluble galactoglucomannans from Norway spruce wood and thermomechanical pulp. *Carbohydr. Polym.* **2003**, *52*, 175–187. [CrossRef]
41. Li, J.; Ma, J.; Chen, S.; He, J.; Huang, Y. Characterization of calcium alginate/deacetylated konjac glucomannan blend films prepared by Ca^{2+} crosslinking and deacetylation. *Food Hydrocoll.* **2018**, *82*, 363–369. [CrossRef]
42. Gómez-Ordóñez, E.; Rupérez, P. FTIR-ATR spectroscopy as a tool for polysaccharide identification in edible brown and red seaweeds. *Food Hydrocoll.* **2011**, *25*, 1514–1520. [CrossRef]
43. Hao, J.; Lu, J.; Xu, N.; Linhardt, R.J.; Zhang, Z. Specific oxidation pattern of soluble starch with TEMPO-NaBr-NaClO system. *Carbohydr. Polym.* **2016**, *146*, 238–244. [CrossRef] [PubMed]
44. Nagel, A.; Conrad, J.; Leitenberger, M.; Carle, R.; Neidhart, S. Structural studies of the arabinogalactans in Mangifera indica L. fruit exudate. *Food Hydrocoll.* **2016**, *61*, 555–566. [CrossRef]
45. Rakhmanberdyeva, R.K.; Filippov, M.P. Infrared spectroscopy study of Gleditsia macracantha seeds. *Chem. Nat. Compd.* **2011**, *47*, 179–181. [CrossRef]
46. Tamaki, Y.; Teruya, T.; Tako, M. The chemical structure of galactomannan isolated from seeds of Delonix regia. *Biosci. Biotechnol. Biochem.* **2010**, *74*, 1110–1112. [CrossRef] [PubMed]
47. Muschin, T.; Yoshida, T. Structural analysis of galactomannans by NMR spectroscopy. *Carbohydr. Polym.* **2012**, *87*, 1893–1898. [CrossRef]
48. Marcon, M.; Carneiro, P.; Wosiacki, G.; Beleski-Carneiro, E.; Petkowicz, C.L.O. Pectins from apple pomace—Characterization by 13C and 1H NMR spectroscopy. *Ann. Magn. Reson.* **2005**, *4*, 56–63.
49. Kartal, F.; Dalbudak, Y.; Özveren, U. Prediction of thermal degradation of biopolymers in biomass under pyrolysis atmosphere by means of machine learning. *Renew. Energy* **2023**, *204*, 774–787. [CrossRef]
50. Saeidy, S.; Nasirpour, A.; Keramat, J.; Desbrières, J.; Cerf, D.L.; Pierre, G.; Delattre, C.; Laroche, C.; Baynast, H.D.; Ursu, A.-V.; et al. Structural characterization and thermal behavior of a gum extracted from *Ferula assa foetida* L. *Carbohydr. Polym.* **2018**, *181*, 426–432. [CrossRef]
51. Lobo, R.E.; Gómez, M.I.; Font de Valdez, G.; Torino, M.I. Physicochemical and antioxidant properties of a gastroprotective exopolysaccharide produced by Streptococcus thermophilus CRL1190. *Food Hydrocoll.* **2019**, *96*, 625–633. [CrossRef]
52. Almeida, E.A.M.S.; Facchi, S.P.; Martins, A.F.; Nocchi, S.; Schuquel, I.T.A.; Nakamura, C.V.; Rubira, A.F.; Muniz, E.C. Synthesis and characterization of pectin derivative with antitumor property against Caco-2 colon cancer cells. *Carbohydr. Polym.* **2015**, *115*, 139–145. [CrossRef]
53. Wang, B.; Lin, C.; Duan, C.; Li, J.; Chen, H.; Xu, J.; Zeng, J.; Gao, W.; Wei, W. Physicochemical characterization of bioactive polysaccharides from three seaweed and application of functional fruit packaging films. *Int. J. Biol. Macromol.* **2024**, *282*, 136765. [CrossRef] [PubMed]
54. Fernandes, P.A.R.; Coimbra, M.A. The antioxidant activity of polysaccharides: A structure-function relationship overview. *Carbohydr. Polym.* **2023**, *314*, 120965. [CrossRef] [PubMed]
55. Sun, Y.; Xu, X.; Wu, Z.; Zhou, H.; Xie, X.; Zhang, Q.; Liu, R.; Pang, J. Structure, Merits, Gel Formation, Gel Preparation and Functions of Konjac Glucomannan and Its Application in Aquatic Food Preservation. *Foods* **2023**, *12*, 1215. [CrossRef]
56. Zannagui, C.; Hassan, A.; Soufian, E.B.; Jilal, I.; Sundman, O.; Salhi, A.; Abou-Salama, M.; El Idrissi, A. Homogeneous succinylation of cellulose acetate: Design, characterization and adsorption study of Pb(II), Cu(II), Cd(II) and Zn(II) ions. *E3S Web Conf.* **2021**, *240*, 02003. [CrossRef]
57. Liu, X.; Hu, Q.; Fang, Z.; Zhang, X.; Zhang, B. Magnetic Chitosan Nanocomposites: A Useful Recyclable Tool for Heavy Metal Ion Removal. *Langmuir: ACS J. Surf. Colloids* **2008**, *25*, 3–8. [CrossRef] [PubMed]
58. Macena, M.; Pereira, H.; Cruz-Lopes, L.; Grosche, L.; Esteves, B. Competitive Adsorption of Metal Ions by Lignocellulosic Materials: A Review of Applications, Mechanisms and Influencing Factors. *Separations* **2025**, *12*. [CrossRef]

59. Persson, I. ChemInform Abstract: Hydrated Metal Ions in Aqueous Solution: How Regular Are Their Structures? *Pure Appl. Chem.* **2010**, *82*, 1901–1917. [CrossRef]
60. Zhao, H.; Ouyang, X.-K.; Yang, L.-Y. Adsorption of lead ions from aqueous solutions by porous cellulose nanofiber–sodium alginate hydrogel beads. *J. Mol. Liq.* **2021**, *324*, 115122. [CrossRef]
61. Alraddadi, H.M.; Fagieh, T.M.; Bakhsh, E.M.; Akhtar, K.; Khan, S.B.; Khan, S.A.; Bahaidarah, E.A.; Homdi, T.A. Adsorptive removal of heavy metals and organic dyes by sodium alginate/coffee waste composite hydrogel. *Int. J. Biol. Macromol.* **2023**, *247*, 125708. [CrossRef]
62. Zhang, P.; Zou, K.; Yuan, L.; Liu, J.; Liu, B.; Qing, T.-P.; Feng, B. A biomass resource strategy for alginate-polyvinyl alcohol double network hydrogels and their adsorption to heavy metals. *Sep. Purif. Technol.* **2022**, *301*, 122050. [CrossRef]
63. Wen, Y.; Xue, C.; Ji, D.; Hou, Y.; Li, K.; Li, Y. Eco-friendly Enteromorpha polysaccharides-based hydrogels for heavy metal adsorption: From waste to efficient materials. *Colloids Surf.: Physicochem. Eng. Asp.* **2023**, *656*, 130531. [CrossRef]
64. Tang, S.; Yang, J.; Lin, L.; Peng, K.; Chen, Y.; Jin, S.; Yao, W. Construction of physically crosslinked chitosan/sodium alginate/calcium ion double-network hydrogel and its application to heavy metal ions removal. *Chem. Eng. J.* **2020**, *393*, 124728. [CrossRef]
65. Nandi, G.; Hasnain, M.S.; Nayak, A.K. Chapter 8—Polysaccharide-based polyelectrolyte complex systems in drug delivery. In *Tailor-Made Polysaccharides in Drug Delivery*; Nayak, A.K., Hasnain, M.S., Eds.; Academic Press: Cambridge, MA, USA, 2023; pp. 177–210.
66. Sibikina, O.V.; Iozep, A.A.; Moskvin, A.V. Polysaccharide complexes with metal cations: Structure and application (a review). *Pharm. Chem. J.* **2009**, *43*, 341–345. [CrossRef]

antioxidants

MDPI

Article

Green Recovery and Identification of Antioxidant and Enzyme Inhibitor Molecules from Pisco Grape Pomace by Targeted Effects Analysis Using Thin-Layer Chromatography, Bioassay, and Mass Spectrometry

Jacqueline Poblete [1], Joaquín Fernández-Martínez [2], Mario Aranda [2] and Issis Quispe-Fuentes [1,*]

[1] Food Engineering Department, Universidad de La Serena, Av. Raúl Bitrán 1305, La Serena 1700000, Chile; j.pobletegalleguillos@gmail.com
[2] Department of Chemistry and Pharmacy, Pontificia Universidad Católica de Chile, Santiago 7810000, Chile
* Correspondence: iquispe@userena.cl; Tel.: +56-512334661

Abstract: The search and identification of inhibitory molecules from novel natural sources, such as pisco grape pomace extract obtained by green techniques, may help to develop agents with therapeutic potential that are beneficial to health with fewer adverse effects than drugs. Many drugs act as enzyme inhibitors, decreasing their activity and thus correcting a metabolic imbalance. This study aims to identify bioactive molecules with antioxidant and inhibitory activity over acetylcholinesterase and cyclooxygenase enzymes present in pisco grape pomace green extracts. Bioactive molecules were detected and identified applying directed effect analysis on planar chromatography coupled to mass spectrometry. For the first time, the presence of antioxidant molecules (quercetin-3-O-glucuronide, quercetin-3-O-glucoside, and gallic acid) and inhibitors of acetylcholinesterase (kaempferol-3-O-glucoside) and cyclooxygenase (gallic acid) enzymes are reported in pisco grape pomace. According to the results, grape pomace could be an alternative to develop novel functional foods and nutraceuticals that provide health benefits and, at the same time, generate a circular economy in the industry.

Keywords: pisco grape pomace; antioxidants; cyclooxygenase; acetylcholinesterase; quercetin; flavonoids

1. Introduction

Isolated and biologically active molecules have generated great interest in the agri-food, pharmaceutical, nutraceutical, and cosmetic industries because they can exert a wide range of biological activities (antioxidant, antimicrobial, anti-inflammatory, wound healing, anti-aging, anti-hyperpigmentation, photoprotective, chemopreventive, anticancer, antidiabetic, immunomodulatory, antitumor, and cardioprotective) [1,2]. These molecules are found in plants and fruit, showing a broad spectrum of pharmacological effects and resulting in a reduction in the risk of suffering chronic non-communicable diseases (CNCDs) [3]. CNCDs are the leading cause of death and disability worldwide, with 41 million people currently dying each year, accounting for 71% of all deaths globally [4].

Grape pomace is an interesting potential source of biologically active molecules. It is a byproduct of the pisco industry that is obtained in large quantities and is mainly discarded as waste or used for composting and animal feed. This residue is composed of seeds and skins that have a high content of (poly) phenols with an important antioxidant capacity [5–7]. Ecological or green extraction techniques are very promising to extract this kind of molecule because they improve the sustainability of extraction by being safe, effective, and economical. Furthermore, these approaches solve the disadvantages of conventional methods, such as low yields, higher solvent volumes, long times, energy, costs, and others [8]. Additionally, in grape pomace, some phenolic compounds are bound to the cell wall or other constituents, which could reduce the extraction yields. To overcome this issue, some novel techniques like enzyme-assisted extraction (EAE) and pressurized liquid extraction (PLE) have been used. The first degrades plant cell wall components using specific enzymes [9] and the second uses water and organic solvents in combination with elevated temperature and pressure [10,11] to improve extraction yields using green solvents like water and ethanol [9,10].

Searching for and identifying inhibitors from novel matrices can be an alternative to developing novel functional foods, ingredients, nutraceuticals, and agents with therapeutic potential that provide health benefits with fewer adverse effects. Enzymatic inhibition is one of the most relevant therapeutic targets, and this kind of molecule is highly distributed in nature. For example, some naturally occurring acetylcholinesterase inhibitors like galantamine are commercially available to treat Alzheimer's disease [12]. Some studies have reported the presence of different types of inhibitors in natural products like cherimoya, fig, ginger, and Tunisian sages [3,13–16]. This inhibitory effect has also been associated with polyphenolic molecules such as flavonoids [17]. Another relevant target is the cyclooxygenase enzyme, whose inhibitors are used as anti-inflammatory agents such as salicylic acid, which gave rise to aspirin [18]. This property has also been reported to come mainly from phenolic acids and flavonoids [19].

One way to evaluate enzyme inhibitors is by high-performance thin-layer chromatography (HPTLC) coupled to biological, biochemical, and micro-chemical assays such as effect-directed analysis (EDA). This coupled system allows for the separation and detection of naturally occurring bioactive molecules in complex samples [3,15]. HPTLC-EDA ensures several benefits, such as simple sample preparation, rapid analysis of several samples at once, and unambiguous identification of the compound(s) responsible for the biological activity by using mass spectrometry (MS) [14,15,20]. In this study, the objective was to identify antioxidant molecules and inhibitors of acetylcholinesterase and cyclooxygenase enzymes present in pisco grape pomace green extracts by HPTLC-EDA-MS.

2. Materials and Methods

2.1. Reagents, Chemicals, Solvents, and Solutions

Ethanol, methanol, acetonitrile, toluene, ultrapure water, formic acid, ethyl acetate, and DPPH (2,2-Diphenyl-1-picrylhydrazyl) were purchased from Merck (Darmstadt, Germany); gallic acid, enzyme cyclooxygenase (COX-2 human; EC1.14.99.1), arachidonic acid (>98.5%), N, N, N', N'-tetramethyl-p-phenylenediamine (TMPD; >98.5%), porcine hematin (>90%), Tris-HCl, enzyme acetylcholinesterase (AChE), Electrophorus electricus (E.C number 3.1.17), 1-naphthyl acetate, fast blue B salt, bovine serum albumin (BSA), α-glucosidase (G5300), 4-nitrophenyl-α-D-glucopiranoside (N1377), and acarbose were purchased from Sigma (St Louis, MO, USA). Diclofenac and donepezil hydrochloride (pharmaceutical secondary standard) were donated from a local pharmaceutical company.

2.2. Raw Material and Preparation

The pisco grape pomace was obtained from the company Pisquera de Chile S.A (Ovalle, Coquimbo Region, Chile). Once the pomace was received, it was dried in a vacuum oven [5].

2.3. Extract Preparation

2.3.1. Extract Obtained by Enzymatic Extraction (EAE)

One gram of dry pomace was weighed and mixed with 20 mL of acetate buffer (pH 4.8; 0.1 M). The enzymes (0.75 U tannase and 40 U cellulase) were added and incubated for 15 min at 20 °C with a speed of 150 rpm (BIOBASE, Singapore, BJPX-2008). Then, 20 mL absolute ethanol was added and shaken for 30 min with a speed of 200 rpm (Boeco, Hamburg, Germany, OS-20) and centrifuged for 5 min at 4193× g. The extract was then frozen at −80 °C for 12 h.

2.3.2. Extract Obtained by Pressurized Liquid Extraction (PLE)

One gram of dry pomace was weighed and mixed with diatomaceous earth (1:1) in a 34 mL stainless steel cell and taken for pressurized liquid extraction (Thermo Scientific Fisher, Dionex ASE 150, Hampton, NH, USA). The extraction conditions were 54% ethanol, 113 °C temperature, and three extraction cycles. The time (3 min) and pressure (103 bar) conditions were fixed. The samples were centrifuged at 4193× g and evaporated at 37 °C. The extract was then frozen at −80 °C for 12 h.

Both extracts were lyophilized (Friologic, Given one 5K, Santiago, Chile). The process conditions were primary drying at −20 to 20 °C with the condenser temperature set at −55 °C and secondary drying at 25 °C and a vacuum of 0.027 kPa for 24 h. The yield in both extracts was approximately 30% from the raw material.

2.4. High-Performance Thin-Layer Chromatography (HPTLC)

The extracts (EAE and PLE) were applied with the HPTLC (CAMAG, Muttenz, Switzerland) TLC autosampler (ATS4) on HPTLC Silica gel F254 plates using the following parameters for 8 tracks per plate: band length 8 mm, track distance 10 mm, dosage rate 50 nL/s, and the first application on the x-axis and the y-axis at 10.0 mm. Application volumes of the sample extracts were 10 μL (100 mg/mL). Chromatography was performed in a 10 × 10 cm twin-trough chamber (without saturation) up to a migration distance of 80 mm using a mobile phase consisting of ethyl acetate/toluene/methanol/water/formic acid (2.5:0.9:0.6:0.6:0.6 $v/v/v/v/v$). Both extracts were applied in duplicate on both sides of the plate to obtain two sections; the first was used for a bioassay and the second was used for mass spectrometry (MS) analysis.

2.5. HPTLC Bioassay for Detection of Antioxidant Molecules in Pisco Grape Pomace Extracts

Bioactive molecules present in EAE and PLE extracts were separated following the methodology described in Section 2.4. After drying the TLC plate to remove the mobile phase for 10 min at 60 °C, it was immersed into a DPPH (1% w/v) methanolic solution using a Chromatogram immersion device 3 (CAMAG, Muttenz, Switzerland).The plate was dried in the dark for 30 min, observing the antioxidant molecules as colorless bands on a purple background. Bands of 10 uL of gallic acid solution (1 mg/mL) were used as positive control.

2.6. HPTLC Bioassay for Detection of Acetylcholinesterase (AChE) Inhibitors in Pisco Grape Pomace Extracts

As described in Section 2.4, extracts and 10 uL of Donepezil solution (1 mg/mL; positive control) were applied on the HPTLC plate. After development, plates were dried in TLC Plate Heater 3 for 10 min at 60 °C to remove the mobile phase. After cooling to room temperature, the plates were nebulized twice with 3 mL of 100 mM Tris buffer (pH = 10) and twice with 100 mM Tris buffer (pH = 8) using a derivatizer (CAMAG, Muttenz, Switzerland) and dried at 60 °C for 10 min. The plate was then nebulized with 3 mL of the enzyme substrate 1-naphthyl acetate (1.5 mg/mL) and dried for 20 min at 60 °C. Then, 3 mL of enzyme solution (1 U/mL with BSA (1 mg/mL) in 50 mM Tris-HCl buffer at pH = 7.8) was uniformly sprayed on the plate. Finally, the plate was incubated at 37 °C for 30 min over a horizontal stand inside a closed moisture chamber containing ca. 50 mL of ultrapure water. Immediately after incubation, a freshly prepared aqueous solution of Fast Blue B salt (1.0 mg/mL) was sprayed onto the plate to obtain a purple background, contrasting with the colorless inhibition zones. The plate image was documented under white light illumination.

2.7. HPTLC Bioassay for Detection of Cyclooxygenase (COX-2) Inhibitors from Pisco Grape Pomace Extracts

Chromatography was carried out as described in Section 2.4. After drying the plate for 10 min at 60 °C, the plates were nebulized twice with 100 mM Tris buffer (pH = 10) and twice with 100 mM Tris buffer (pH = 8) using a derivatizer (CAMAG, Muttenz, Switzerland). After the drying step (60 °C for 10 min), plates were nebulized with the substrates (0.1 mM arachidonic acid and 50 mM TMPD) and dried for 20 min at 60 °C. Then, 3 mL of enzyme solution (0.15 U/mL COX-2 plus 0.60 µg/mL hematin in 100 mM Tris-HCl buffer, pH = 8) was uniformly sprayed onto the plate. Finally, the plate was incubated at 37 °C over a horizontal stand inside a closed moisture chamber containing ca. 50 mL of ultra-pure water. After incubation, the plate was dried at 60 °C and the COX-2 inhibitors were observed as colorless bands on a purple/blue background (TMPD oxidation product). The plate image was documented under white light illumination.

2.8. Bioactive Compound Identification by Mass Spectrometry

Bioactive bands were cut out from the plates and analyzed by liquid chromatography (LC) mass spectrometry (MS) by means of a Waters (Milford, MA, USA) system composed of ACQUITY UPLC I-Class PLUS coupled to Synapt XS HRMS (Table 1). Separation was carried out into a BEH C18 1.7 µm 2.1 × 100 mm column (for bands 1, 2, and 3) and Phenomenex (Torrance, CA, USA) Kinetex XB-C18 3.5 µm 4.1 × 150 mm column (for band 4) using a binary mobile phase composed of water (A) and acetonitrile (B), both acidified with 0.1% (*v*/*v*) formic acid, which was used for applying the following gradient program at a flow rate of 0.4 mL/min: 0–2 min 20% B (isocratic step), 2–6 min 20–30% B, 6–14 min 30–70% B, 14–15 min 70% B (isocratic step), and 15–18 min 70–20%, plus 2 min for column conditioning for a total of 20 min per run.

Table 1. Identification of phenolic compounds with antioxidant capacity and enzymatic inhibition of acetylcholinesterase and cyclooxygenase in pisco grape pomace extracts.

Band	Compounds	$[M]^+$ $[M\text{-}H]^-$ m/z [a]	MS/MS Fragments m/z [a]
1	Quercetin 3-O-glucuronide	477	301
2	Quercetin 3-O-glucoside	463	300
3	Kaempferol-3-O-glucoside	447	284
4	Gallic acid	169	125
5	Not detected	-	-

m/*z*: mass divided by charge number; [a] positive ion mode.

Mass spectrometry analysis was performed using the following parameters: ESI (-), MS ^{e} in continuum mode, m/z range of 100–600, capillary voltage of 1.0–3.0 kV, and scan speed of 0.3 s. A collision energy ramp was set from 15 to 45 eV for high energy and 4 eV for low energy. The cone voltage was 40 V and argon was used as the collision gas. For m/z correction, a lock mass solution of Leucine enkephalin ($[M\text{-}H]^-$ m/z: 556.2771) was applied in 30 s intervals during the sample acquisition, and mass calibration was performed using sodium formate (100−1500 m/z). Data were acquired and analyzed by Waters Masslynx software (version 4.2).

2.9. Characterization of EAE and PLE Extracts

The total polyphenol content of EAE and PLE extracts was determined according to the method described by Singleton et al. [21] with some modifications in a multiplate reader (Perkin-Elmer, Waltham, MA, USA, Vitor TM X3) at a 750 nm wavelength. The results were expressed in mg of gallic acid per gram of dry extract. The total flavonoid content was determined according to the method reported by Dini et al. [22] with some modifications using a UV/VIS spectrophotometer (Thermo Scientific, ORION AQUAMATE 8000, Madison, WI, USA) at 415 nm; the results were expressed in mg of quercetin per gram of dry extract. The antioxidant capacity was evaluated using different assays, namely DPPH (2,2-diphenyl-1-picrylhydrazyl) [23], ORAC (oxygen radical absorbance capacity) [24], FRAP (ferric reducing agent) [25], and ABTS (2,2′-azino-bis (3-ethylbenz-thiazoline-6-sulfonic acid)) [26]. These results were expressed in μmol trolox equivalent (TE) per gram of dry extract. In addition, the phenol profile was determined by mass spectrometry according to the conditions described in Section 2.8. UV detection was performed at 280, 320, and 360 nm.

2.10. Statistical Analysis

The effect of extraction methods was assessed using analysis of variance (Statgraphics Centurion XV (Statistical Graphics Corp., Herndon, VA, USA). Differences were determined to be statistically significant when $p < 0.05$.

3. Results and Discussion

3.1. Detection of Molecules with Antioxidant Capacity in Pisco Grape Pomace Extracts

Oxidative stress plays an essential role in the development and progression of neurodegenerative diseases, and the free radical scavengers present in foods known as antioxidants are attributed to health benefits [27].

Figure 1 shows the HPTLC separation and the antioxidant capacity bioassay (DPPH) for the EAE and PLE extracts. The reaction of the bands with the DPPH reagent generated three yellow-colored bands representing the antioxidant compounds in both extracts. These bands demonstrate antioxidant potential by reducing the DPPH radical in the presence of an antioxidant fraction by hydrogen or electron exchange. Visually, the color of the test medium changes from purple (DPPH) to yellow (DPPH-H) [28].

Compounds present in each band were identified by mass spectrometry (Table 1). In band 1, the presence of quercetin-3-O-glucuronide was established by the presence of a parent ion at m/z 477 $[M\text{-}H]^-$ and a main fragment at m/z 301. In bands 2 and 4, the compounds determined were quercetin-3-O-glucoside (m/z 463 and fragment at 300) and gallic acid (m/z 169 and fragment at 125).

Several studies have reported the presence of these compounds in different varieties of grape pomace [29–32]. Jara-Palacios et al. [33] mentioned that quercetin-3-O-glucoside is the most abundant in different types of grape pomace in Moscatel varieties, while the concentration of quercetin-3-O-glucuronide is higher in Pedro Ximénez varieties, both varieties that make up the pomace used in this work.

In the pisco pomace extracts obtained by EAE and PLE in a previous study, a concentration of 2.5 mg/g d.w. of quercetin-3-rutinoside hydrate was determined by liquid chromatography. Quercetin is a potent antioxidant due to its ability to trap free radicals and bind metal ions [34,35]; this correlates with the values obtained in the antioxidant capacity. Among the properties possessed by the identified compounds are quercetin-3-O-glucuronide, which is a primary functional flavonoid with anti-inflammatory, antiatherogenic, and antioxidant properties [36]; quercetin-3-O-glucoside exerts antioxidant, anti-inflammatory, antihypertensive, and cytoprotective activities [35]; and gallic acid is a phenolic substance widely distributed in various plants that has potent antioxidant and antimicrobial activity and is more effective in scavenging free radicals than other phenolic compounds [37,38].

These results correlate with the total polyphenol and flavonoid content in both extracts, whose values varied between 33 and 63 mg GAE/g extract and 134–254 mg QE/g extract, respectively (Table 2). The flavonoid values were higher, which is related to the compounds obtained in the phenolic profile, which mainly belong to this group of compounds (Figure 2).

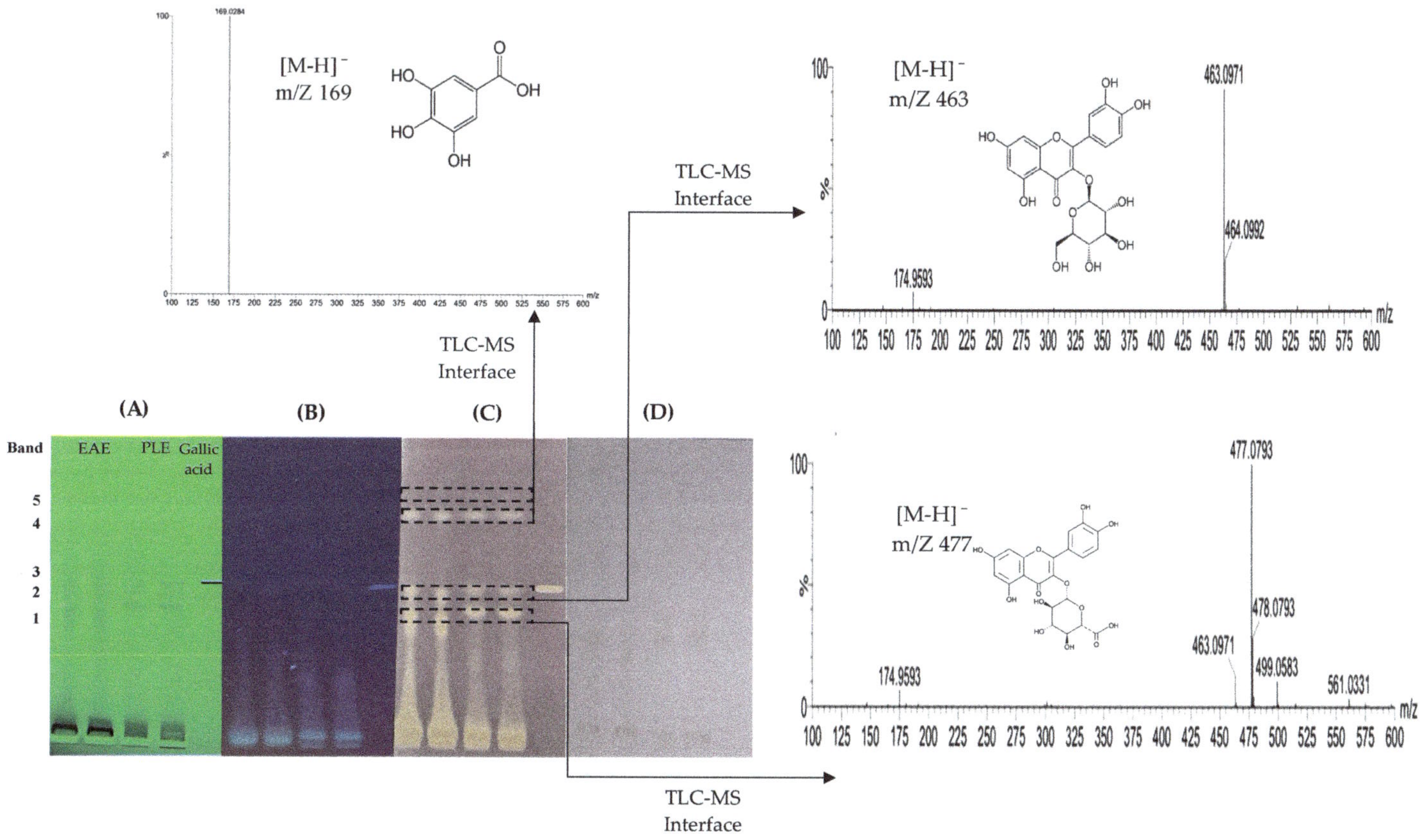

Figure 1. HPTLC chromatogram of pisco grape pomace extracts in TLC silica gel 60G F_{254}. Detection of polyphenolic compounds with antioxidant capacity by bioassay. Photo-documentation at 366 nm, green background (**A**). Photo-documentation at 254 nm, blue background (**B**). Antioxidant capacity by DPPH, as colorless bands on a purple background (**C**). Photo documentation under white light (**D**), positive control gallic acid, and HPTLC-ESI-MS of selected bands. EAE: enzyme-assisted extraction; PLE: pressurized liquids extraction.

Table 2. Determination of total phenolic compounds and antioxidant capacity of EAE and PLE extracts.

	Extraction Method	
Analysis	**EAE**	**PLE**
TPC (mg GAE)/g dry extract)	33.21 ± 1.54 [a]	62.68 ± 3.62 [b]
TFC (mg QE/g dry extract)	134.2 ± 2.80 [a]	254.17 ± 2.10 [b]
DPPH (μmol TE/g dry extract)	399.4 ± 23.54 [a]	557.61 ± 25.03 [b]
ORAC (μmol TE/g dry extract)	668.2 ± 26.3 [a]	871.28 ± 12.03 [b]
FRAP (μmol TE/g dry extract)	414.4 ± 11.73 [a]	668.22 ± 22.45 [b]
AATS (μmol TE/g dry extract)	317.4 ± 10.67 [a]	469.05 ± 14.65 [b]

Values with different letters (a,b) in the same row are significantly different ($p < 0.05$). TPC: total polyphenol content; TFC: total flavonoid content; EAE: enzyme-assisted extraction; PLE: pressurized liquids extraction.

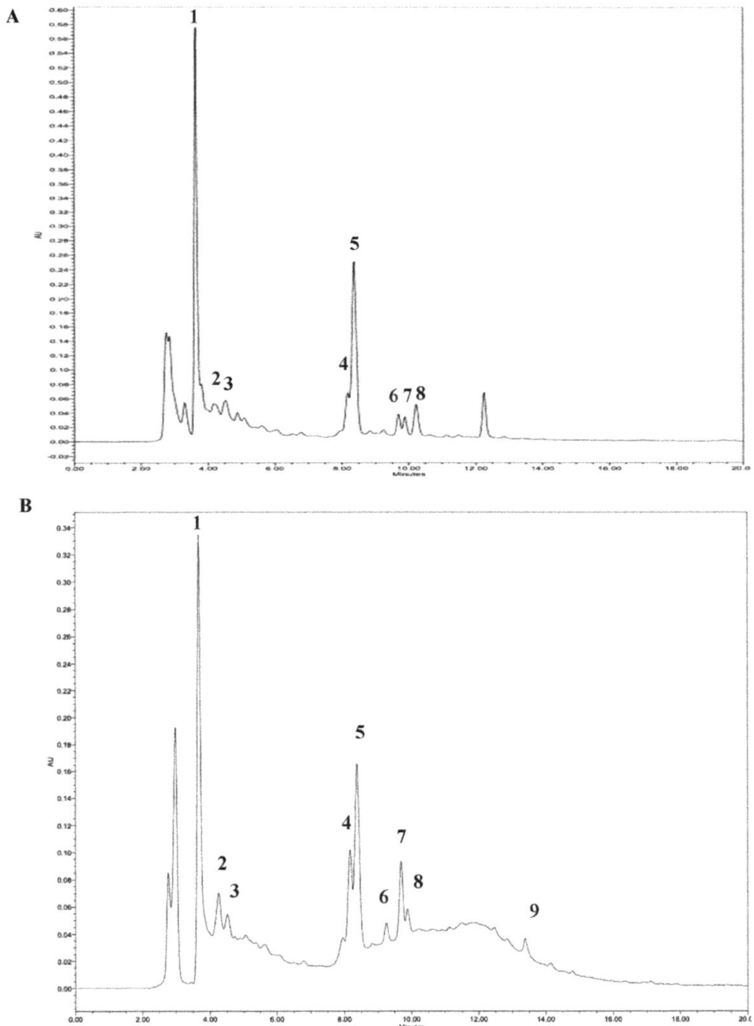

Figure 2. LC-MS data of phenolic compound profile of pisco grape pomace extracts. (**A**) Enzyme-assisted extraction; (**B**) extraction by pressurized liquids. (1) Gallic acid; (2) catechin; (3) (−) epi-catechin; (4) quercetin-3-O-glucoside; (5) quercetin-3-O-galactoside; (6) quercetin-3-O-glucoronide; (7) quercetin-3-O-rhamnoside; (8) kaempferol-3-O-glucoside; and (9) quercetin.

The antioxidant capacity evaluated by the different assays showed that PLE obtained the highest values compared to EAE. However, both extracts showed the same biological activities. The results obtained by ORAC showed the highest values between 668 and 871 μmol TE/g extract, while between DPPH, FRAP, and ABTS, the values varied between 317 and 668 μmol TE/g extract; these variations between the assays are attributed to the mechanism of action, as they are based on redox reactions, where a reagent changes color due to its oxidative state and ability to deplete the amount of radicals [39], while ORAC uses peroxyl radicals as reagents with redox potential and a reaction mechanism similar to physiological oxidants [40] (Table 2). In addition, the IC50 of DPPH was determined, obtaining values of 1.63 ± 0.15 mg/mL extract and 1.00 ± 0.01 mg/mL extract for EAE and PLE, respectively. The IC50 was evaluated to determine the extract concentration necessary to inhibit 50% free radical scavenging; low IC50 values indicate greater antioxidant efficacy [41].

It is important to highlight that grape pomace extracts can be used in the food sector as natural food additives due to their beneficial biological properties and in the pharmaceutical industry as dietary supplements and complementary medicines, considering the content of polyphenols and how they participate in health [42].

3.2. Detection of Acetylcholinesterase Inhibitors in Pomace Extracts

Alzheimer's disease (AD) is a chronic degenerative disease of the central nervous system that usually appears in older populations [43].

Acetylcholinesterase (AChE) has been used as a drug for the treatment of neurodegenerative disorders [44]. Its inhibition may be beneficial in patients with AD who suffer cognitive impairment [27].

The AChE assay is based on the enzymatic activity converting the substrate (α-naphthyl acetate to naphthol) and its azo bonding with Fast Blue B salt (FBB) to form the purple diazonium dye (bottom of the plate) [14] (Figure 3). The HPTLC–bioassay for acetylcholinesterase inhibitor detection shows colorless bands on a purple background (EAE and PLE). MS identified the compound as kaempferol 3-O-glucoside, corresponding to band 3 with a parent ion of m/z 447 $[M\text{-}H]^-$ and the main fragment at m/z 284 (Table 1). This compound has been previously reported in grape pomace [30,31,33,45]. This compound belongs to the group of flavonoids and is commonly found as glucosides, which are the most relevant. In addition, it was reported as a potential therapeutic agent because it exhibits neuroprotective activities [46,47]; the presence of this compound is related to the high values of total flavonoids obtained (Table 2).

Grape plant extracts and metabolites are an alternative in preventing and treating AD. As mentioned above, the flavonoids possess AChE inhibitory activity [48].

The HPTLC-AChE tool has been used to identify inhibitors of this enzyme in fig leaf, custard apple, sage, ginger, hemp seed oil, flax, and canola samples [3,15,16,49], but no report was found in grapes.

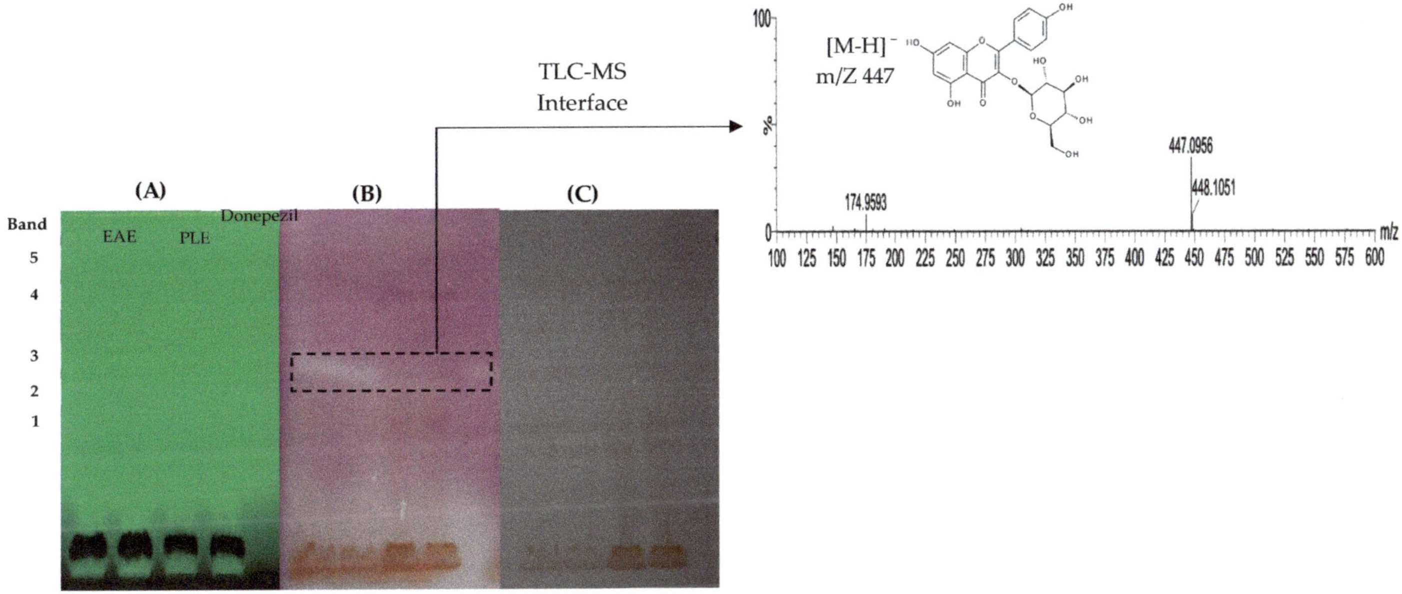

Figure 3. HPTLC chromatogram of pisco grape pomace extracts in TLC silica gel 60G F_{254}. Detection of polyphenolic compounds inhibiting acetylcholinesterase by bioassay. Photo documentation at 366 nm, green background (**A**). Acetylcholinesterase inhibition, purple background (**B**). Photo documented under white light (**C**), positive control Donepezil, and HPLC-ESI-MS of selected bands. EAE: enzyme-assisted extraction; PLE: pressurized liquids extraction.

3.3. Detection of Cyclooxygenase (COX-2) Inhibitors from Pisco Grape Pomace Extracts

Selective cyclooxygenase-2 inhibitor anti-inflammatory drugs are a specific category of non-steroidal anti-inflammatory drugs designed to reduce the production of inflammatory substances [50].

Cyclooxygenase (COX-2) is involved in the inflammatory process, triggering complex physiological responses in injured tissues [20].

Bands with inhibitory molecules were observed as colorless bands on a purple background (Figure 4) By MS, the COX-2 inhibitor was identified as gallic acid with a parent ion at m/z 169 $[M\text{-}H]^-$ and a main fragment at m/z 125 (Table 1). This phenolic compound also exhibited antioxidant activity, as mentioned in Section 3.1. This finding is in agreement with the phenolic profile, where gallic acid was observed at 3.6 min (Figure 2). The presence of gallic acid has been reported in various types of grape pomaces [30,31,51–53], as well as its anti-inflammatory potential, antioxidant, and anti-diabetic agents [54,55].

The search and identification of antioxidant and inhibitory molecules beneficial for health from Pisco grape pomace obtained through green extraction techniques (EAE and PLE) can be an alternative to develop novel functional foods and, at the same time, generate a circular economy. HPTLC–bioassay–MS is a simple technique that can be used to obtain an early estimate of the biological effects of the compounds present in the extracts. The results obtained in this work reported, for the first time, molecules in pisco grape pomace with antioxidant activity, such as quercetin-3-O-glucuronide, quercetin-3-O-glucoside, and gallic acid. In addition, kaempferol-3-O-glucoside proved to be an inhibitor of acetylcholinesterase and gallic acid of cyclooxygenase. Therefore, pomace is a valuable subproduct and an agent with therapeutic potential for the food and pharmaceutical industries. In addition, these green extraction methods are sustainable processes. Although the results obtained are interesting, this study is not free of limitations. In particular, the techniques used to face scaled-up challenges make further research necessary to optimize these technological processes both at the pilot level and on an industrial scale.

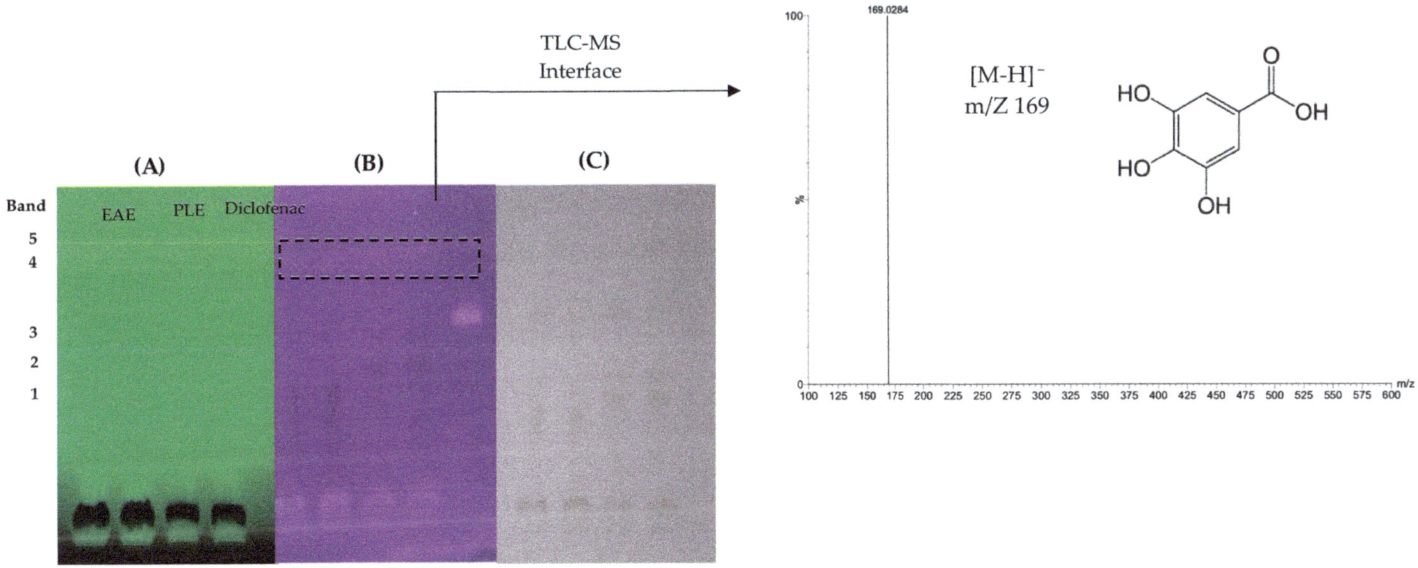

Figure 4. HPTLC chromatogram of pisco grape pomace extracts TLC silica gel 60G F_{254}. Detection of polyphenolic compounds inhibiting cyclooxygenase by bioassay. Photo documentation at 366 nm, green background (**A**). Cyclooxygenase inhibition, purple/blue background (**B**). Photo documentation under white light (**C**), positive control Diclofenac, and HPTLC-ESI-MS of selected bands. EAE: enzyme-assisted extraction; PLE: pressurized liquids extraction.

Author Contributions: Conceptualization, I.Q.-F.; data curation, J.P. and J.F.-M.; formal analysis, J.P. and I.Q.-F.; funding acquisition, M.A. and I.Q.-F.; investigation, J.P.; methodology, J.P., J.F.-M. and I.Q.-F.; project administration, I.Q.-F.; resources, I.Q.-F.; supervision, M.A. and I.Q.-F.; validation, I.Q.-F.; writing—original draft, J.P.; writing—review and editing, J.P., M.A. and I.Q.-F. All authors have read and agreed to the published version of the manuscript.

Funding: This research was funded by the National Agency for Research and Development (ANID) for the Initiation Fondecyt N° 11230866 and Regular Fondecyt N° 1211803. ANID also supported the national doctoral scholarship (Folio: 21221853) and Fondequip EQM 210203 and EQY220018.

Institutional Review Board Statement: Not applicable.

Informed Consent Statement: Not applicable.

Data Availability Statement: The data used or analyzed during the current study are available from the corresponding author upon reasonable request.

Acknowledgments: The authors thank the Department of Food Engineering of the University of La Serena and the Department of Chemistry and Pharmacy, Pontificia Universidad Católica de Chile. The company Pisco de Chile S.A. (CCU-Control), Sotaqui Plant, IV region, facilitated access to raw materials.

Conflicts of Interest: The company Pisquera de Chile S.A., which provided the pisco grape pomace product, had no role in the design of the study; in the collection, analysis, or interpretation of data; in the writing of the manuscript; or in the decision to publish the results.

References

1. Di Prima, G.; Belfiore, E.; Angellotti, G.; De Caro, V. Green Next-Generation Excipients Enriched in Polyphenols from Recovery of Grape Processing Waste Black Bentonite: Influence of Unconventional Extraction Solvents on Antioxidant Properties and Composition. *Sustain. Chem. Pharm.* **2024**, *37*, 101414. [CrossRef]
2. Kato-Schwartz, C.G.; Corrêa, R.C.G.; de Souza Lima, D.; de Sá-Nakanishi, A.B.; de Almeida Gonçalves, G.; Seixas, F.A.V.; Haminiuk, C.W.I.; Barros, L.; Ferreira, I.C.F.R.; Bracht, A.; et al. Potential Anti-Diabetic Properties of Merlot Grape Pomace Extract: An in Vitro, in Silico and in Vivo Study of α-Amylase and α-Glucosidase Inhibition. *Food Res. Int.* **2020**, *137*, 109462. [CrossRef] [PubMed]
3. Galarce-Bustos, O.; Pavón, J.; Henríquez-Aedo, K.; Aranda, M. Detection and Identification of Acetylcholinesterase Inhibitors in Annona Cherimola Mill. by Effect-Directed Analysis Using Thin-Layer Chromatography-Bioassay-Mass Spectrometry. *Phytochem. Anal.* **2019**, *30*, 679–686. [CrossRef] [PubMed]
4. World Health Organization. Available online: https://www.who.int/ (accessed on 27 June 2024).
5. Poblete, J.; Quispe-Fuentes, I.; Aranda, M.; Vega-Gálvez, A. Application of Vacuum and Convective Drying Processes for the Valorization of Pisco Grape Pomace to Enhance the Retention of Its Bioactive Compounds. *Waste Biomass Valorization* **2023**, *15*, 3093–3107. [CrossRef]
6. Rodríguez-Ramos, F.; Cañas-Sarazúa, R.; Briones-Labarca, V. Pisco Grape Pomace: Iron/Copper Speciation and Antioxidant Properties, towards Their Comprehensive Utilization. *Food Biosci.* **2022**, *47*, 101781. [CrossRef]
7. Allcca-Alca, E.E.; León-Calvo, N.C.; Luque-Vilca, O.M.; Martínez-Cifuentes, M.; Pérez-Correa, J.R.; Mariotti-Celis, M.S.; Huamán-Castilla, N.L. Hot Pressurized Liquid Extraction of Polyphenols from the Skin and Seeds of Vitis Vinifera L. Cv. Negra Criolla Pomace a Peruvian Native Pisco Industry Waste. *Agronomy* **2021**, *11*, 866. [CrossRef]
8. Verma, N. Himshweta Extraction of Phenolic Compounds by Conventional and Green Innovative Techniques. In *Green Sustainable Process for Chemical and Environmental Engineering and Science: Green Solvents and Extraction Technology*; Elsevier: Amsterdam, The Netherlands, 2023; pp. 355–394. ISBN 9780323951562.
9. Pimpley, V.A.; Murthy, P.S. Influence of Green Extraction Techniques on Green Coffee: Nutraceutical Compositions, Antioxidant Potential and in Vitro Bio-Accessibility of Phenolics. *Food Biosci.* **2021**, *43*, 101284. [CrossRef]
10. Machado, T.O.X.; Portugal, I.; de Kodel, H.A.C.; Droppa-Almeida, D.; Dos Santos Lima, M.; Fathi, F.; Oliveira, M.B.P.P.; de Albuquerque-Júnior, R.L.C.; Dariva, C.; Souto, E.B. Therapeutic Potential of Grape Pomace Extracts: A Review of Scientific Evidence. *Food Biosci.* **2024**, *60*, 104210. [CrossRef]
11. García, P.; Fredes, C.; Cea, I.; Lozano-Sánchez, J.; Leyva-Jiménez, F.J.; Robert, P.; Vergara, C.; Jimenez, P. Recovery of Bioactive Compounds from Pomegranate (*Punica Granatum* L.). *Foods* **2021**, *10*, 203. [CrossRef]
12. Cheng, B.; Wang, Q.; An, Y.; Chen, F. Recent Advances in the Total Synthesis of Galantamine, a Natural Medicine for Alzheimer's Disease. *Nat. Prod. Rep.* **2024**, *41*, 1060–1090. [CrossRef]
13. Galarce-Bustos, O.; Pavón-Pérez, J.; Henríquez-Aedo, K.; Aranda, M. An Improved Method for a Fast Screening of α-Glucosidase Inhibitors in Cherimoya Fruit (Annona Cherimola Mill.) Applying Effect-Directed Analysis via High-Performance Thin-Layer Chromatography-Bioassay-Mass Spectrometry. *J. Chromatogr. A* **2019**, *1608*, 460415. [CrossRef] [PubMed]

14. Agatonovic-Kustrin, S.; Wong, S.; Dolzhenko, A.V.; Gegechkori, V.; Ku, H.; Tucci, J.; Morton, D.W. Evaluation of Bioactive Compounds from Ficus Carica L. Leaf Extracts via High-Performance Thin-Layer Chromatography Combined with Effect-Directed Analysis. *J. Chromatogr. A* **2023**, *1706*, 464241. [CrossRef] [PubMed]
15. Reguigui, A.; Ott, P.G.; Darcsi, A.; Bakonyi, J.; Romdhane, M.; Móricz, Á.M. Nine-Dimensional Bioprofiles of Tunisian Sages (Salvia Officinalis, *S. Aegyptiaca* and *S. Verbenaca*) by High-Performance Thin-Layer Chromatography—Effect-Directed Analyses. *J. Chromatogr. A* **2023**, *1688*, 463704. [CrossRef] [PubMed]
16. Krüger, S.; Bergin, A.; Morlock, G.E. Effect-Directed Analysis of Ginger (Zingiber Officinale) and Its Food Products, and Quantification of Bioactive Compounds via High-Performance Thin-Layer Chromatography and Mass Spectrometry. *Food Chem.* **2018**, *243*, 258–268. [CrossRef]
17. Khan, H.; Marya; Amin, S.; Kamal, M.A.; Patel, S. Flavonoids as Acetylcholinesterase Inhibitors: Current Therapeutic Standing and Future Prospects. *Biomed. Pharmacother.* **2018**, *101*, 860–870. [CrossRef]
18. Patrignani, P.; Patrono, C. Cyclooxygenase Inhibitors: From Pharmacology to Clinical Read-Outs. *Biochim. Biophys. Acta Mol. Cell Biol. Lipids* **2015**, *1851*, 422–432. [CrossRef]
19. Paun, G.; Neagu, E.; Moroeanu, V.; Albu, C.; Ursu, T.M.; Zanfirescu, A.; Negres, S.; Chirita, C.; Radu, G.L. Anti-Inflammatory and Antioxidant Activities of the Impatiens Noli-Tangere and Stachys Officinalis Polyphenolic-Rich Extracts. *Rev. Bras. Farmacogn.* **2018**, *28*, 57–64. [CrossRef]
20. Oyarzún, P.; Carrasco, J.; Peterssen, D.; Tereucan, G.; Aranda, M.; Henríquez-Aedo, K. A High Throughput Method for Detection of Cyclooxygenase-2 Enzyme Inhibitors by Effect-Directed Analysis Applying High Performance Thin Layer Chromatography-Bioassay-Mass Spectrometry. *J. Chromatogr. A* **2023**, *1711*, 464426. [CrossRef]
21. Singleton, V.L.; Rossi, J.A. Colorimetry of Total Phenolics with Phosphomolybdic-Phosphotungstic Acid Reagents. *Am. J. Enol. Vitic.* **1965**, *16*, 144–158. [CrossRef]
22. Dini, I.; Tenore, G.C.; Dini, A. Antioxidant Compound Contents and Antioxidant Activity before and after Cooking in Sweet and Bitter Chenopodium Quinoa Seeds. *LWT* **2010**, *43*, 447–451. [CrossRef]
23. Grajeda-Iglesias, C.; Salas, E.; Barouh, N.; Baréa, B.; Panya, A.; Figueroa-Espinoza, M.C. Antioxidant Activity of Protocatechuates Evaluated by DPPH, ORAC, and CAT Methods. *Food Chem.* **2016**, *194*, 749–757. [CrossRef] [PubMed]
24. Zhang, L.; Li, J.; Hogan, S.; Chung, H.; Welbaum, G.E.; Zhou, K. Inhibitory Effect of Raspberries on Starch Digestive Enzyme and Their Antioxidant Properties and Phenolic Composition. *Food Chem.* **2010**, *119*, 592–599. [CrossRef]
25. Briones-Labarca, V.; Plaza-Morales, M.; Giovagnoli-Vicuña, C.; Jamett, F. High Hydrostatic Pressure and Ultrasound Extractions of Antioxidant Compounds, Sulforaphane and Fatty Acids from Chilean Papaya (Vasconcellea Pubescens) Seeds: Effects of Extraction Conditions and Methods. *LWT* **2015**, *60*, 525–534. [CrossRef]
26. Re, R.; Pellegrini, N.; Proteggente, A.; Pannala, A.; Yang, M.; Rice-Evans, C. Antioxidant activity applying an improved ABTS radical cation decolorization assay. *Free. Radic. Biol. Med.* **1999**, *26*, 1231–1237. [CrossRef] [PubMed]
27. Cabezudo, I.; Salazar, M.O.; Ramallo, I.A.; Furlan, R.L.E. Effect-Directed Analysis in Food by Thin-Layer Chromatography Assays. *Food Chem.* **2022**, *390*, 132937. [CrossRef]
28. Roussel, J.M.; Bardot, V.; Berthomier, L.; Cotte, C.; Dubourdeaux, M.; Holowacz, S.; Bernard-Savary, P. Application of the Life Cycle Management of Analytical Methods Concept to a HPTLC-DPPH Assay Method for Acteoside Content in Industrial Extracts of Plantago Lanceolata L. *J. Chromatogr. B Analyt Technol. Biomed. Life Sci.* **2021**, *1181*, 122923. [CrossRef]
29. Kadouh, H.C.; Sun, S.; Zhu, W.; Zhou, K. α-Glucosidase Inhibiting Activity and Bioactive Compounds of Six Red Wine Grape Pomace Extracts. *J. Funct. Foods* **2016**, *26*, 577–584. [CrossRef]
30. Abouelenein, D.; Mustafa, A.M.; Caprioli, G.; Ricciutelli, M.; Sagratini, G.; Vittori, S. Phenolic and Nutritional Profiles, and Antioxidant Activity of Grape Pomaces and Seeds from Lacrima Di Morro d'Alba and Verdicchio Varieties. *Food Biosci.* **2023**, *53*, 102808. [CrossRef]
31. Serni, E.; Tomada, S.; Haas, F.; Robatscher, P. Characterization of Phenolic Profile in Dried Grape Skin of Vitis Vinifera L. Cv. Pinot Blanc with UHPLC-MS/MS and Its Development during Ripening. *J. Food Compos. Anal.* **2022**, *114*, 104731. [CrossRef]
32. Kammerer, D.; Claus, A.; Carle, R.; Schieber, A. Polyphenol Screening of Pomace from Red and White Grape Varieties (*Vitis Vinifera* L.) by HPLC-DAD-MS/MS. *J. Agric. Food Chem.* **2004**, *52*, 4360–4367. [CrossRef]
33. José Jara-Palacios, M.; Hernanz, D.; Escudero-Gilete, M.L.; Heredia, F.J. Antioxidant Potential of White Grape Pomaces: Phenolic Composition and Antioxidant Capacity Measured by Spectrophotometric and Cyclic Voltammetry Methods. *Food Res. Int.* **2014**, *66*, 150–157. [CrossRef]
34. D'Andrea, G. Quercetin: A Flavonol with Multifaceted Therapeutic Applications? *Fitoterapia* **2015**, *106*, 256–271. [CrossRef] [PubMed]
35. Septembre-Malaterre, A.; Boumendjel, A.; Seteyen, A.L.S.; Boina, C.; Gasque, P.; Guiraud, P.; Sélambarom, J. Focus on the High Therapeutic Potentials of Quercetin and Its Derivatives. *Phytomedicine Plus* **2022**, *2*, 100220. [CrossRef] [PubMed]
36. Vollmannová, A.; Bojňanská, T.; Musilová, J.; Lidiková, J.; Cifrová, M. Quercetin as One of the Most Abundant Represented Biological Valuable Plant Components with Remarkable Chemoprotective Effects—A Review. *Heliyon* **2024**, *10*, e33342. [CrossRef]
37. Xiang, Z.; Guan, H.; Zhao, X.; Xie, Q.; Xie, Z.; Cai, F.; Dang, R.; Li, M.; Wang, C. Dietary Gallic Acid as an Antioxidant: A Review of Its Food Industry Applications, Health Benefits, Bioavailability, Nano-Delivery Systems, and Drug Interactions. *Food Res. Int.* **2024**, *180*, 114068. [CrossRef]

38. Rahmawati, I.; Pratama, A.W.; Pratama, S.A.; Khozin, M.N.; Firmanda, A.; Irawan, F.H.; Asranudin; Ansori, A.N.M.; Sucipto, T.H. Gallic Acid: A Promising Bioactive Agent for Food Preservation and Sustainable Packaging Development. *Case Stud. Chem. Environ. Eng.* **2024**, *10*, 100776. [CrossRef]
39. Pohanka, M. Assays of Antioxidant Capacity: Optics and Voltammetry. *Int. J. Electrochem. Sci.* **2023**, *18*, 100276. [CrossRef]
40. Bisby, R.H.; Brooke, R.; Navaratnam, S. Effect of Antioxidant Oxidation Potential in the Oxygen Radical Absorption Capacity (ORAC) Assay. *Food Chem.* **2008**, *108*, 1002–1007. [CrossRef]
41. Bello, A.A.; Katta, A.; Obaydo, R.H.; Jazmati, A. Phytochemical Analysis and Antioxidant Efficacy of Chrysojasminum fruticans (L.) Banfi in Syrian Flora. *Heliyon* **2024**, *10*, e37322. [CrossRef]
42. Wang, C.; You, Y.; Huang, W.; Zhan, J. The High-Value and Sustainable Utilization of Grape Pomace: A Review. *Food Chem. X* **2024**, *24*, 101845. [CrossRef]
43. Dai, R.; Sun, Y.; Su, R.; Gao, H. Anti-Alzheimer's Disease Potential of Traditional Chinese Medicinal Herbs as Inhibitors of BACE1 and AChE Enzymes. *Biomed. Pharmacother.* **2022**, *154*, 113576. [CrossRef] [PubMed]
44. Surya Prakash, V.; Radhakrishnan, N.; Vasantha-Srinivasan, P.; Veeramani, C.; El Newehy, A.S.; Alsaif, M.A.; Al-Numair, K.S. In Silico Analysis of Selected Nutrition Rich Fruit of Bunch Berry (*Lantana camara*) Constituents as Human Acetylcholinesterase (HAchE), Carbonic Anhydrase II (HCA-II) and Carboxylesterase 1 (HCES-1) Inhibitory Agents. *Saudi J. Biol. Sci.* **2023**, *30*, 103847. [CrossRef] [PubMed]
45. Huamán-Castilla, N.L.; Campos, D.; García-Ríos, D.; Parada, J.; Martínez-Cifuentes, M.; Mariotti-Celis, M.S.; Pérez-Correa, J.R. Chemical Properties of Vitis Vinifera Carménère Pomace Extracts Obtained by Hot Pressurized Liquid Extraction, and Their Inhibitory Effect on Type 2 Diabetes Mellitus Related Enzymes. *Antioxidants* **2021**, *10*, 472. [CrossRef] [PubMed]
46. Jin, S.; Zhang, L.; Wang, L. Kaempferol, a Potential Neuroprotective Agent in Neurodegenerative Diseases: From Chemistry to Medicine. *Biomed. Pharmacother.* **2023**, *165*, 115215. [CrossRef]
47. Hussain, M.S.; Altamimi, A.S.A.; Afzal, M.; Almalki, W.H.; Kazmi, I.; Alzarea, S.I.; Gupta, G.; Shahwan, M.; Kukreti, N.; Wong, L.S.; et al. Kaempferol: Paving the Path for Advanced Treatments in Aging-Related Diseases. *Exp. Gerontol.* **2024**, *188*, 112389. [CrossRef]
48. Dwibedi, V.; Jain, S.; Singhal, D.; Mittal, A.; Rath, S.K.; Saxena, S. Inhibitory Activities of Grape Bioactive Compounds against Enzymes Linked with Human Diseases. *Appl. Microbiol. Biotechnol.* **2022**, *106*, 1399–1417. [CrossRef]
49. Teh, S.S.; Morlock, G.E. Effect-Directed Analysis of Cold-Pressed Hemp, Flax and Canola Seed Oils by Planar Chromatography Linked with (Bio)Assays and Mass Spectrometry. *Food Chem.* **2015**, *187*, 460–468. [CrossRef]
50. Biase, T.M.M.A.; Rocha, J.G.M.; Silva, M.T.; Ribeiro-Vaz, I.; Galvao, T.F. Renal Effects of Selective Cyclooxygenase-2 Inhibitor Anti-Inflammatory Drugs: A Systematic Review and Meta-Analysis. *Explor. Res. Clin. Soc. Pharm.* **2024**, *15*, 100475. [CrossRef]
51. Majeed, U.; Shafi, A.; Majeed, H.; Akram, K.; Liu, X.; Ye, J.; Luo, Y. Grape (*Vitis Vinifera* L.) Phytochemicals and Their Biochemical Protective Mechanisms against Leading Pathologies. *Food Chem.* **2023**, *405*, 134762. [CrossRef]
52. Bilawal, A.; Ishfaq, M.; Gantumur, M.A.; Qayum, A.; Shi, R.; Fazilani, S.A.; Anwar, A.; Jiang, Z.; Hou, J. A Review of the Bioactive Ingredients of Berries and Their Applications in Curing Diseases. *Food Biosci.* **2021**, *44*, 101407. [CrossRef]
53. Venkidasamy, B.; Samynathan, R.; Ramasamy, P.; Santhosh Kumar, M.P.; Thiruvengadam, M.; Khayrullin, M.; Shariati, M.A.; Nile, A.S.; Nile, S.H. Unveiling Novel Applications of Fruit Pomace for Sustainable Production of Value-Added Products and Health Benefits: A Review. *Food Biosci.* **2024**, *61*, 104533. [CrossRef]
54. Zhao, X.; Zhang, L.; Wu, N.; Liu, Y.; Xie, J.; Su, L.; Zhu, Q.; Unger, B.H.; Altaf, F.; Hu, Y.; et al. Gallic Acid Acts as an Anti-Inflammatory Agent via PPARγ-Mediated Immunomodulation and Antioxidation in Fish Gut-Liver Axis. *Aquaculture* **2024**, *578*, 740142. [CrossRef]
55. Jajo, H.; Baishya, T.; Das, P.; Ashraf, G.J.; Dua, T.K.; Paul, P.; Nandi, G.; Sahu, R. GC-MS and HPTLC Bioautography-Based Phytochemical Profiling and Evaluation of Biological Activity Neptunia Prostrata Linn Whole Plant and Leaves. *Pharmacol. Res.—Nat. Prod.* **2024**, *2*, 100013. [CrossRef]

antioxidants

MDPI

Article

Metabolome and Metagenome Integration Unveiled Synthesis Pathways of Novel Antioxidant Peptides in Fermented Lignocellulosic Biomass of Palm Kernel Meal

Hammad Qamar [1], Rong He [1], Yuanfei Li [2], Min Song [1], Dun Deng [1], Yiyan Cui [1], Miao Yu [1,3] and Xianyong Ma [1,3,*]

1 Institute of Animal Science, Guangdong Academy of Agricultural Sciences, State Key Laboratory of Livestock and Poultry Breeding, Key Laboratory of Animal Nutrition and Feed Science in South China, Ministry of Agriculture and Rural Affairs, Guangdong Provincial Key Laboratory of Animal Breeding and Nutrition, Guangzhou 510640, China; drhammadqamar@gmail.com (H.Q.); herong4747@163.com (R.H.); song-min@gdaas.cn (M.S.); dengdun@gdaas.cn (D.D.); cuiyiyan@gdaas.cn (Y.C.); yumiao@gdaas.cn (M.Y.)

2 Institute of Biological Technology, Jiangxi Provincial Key Laboratory of Poultry Genetic Improvement, Nanchang Normal University, Nanchang 330032, China; li-yuan-fei@outlook.com

3 Maoming Branch, Guangdong Laboratory for Lingnan Modern Agricultural, Maoming 525000, China

* Correspondence: maxianyong@gdaas.cn

Abstract: Approximately one-third of the entire world's food resources are deemed to be wasted. Palm kernel meal (PKM), a product that is extensively generated by the palm oil industry, exhibits a unique nutrient-rich composition. However, its recycling is seldom prioritized due to numerous factors. To evaluate the impact of enzymatic pretreatment and *Lactobacillus plantarum* and *Lactobacillus reuteri* fermentation upon the antioxidant activity of PKM, we implemented integrated metagenomics and metabolomics approaches. The substantially enhanced ($p < 0.05$) property of free radicals scavenging, as well as total flavonoids and polyphenols, demonstrated that the biotreated PKM exhibited superior antioxidant capacity. Non-targeted metabolomics disclosed that the *Lactobacillus* fermentation resulted in substantial ($p < 0.05$) biosynthesis of 25 unique antioxidant biopeptides, along with the increased ($p < 0.05$) enrichment ratio of the isoflavonoids and secondary metabolites biosynthesis pathways. The 16sRNA sequencing and correlation analysis revealed that *Limosilactobacillus reuteri, Pediococcus acidilactici, Lacticaseibacillus paracasei, Pediococcus pentosaceus, Lactiplantibacillus plantarum, Limosilactobacillus fermentum*, and polysaccharide lyases had significantly dominated ($p < 0.05$) proportions in PMEL, and these bacterial species were strongly ($p < 0.05$) positively interrelated with antioxidants peptides. Fermented PKM improves nutritional value by enhancing beneficial probiotics, enzymes, and antioxidants and minimizing anti-nutritional factors, rendering it an invaluable feed ingredient and gut health promoter for animals, multifunctional food elements, or as an ingredient in sustainable plant-based diets for human utilization, and functioning as a culture substrate in the food sector.

Keywords: enzymes; fermentation; metabolomics; metagenomics; antioxidants; *Lactobacillus*; palm kernel meal

1. Introduction

Processing foods generates around 33% of the worldwide wasted food, therefore giving rise to substantial economic, environmental, and climatic concerns [1]. Across the world, extraction and utilization of primary vegetable oils have seen substantial growth, with palm oil emerging as the dominating processed plant-based edible oil [2]. Currently, palm oil accounts for roughly 33% of the worldwide need for vegetable oils. Its annual supply is expected to reach 80.19 million tons by 2024. Consequently, the edible oil sector is expected to generate around 10.87 million tons of palm kernel meal (PKM) in 2024 [3,4]. Overall, PKM is regarded as a major protein and energy resource with a composition of 91% dry matter, 13–22% crude protein, 16–34% crude fiber, 11% oil, and 5.5% ash [5]. Furthermore, it serves as a reservoir of certain macronutrients, including calcium and phosphorus [6,7]. The difficult-to-digest components consist of non-starch polysaccharides (NSP), including 3% water-insoluble glucoxylans, 3% arabinoxylans, 12% cellulose, and 78% β-mannan [8]. Although the PKM has significant potential for exploitation, it remains underused despite its affordability and widespread availability [7]. The current application of PKM is primarily limited to using it for organic fertilizer, feed for animals, or energy production using biogas or combustion [9]. Utilizing PKM as fertilizer is restricted by its high oil percentage, which may hinder soil aeration and microbial growth [10]. As animal feed, its substantial amount of fiber may reduce digestibility in monogastric species like chickens [11]. Energy production techniques such as generating biogas are obstructed by the complicated lignocellulosic structure, which challenges microbial digestion [8]. In addition, PKM is utilized for power and steam production or is discarded at specified dumps, contributing to environmental pollution and greenhouse gas emissions (0.25–0.5 tons of CH_4 is estimated to be produced per ton of PKM discarded) [12,13]. Therefore, while these methods offer possibilities, they do not come without challenges, necessitating additional investigation for efficient and sustainable ways of disposing.

The extraction of vital nutrients from PKM exhibits a substantial challenge since nutrients and proteins are strongly embedded into cellulose and hemicellulose, i.e., mannans and arabinoxylans [14,15]. To surmount this difficulty, of all the available alternatives, biologically pretreatment processes stand out as a more environmentally sustainable solution [16]. The enzymatic approach for hydrolysis of PKM is preferred due to its moderate conditions for operation, substrate precision, and environmental sustainability. Contrary to chemical hydrolysis, enzymatic procedures run at lower temperatures and neutral pH levels, which assists in maintaining the nutritional integrity and bioactive characteristics of the resulting hydrolysates and avoids the formation of harmful by-products [16,17]. Enzymes including cellulases and hemicellulases especially target the cellulose and hemicellulose portions, enabling the targeted emission of bioactive peptides, amino acids, or fermentable carbohydrates without creating harmful compounds [18]. This method additionally improves the digestibility of PKM, making it ideal for use in feeds for animals by improving nutrient accessibility [11]. The enzymatic operation can be tailored and scaled up, providing a more sustainable and economically feasible strategy for PKM utilization in different industries.

The beneficial effects of *Lactobacillus* fermentation for augmenting the antioxidant properties of numerous agricultural industries' secondary products are becoming more prevalent. The process of fermentation initiated by *Lactobacillus* spp. assists in the dispensation and formation of antioxidant compounds, for instance, flavonoids, isoflavonoids, and polyphenolic compounds, from plant-derived ingredients [19]. *Lactobacillus* fermentation improves antioxidant capability through enzyme-mediated hydrolysis, creating bioactive compounds and enhancing the activity of antioxidant enzymes. This procedure boosts the bioavailability of antioxidants and reduces chemical elements that inhibit the action of antioxidants, consequently boosting the overall antioxidant potential of the final fermentation product [20]. The *Lactobacillus* fermentation efficiently degrades anti-nutritional elements, like phytic acid and tannins, that naturally occur in PKM and improves the bioavailability of important nutrients, including amino acids and minerals [21]. The creation of lactic acid by lactic acid bacteria decreases the pH, hindering the growth of spoilage bacteria and enhancing the meal's shelf life [22]. Furthermore, *Lactobacilli* fermentation may supplement PKM with health-promoting probiotics, enhancing the gut functioning of livestock [23]. This approach is ecologically friendly, demanding no chemical compounds and producing less waste than conventional chemical treatments, rendering it a desirable choice for sustainable processing of PKM [24].

The overall nutritional value of fermented food is primarily governed by the microbial population and their main and secondary products [25]. Modern sequencing research, "omics" technology, and the integration of microbiome and metabolomic approaches have been demonstrated to be excellent tools for undertaking comprehensive studies of microbial populations and analyzing the metabolites of fermented substrates [26]. Metabolomics enables the analysis of several individual metabolites simultaneously rather than in groupings. Metabolomics is a useful tool for assessing the different metabolites involved in PKM fermentation. These approaches seem to be effective in determining how the complete microbial population and microbial secondary metabolites affect the nutritional content and functional parameters of fermented foods [27,28].

Very little emphasis has been paid to reducing the anti-nutritional components and increasing the nutritive value of PKM using a mix of enzymatic degradation and fermentation by microbes. The objective of the proposed study is to assess how enzymatic treatment can boost *Lactobacilli* fermentation and improve the nutritive value and antioxidant properties of PKM, as well as to investigate the metabolic changes and mechanisms involved during the fermentation process.

2. Materials and Methods

2.1. Enzymes and Microbial Strains

Kangyuan Hezhong Biotech (Zibo, China) and Xihai Biotech (Xi'an, China) provided the commercially prepared probiotics, *Lactobacillus plantarum* and *Lactobacillus reuteri*, respectively. The bacterial strains used for fermenting substrate possessed 10^{10} CFU/g of concentration. The two commercial enzymes (pectinase and cellulase) were gratefully delivered by VTR Biotechnology (Zhuhai, China), having an activity of 750,000 U/g and 50,000 U/g, respectively. The normal dosage amounts for cellulase and pectinase were 0.5 g and 0.3 g per kilogram, respectively. The commercially available enzyme (xylanase) was acquired from Sunson Biotechnology Limited (Beijing, China), having an activity of 30,000 U/g, and the dosage rate was 0.3 g per kilogram. Pectinase was selected because it targets pectin and helps in breaking the plant cell wall structure, encouraging the accessibility of other enzymes to cellulose and hemicellulose [29]. PKM is enriched with cellulose, and cellulase is particularly successful at converting this cellulose into fermentable sugars [30]. Xylanase was chosen since it hydrolyzes hemicellulose, specifically xylan, which is an essential element of the PKM cell wall [31].

After weighing precisely, all enzymes were dissolved in 3 mL of deionized water with the aid of an oscillator for around 10 min. Following that, the centrifugation of the enzyme mixture was performed at 5000× *g* for 10 min, and PKM was hydrolyzed using supernatant in subsequent experiments.

2.2. Palm Kernel Meal Hydrolysis and Lactobacilli Fermentation

The PKM was bought from the local feed mill. Prior to the experiment, PKM was originally processed by crushing it using a centrifugal grinder (Ultra-speed grinder, ZX Instrument, Beijing, China) to decrease the particle size. After crushing, the material was passed using a 40-mesh screen, which functioned to separate the larger particles from the finer ones. The larger particles that failed to go through the sieve were taken apart, re-ground utilizing the same centrifugal crusher, and screened once again via the 40-mesh sieve to ensure uniformity in particle size. This process was continued until the particles were adequately homogeneous. Eventually, only the fine, unified fragments that passed easily across the 40-mesh sieve were chosen for further experimentation for better consistency in the following analyses and procedures.

The PKM was fermented in 150 mL conical flasks sealed with aluminum foil and butyl rubber corks. There were three treatments, namely, the control group, with PKM only (CON); PKM with enzymes only (PME); and PKM with enzymes and *Lactobacilli* bacteria (PMEL). The QC01, QC02, and QC03 denote quality control samples. The H1, H2, and H3 represent the CON group. The H8, H9, and H10 symbolize the PME group, and H30, H31, and H32 denote the PMEL treatment.

Each treatment was run with six replicates. The 10 g of PKM and 37 mL of distilled water were added to each flask. At the beginning of the trial, 3 mL enzyme solution was poured into both the PME and PMEL treatments, while 3 mL of distilled water was poured into the CON treatment only. All the flasks were then transferred onto a shaker that was set to operate at 50 °C and 150 rpm for 28 h. Following 28 h, 2 g of the *Lactobacilli* mixture (1 g each of *L. plantarum* and *L. reuteri*) was incorporated into the PMEL alone, and the shaker was once again utilized for a further 48 h. The commercial enzymes exhibit peak efficiency around 50 °C. This temperature guarantees that the enzymes retain their catalytic activity without being denatured, thereby increasing the conversion rate of the substrate into wanted products [32]. The duration of 28 h is preferred to allow appropriate time for the enzymes to act completely on the substrate, warranting complete hydrolysis with no excessive degradation or loss of functional elements [33]. A 48 h period is perfect because it offers enough time for the *Lactobacillus* to multiply and generate lactic acid and other secondary metabolites, which is crucial to the fermentation process [19].

For further examination, the PKM samples were procured and saved at −80 °C after the experiment.

2.3. Metabolome Analysis

The methods previously mentioned were used to perform the metabolome analysis [34–36]. In short, 50 mg of PKM specimen and 6 mm crushing beads were added to a 2 mL centrifugal tube. Metabolite extraction was carried out employing a 400 µL retrieval solution (methanol–water = 4:1 v/v) comprising 0.02 mg/mL of an internal reference compound (L-2-chlorophenylalanine). All the samples were ground for 6 min (−10 °C, 50 Hz) utilizing a Wonbio-96c frozen specimen grinding machine (Shanghai Wanbo Biotechnology, Shanghai, China). After that, they were purified for 30 min at a low temperature (5 °C, 40 kHz) utilizing ultrasonic purifying technology. The collected samples were then centrifuged for 15 min at 4 °C and 13,000× g after being held at −20 °C for 30 min. The supernatant was subsequently shifted to an infusion tube for the analysis of metabolites via the high-performance liquid chromatography-mass spectrometry (HPLC-MS) technique utilizing a ThermoFisher HPLC apparatus (Thermo Fisher Scientific Inc., Waltham, MA, USA). This HPLC system contained a particular ACQUITY HSS T3 column (100 mm × 2.1 mm, 1.8 µm; Waters, Milford, MA, USA). For the separation process, two mobile phases were utilized: one made of 0.1% formic acid in a solution of water and acetonitrile (95:5, v/v), and the other was developed of 0.1% formic acid in a combination of acetonitrile, isopropanol, and water (47.5:47.5, v/v). The system functioned at a flow rate of 0.40 mL per minute, with the column maintained at a constant temperature of 40 °C. Every specimen injection volume was 3 µL. The mass spectrometric data were acquired employing the same HPLC system, packed with an electrospray ionization (ESI) source that operated in positive as well as negative ion modes. The analysis governed a mass range of 70 to 1050 m/z, offering comprehensive identification of the compounds found in the sample.

The metabolites were recognized by exploring the HMDB database (http://www.hmdb.ca/ (accessed on 15 May 2024)) and the KEGG database (http://www.genome.jp/kegg (accessed on 15 May 2024)). The distinct metabolites between the two groups were tracked into their biological pathways via metabolite enrichment and pathways analysis following the KEGG database of metabolites (https://www.genome.jp/kegg/pathway.html (accessed on 25 May 2024)). The Python package "scipy.stats" (https://docs.scipy.org/doc/scipy/ (accessed on 25 June 2024)) was utilized to carry out the analysis of principal components (PCA) and orthogonal least partial squares discriminant assessment (OPLS-DA), the reliability of the model (assessed using a 7-cycle interactive validation process), and to carry out an enrichment assessment to identify the most pertinent biological processes for different treatments. The metabolites with a VIP value of more than 1 and a p-value of less than 0.05 were identified as substantially distinct metabolites employing the variable significance in the projection (VIP) derived from the OPLS-DA model and the p-value generated through the Student's t-test.

2.4. Metagenome Analysis

Following the previously explained methodologies, the extraction of DNA, metagenome sequencing process, and functional interpretations were carried out for metagenomic analysis [37]. The PKM quantity utilized for extracting DNA was 1 mL. In short, the Mag-Bind® Soil DNA Kit was utilized to acquire entire genomic DNA from PKM samples following the directions provided by the manufacturer (Omega Biotek, Norcross, GA, USA). The NEXTFLEX Rapid DNA-Seq (Bio Scientific, Austin, TX, USA) was used for library generation, and DNA sequencing was accomplished on Illumina NovaSeq 6000 (Illumina Inc., San Diego, CA, USA) following the producer's directions (www.illumina.com (accessed on 28 May 2024)).

After eliminating adaptors from the paired-end Illumina reads, poor-quality reads (less than 50 bp in length, less than 20 quality points, or N bases) were eliminated using fastp [38] (https://github.com/OpenGene/fastp (accessed on 30 May 2024)). Megahit (v 1.1.2) was employed for assembling the metagenomics dataset [39]. The eventual assembly outcome was contigs having a length of ≥300 bp, which were later utilized to carry out gene identification and annotations. The prodigal program was utilized for the prediction of the open reading frames (ORFs) across each assembled contig [40] (https://github.com/hyattpd/Prodigal/ (accessed on 12 June 2024)). Utilizing CD-HIT, a non-redundant genetic catalog was developed [41] (http://www.bioinformatics.org/cd-hit/ (accessed on 15 June 2024)). By employing SOAPaligner, the top-quality reads were correlated to non-redundant genetic catalogs to determine gene proportions possessing 95% similarity [42] (https://github.com/ShujiaHuang/SOAPaligner (accessed on 22 June 2024)). The Diamond was implemented to align the selected sequences from the non-redundant (NR) gene catalog to the NR dataset for assessing the taxonomic annotations [43] (http://diamondsearch.org/forums/index.php (accessed on 22 June 2024)). The Diamond tool was employed [43] (http://diamondsearch.org/forums/index.php (accessed on 22 June 2024)) for performing the KEGG annotation via mapping on the KEGG database (https://www.genome.jp/kegg/annotation/ (accessed on 22 June 2024)). The carbohydrate-active enzyme (CAZy) annotation was performed utilizing hmmscan (https://www.ebi.ac.uk/Tools/hmmer/search/hmmscan (accessed on 22 June 2024)) by mapping on the CAZy database (http://www.cazy.org/ (accessed on 22 June 2024)).

2.5. Antioxidant Capacity of PKM and Polyphenols and Flavonoids Content

The ABTS and the DPPH methods were implemented to assess the overall antioxidant activity of PKM, and commercially available kits were utilized for the determination of the total flavonoid and polyphenol percentage under the manufacturer's guidelines [44,45]. All the commercial kits were procured from Michy Bioengineering (Suzhou, China).

The antioxidant capability of the PKM sample was determined via the ABTS method by detecting the modification in absorption at 734 nm against the control substance Trolox. A 1 mL extracting reagent was added after weighing a 0.10 g PKM sample. Post ice-based homogenization, centrifugation of PKM samples was carried out for 10 min at 10,000× g and 4 °C. For further experiments, the fluid portion was collected.

To determine the specimen's antioxidant capacity via the DPPH method, the measurement of absorption variation was taken at 515 nm against the control substance Trolox. After weighing 0.10 g of PKM, 1 mL of extracting reagent was introduced. After homogenizing the specimens on ice, they went through a centrifuge under a temperature of 4 °C at 10,000× g for 10 min. A supernatant was acquired for further assessment.

To determine the flavonoid percentage, the PKM samples were desiccated to a constant mass, grounded, and sifted through a 40-mesh screen. The 60% ethanol was poured into 0.05 g of PKM. The solution was subsequently blended at 60 °C for 2 h before being centrifuged at 25 °C for 10 min at 10,000× g. The resulting solution above the sediment was collected for further analysis. The flavonoid percentage was determined by finding the absorption value of the isolate from the specimen at 510 nm.

The overall polyphenol concentration of the PKM material was determined by homogenizing it after drying to a uniform weight and then screening it over a 40-mesh sieve.

A precisely weighed 0.02 g PKM sample was mixed with 1 mL of the extracting solvent. At 60 °C, a two-hour agitation was performed, and the resultant mixture was subjected to a ten-minute centrifugal rotation with force of 10,000× *g* at 25 °C temperature. The liquid phase was extracted to do further tests. The total amount of phenol in this specimen was estimated via measurement of its absorption at 765 nm. All the above specimens were analyzed for absorption employing the microplate reader.

2.6. Statistical Analysis

In the present research, we employed statistical tools from the Python module "scipy.stats" (https://docs.scipy.org/doc/scipy/ (accessed on 25 June 2024)) for assessing microorganisms' taxonomic classification, CAZymes, KEGG pathways, and metabolites of interest. The Mann–Whitney test was utilized to determine variances among groups. To evaluate the interactions between microbial species and compounds, we applied Spearman correlation modeling, considering solely correlations possessing an absolute value of $|r|$ above 0.6 and a p-value below 0.05 as having significance. Significant variances were noted at $p < 0.05$, and to further regulate for false positives, we customized the p-values employing the false discovery rate (FDR), with the ultimate significance threshold established at an adjusted p-value of ≤ 0.05. The Gephi application was subsequently employed to visualize the correlation networks (https://gephi.org/ (accessed on 25 June 2024)).

3. Results

3.1. Multi-Statistical Analysis

The coefficients related to the correlation of H30 with H31 and H32 were 0.97 and 0.98, correspondingly. The coefficients related to the correlation of H1 with H2 and H3 were 0.96 and 0.97, respectively. The CON samples showed a stronger correlation with the PME samples than the PMEL samples (Figure 1A). The PMEL asserted the greatest number of entire metabolites (1355), followed by PME (1335) and CON (1333) (Figure 1B).

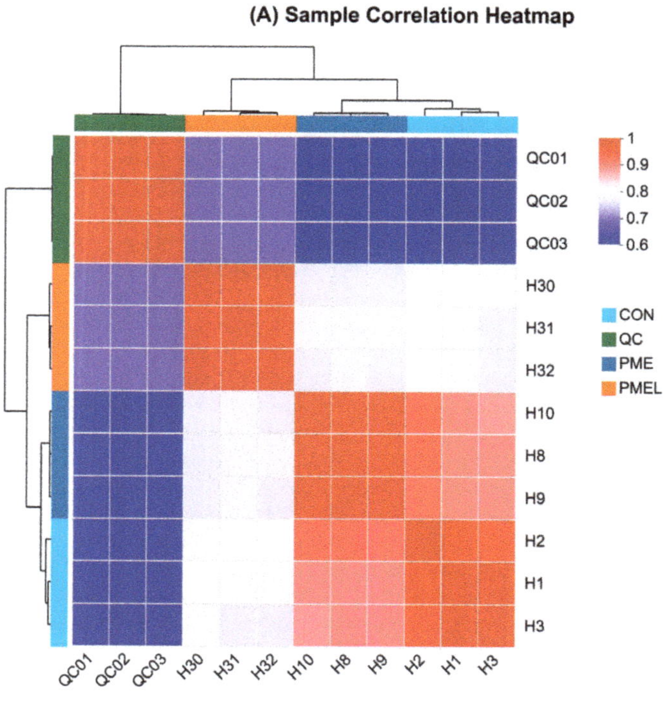

Figure 1. *Cont.*

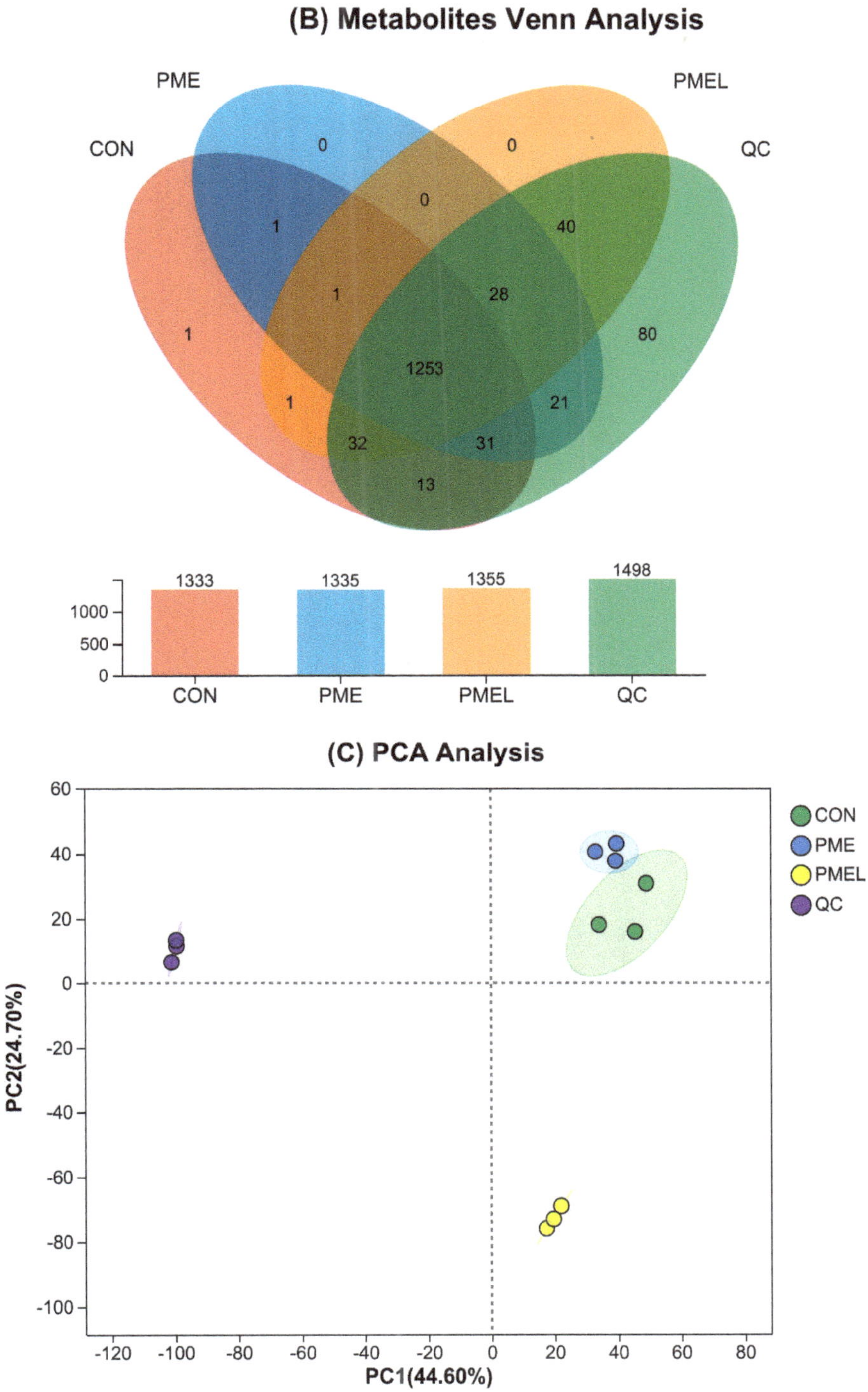

Figure 1. *Cont.*

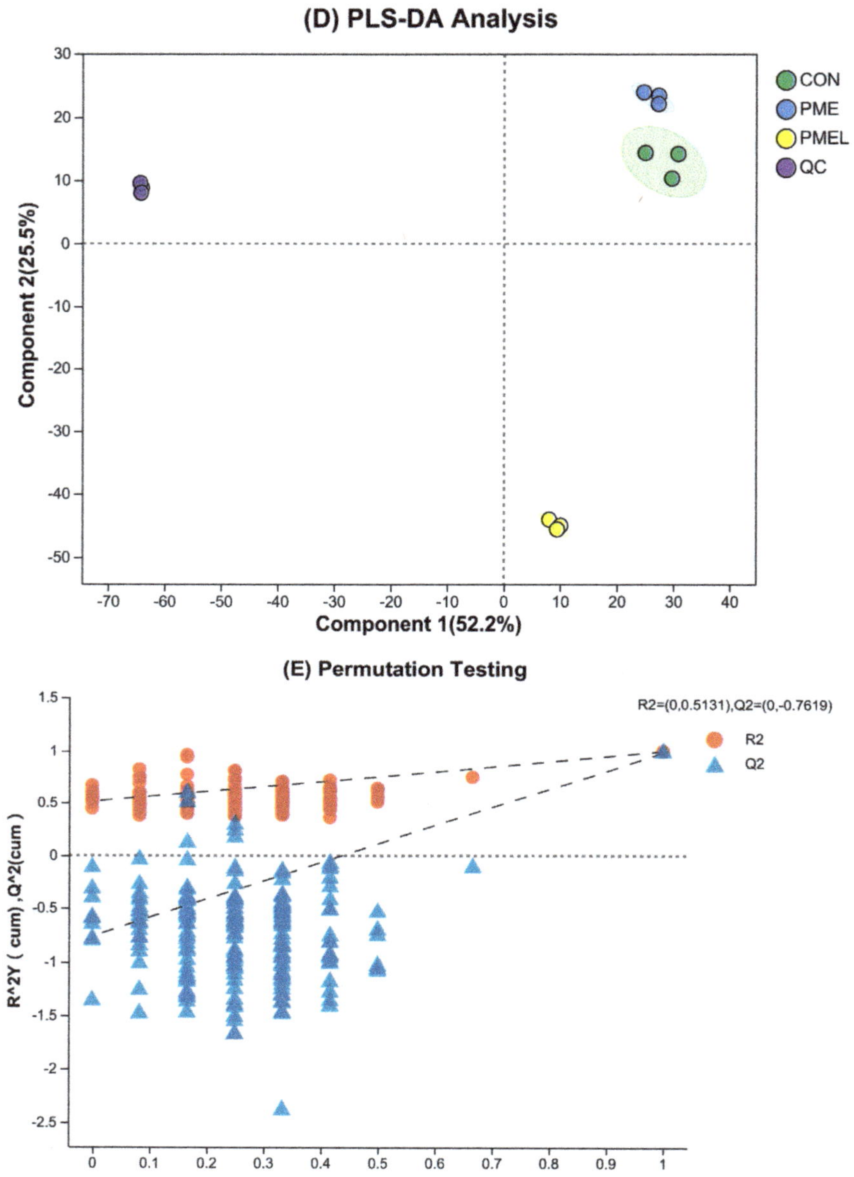

Figure 1. Samples Analyses: (**A**) Samples correlation heatmap analysis; (**B**) Metabolites Venn diagram analysis; (**C**) PCA assessment; (**D**) PLS-DA assessment; and (**E**) Permutation Testing of PLS-DA plot. Every box in the heatmap illustration represents the correlation between the two samples. The various hues correspond to the corresponding value of the correlation factor amongst samples, which ranged from 0.6 to 1. The different hues of Venn diagrams indicate distinct groupings. The Venn diagram's sections that overlap show how many similar compounds there are in each group, while the non-overlapping sections show how many unique metabolites there are in each group, indicating variations in the groups' metabolic profiles.

Principal component (PC) 1 stipulated 44.60% of the variation, although PC 2 was accountable for 24.70% of the variation, according to the PCA scoring assessment. Collectively, PCs 1 and 2 had possession of 69.30% of the data's fluctuation (Figure 1C). Moreover, the PLS-DA evaluation of the data demonstrated that Component 1 might explain 52.2% of the data deviation, whereas Component 2 was able to forecast 25.5% (Figure 1D). This proposed that there was an important distinction existing between CON and PMEL in metabolite diversity, which was additionally apparent from the PCA assessment. The Permutation Testing is designed to examine PLS-DA's modeling and prediction performance. The R2 value of the PLS-DA plot was equivalent to 0.5131, and the upward trajectory of the regression line, coupled with reductions in R2 and Q2, revealed that the regression model did not appear overestimated and was trustworthy (Figure 1E).

3.2. Expression of Distinct Metabolites

We established Volcano plots to assess the metabolites across different treatments. These plots enabled us to identify those molecules that were upregulated, downregulated or demonstrated no significant alteration between the groups. There were 224 upregulated and 111 downregulated metabolites detected in the PME group in comparison between CON and PME (Figure 2A). In comparison between CON and PMEL, 181 upregulated and 138 downregulated compounds were noticed in PMEL (Figure 2B). However, in comparison between PME and PMEL, increased and decreased metabolites were 179 and 241, respectively, in PMEL (Figure 2C). The Venn plot assessment of the differentially expressed metabolites (DEM) in the several groups being compared demonstrated that each group showcased distinct metabolites, indicating variations in metabolic profile among the various groups. There were 88, 24, and 62 distinct (DEM) detected in CON versus PME, CON vs. PMEL, and PME vs. PMEL, correspondingly (Figure 2D).

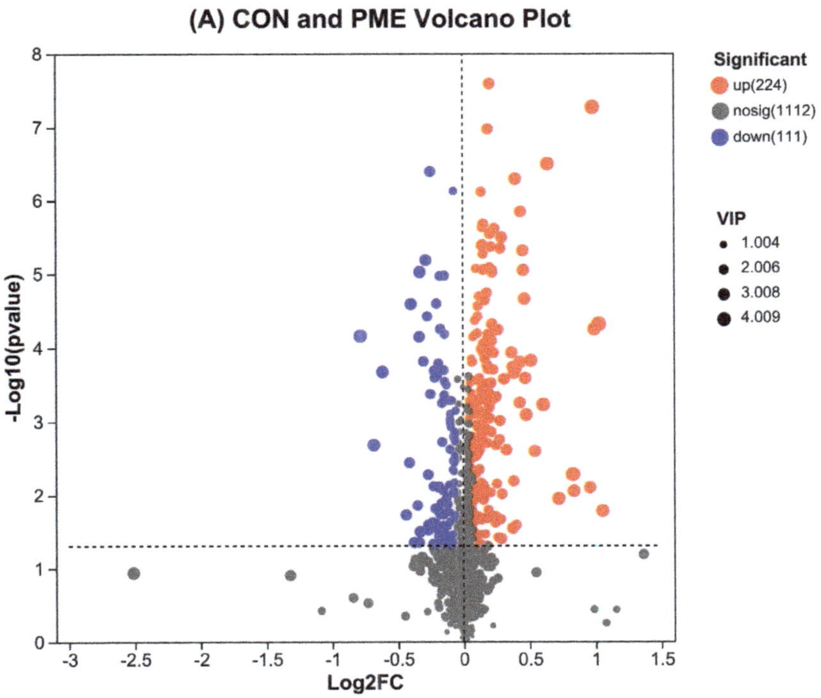

Figure 2. *Cont.*

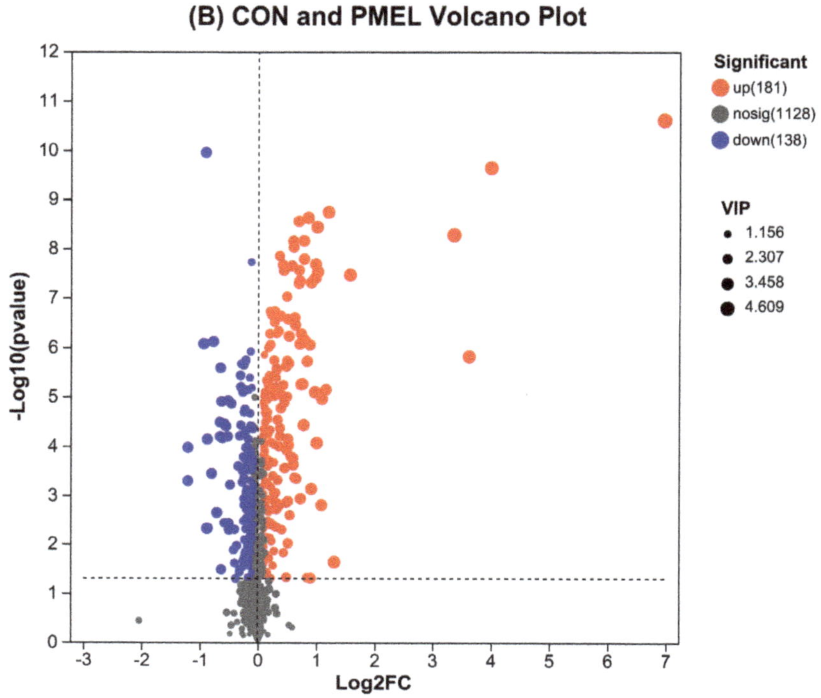

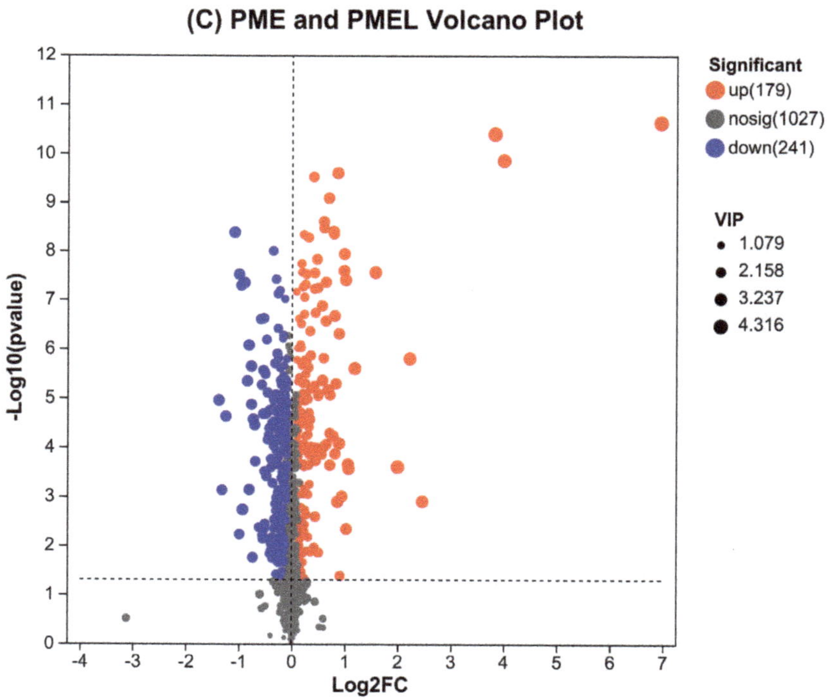

Figure 2. *Cont.*

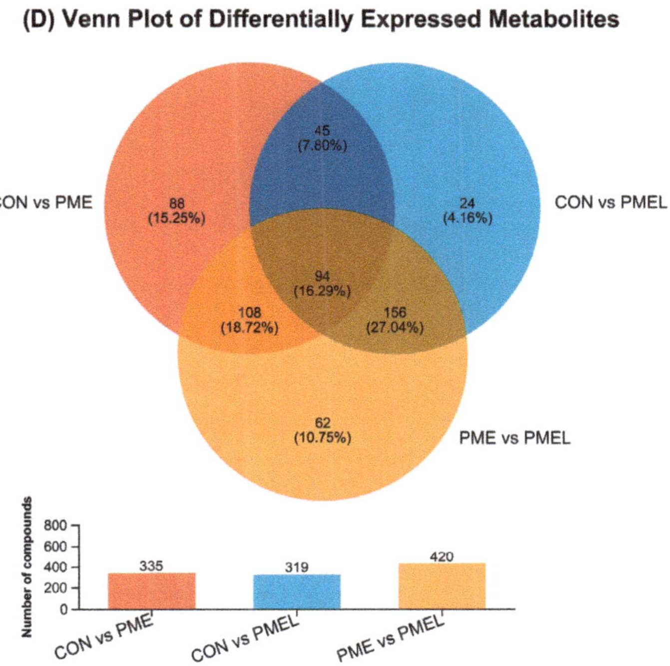

Figure 2. Summary of expression of distinct metabolites: (**A**) CON and PME Volcano plot. The top three upregulated metabolites in PME were trehalose, L-methionine, and glucoside. The top three downregulated metabolites in PME were glucosinolate, aminopicolinic acid, and salicylic acid; (**B**) CON and PMEL Volcano plot. The top three upregulated metabolites in PMEL were kaempferol 3-neohesperidin, tectorigenin, and kaempferol 3-neohesperidoside. The top three downregulated metabolites in PMEL were butyric acid, glyceric acid, and loperamide; (**C**) PME and PMEL Volcano plot. The top three upregulated metabolites in PMEL were kaempferol 3-neohesperidin, sesamolinol glucoside, and tectorigenin. The top three downregulated metabolites in PMEL were loperamide, tetraethylene glycol, and oxoundecylcarnitine, and (**D**) Venn plot of differentially expressed metabolites.

3.3. Compound Categorization in Accordance with KEGG and HMDB Databases

To make categorization sense, the KEGG and HMDB indexes were employed for the categorizing of DEM. An aggregate of 224 metabolites was discovered based on the KEGG categorization. An aggregate of 72 metabolites was found conforming to the KEGG categorization of compounds having biological functions. The most varied and prevalent substances consisted of carbohydrates, which were followed by phospholipids, amino acids, monosaccharides, and oligosaccharides (Figure 3A). There were 60 biochemicals identified in the KEGG repository of phytochemical substances. Between them, flavonoids were the most varied and prevalent biomolecules, followed by monolignols, alkaloids, monoterpenoids, and isoflavonoids (Figure 3B). An aggregate of 92 metabolites was identified using the KEGG grouping of lipid substances. The most varied and prevalent substances were the conjugates and fatty acids, which were succeeded by steroids, flavonoids, octadecanoids, and isoprenoids (Figure 3C).

In accordance with the HMDB categorization of substances, 810 biochemicals were discovered. The most ample and diverse compounds were carboxylic acids and derivatives (26.91%), followed by organooxygen compounds (14.69%), fatty acyls (11.23%), benzene and substituted derivatives (3.83%), pyridines and derivatives (2.59%), prenol lipids (2.59%), and flavonoids (2.10%) (Figure 3D).

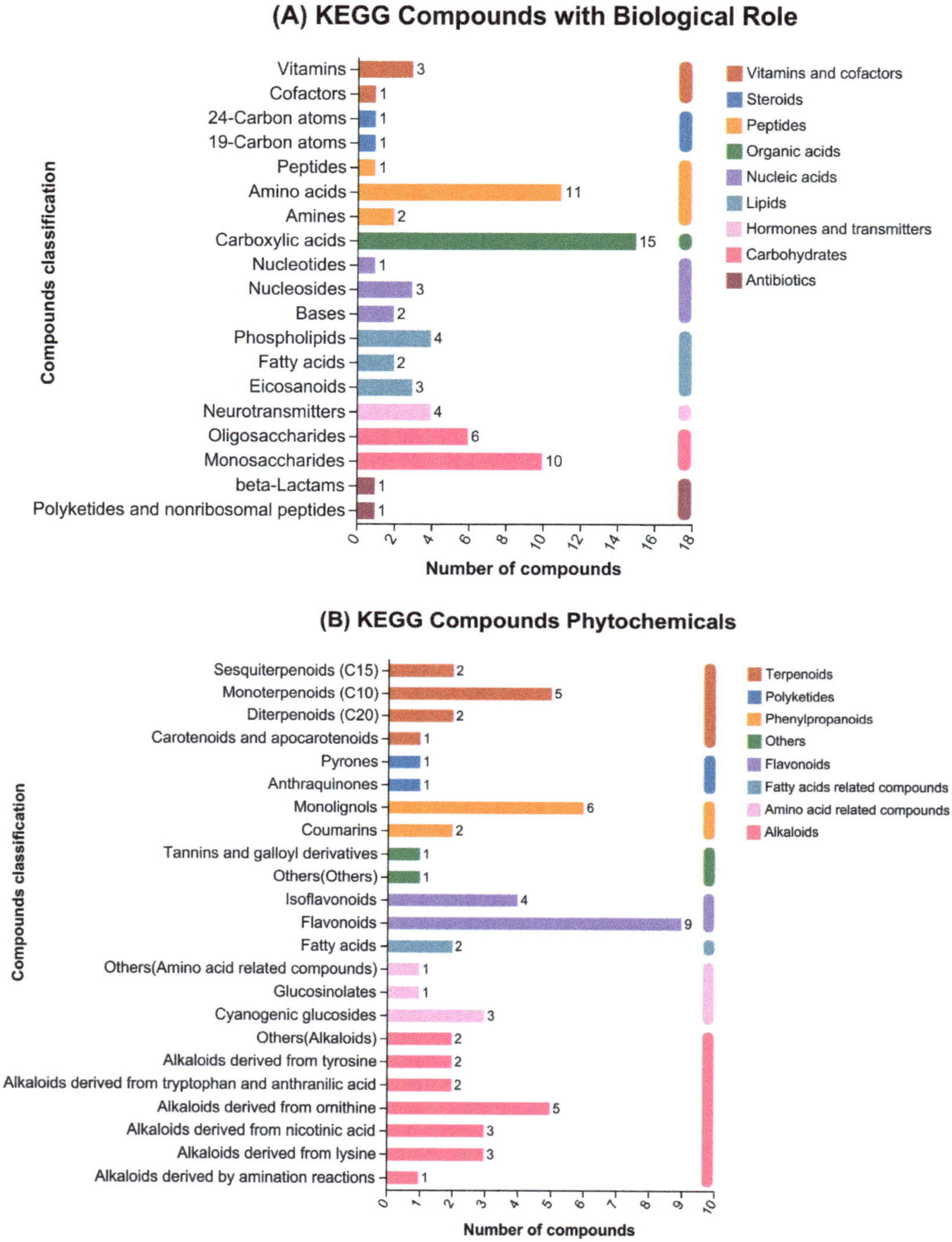

Figure 3. *Cont.*

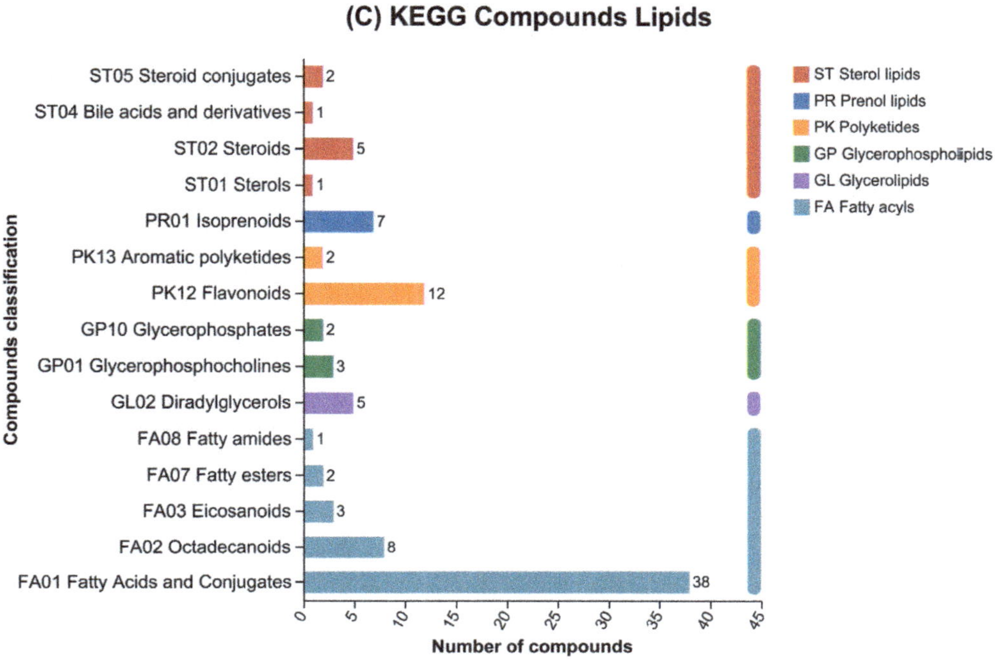

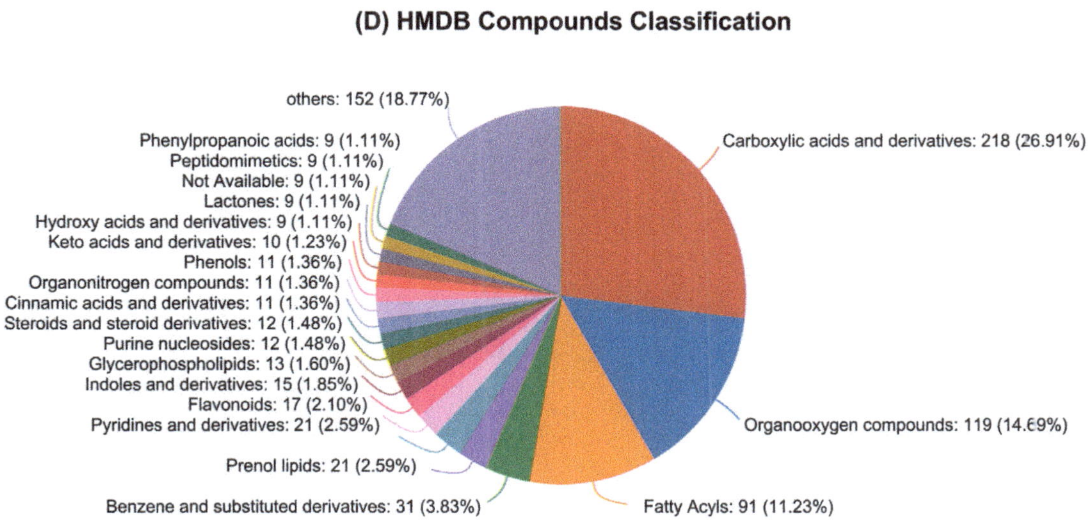

Figure 3. *Cont.*

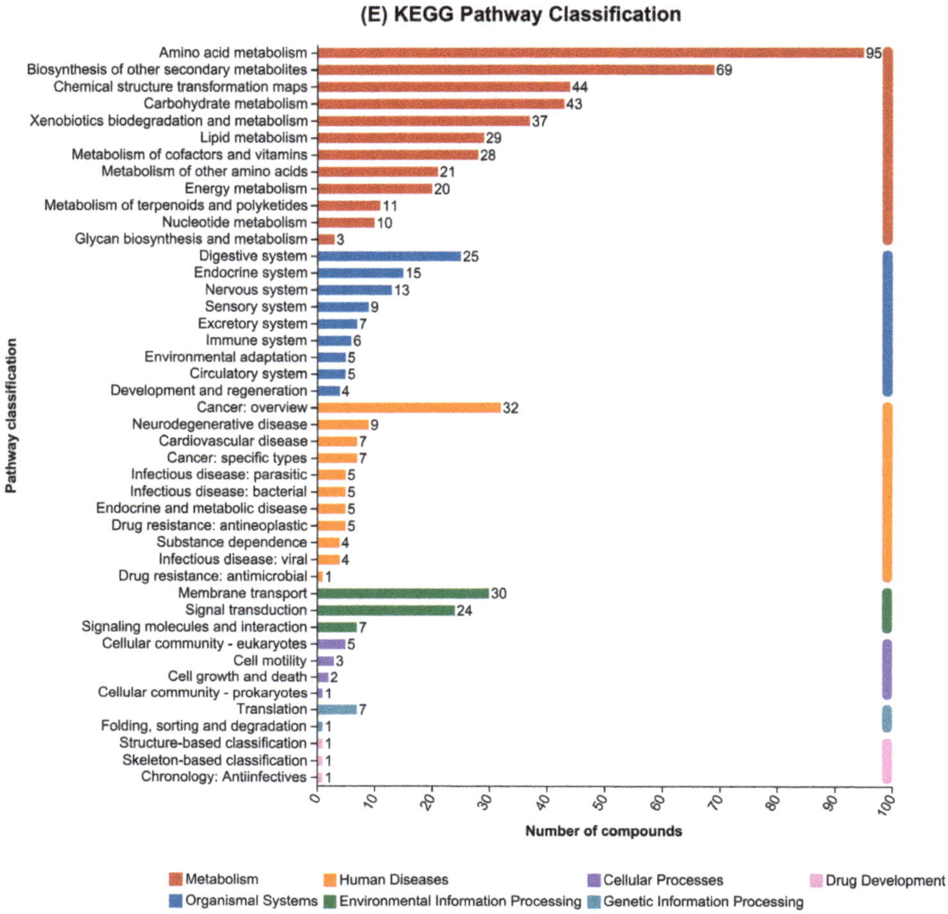

Figure 3. KEGG and HMDB compounds organization: (**A**) KEGG Compounds classification with biological role; (**B**) KEGG compounds classification of phytochemicals; (**C**) KEGG compounds classification of lipids; (**D**) HMDB compounds classification, and (**E**) KEGG compounds pathway classification.

The different metabolites were assigned to the KEGG pathway record to find the relevant pathways of metabolism. The KEGG pathway classification revealed that a total of 666 DEMs were mapped on different pathways. The amino acid metabolism pathway grasped the largest number of compounds, succeeded by the production of other secondary metabolites, xenobiotics biodegradation, chemical structure alteration maps, carbohydrate metabolism, and membrane transport (Figure 3E).

3.4. Enrichment Analysis of KEGG Pathway and Differential Abundance Score

The distinct metabolites were tracked on the KEGG repository, and the KEGG pathway enrichment assessment was performed to examine the biological functions of molecules of various treatments. The differential abundance score (DAS) was utilized to assess the rise or reduction of metabolites in certain metabolism pathways. The enrichment ratio for the most promising 20 KEGG pathways was examined.

In CON vs. PMEL, the highest enrichment ratio ($p < 0.001$) was noted in PME for arginine biosynthesis, followed by C5-Bracnhed dibasic acid metabolism, galactose metabolism, glycerolipid metabolism, and glycerophospholipid metabolism. Apart from this, pathways related to the biosynthesis of isoflavonoid and various other secondary metabolites were also significantly enriched pathways ($p < 0.05$); however, flavone and flavonol biosynthesis pathways tended to be enriched pathways in PMEL (Figure 4A). The DAS of CON vs. PMEL revealed that the glycerolipid metabolism pathway was significantly downregulated ($p < 0.001$), whereas the pathways having significantly upregulated ($p < 0.001$) compounds were concerned with arginine biosynthesis and galactose metabolism. The isoflavonoid biosynthesis and biosynthesis of other various secondary metabolites were also significantly upregulated pathways ($p < 0.05$); however, flavone and flavonol biosynthesis pathways tended to be upregulated pathways in PMEL (Figure 4B).

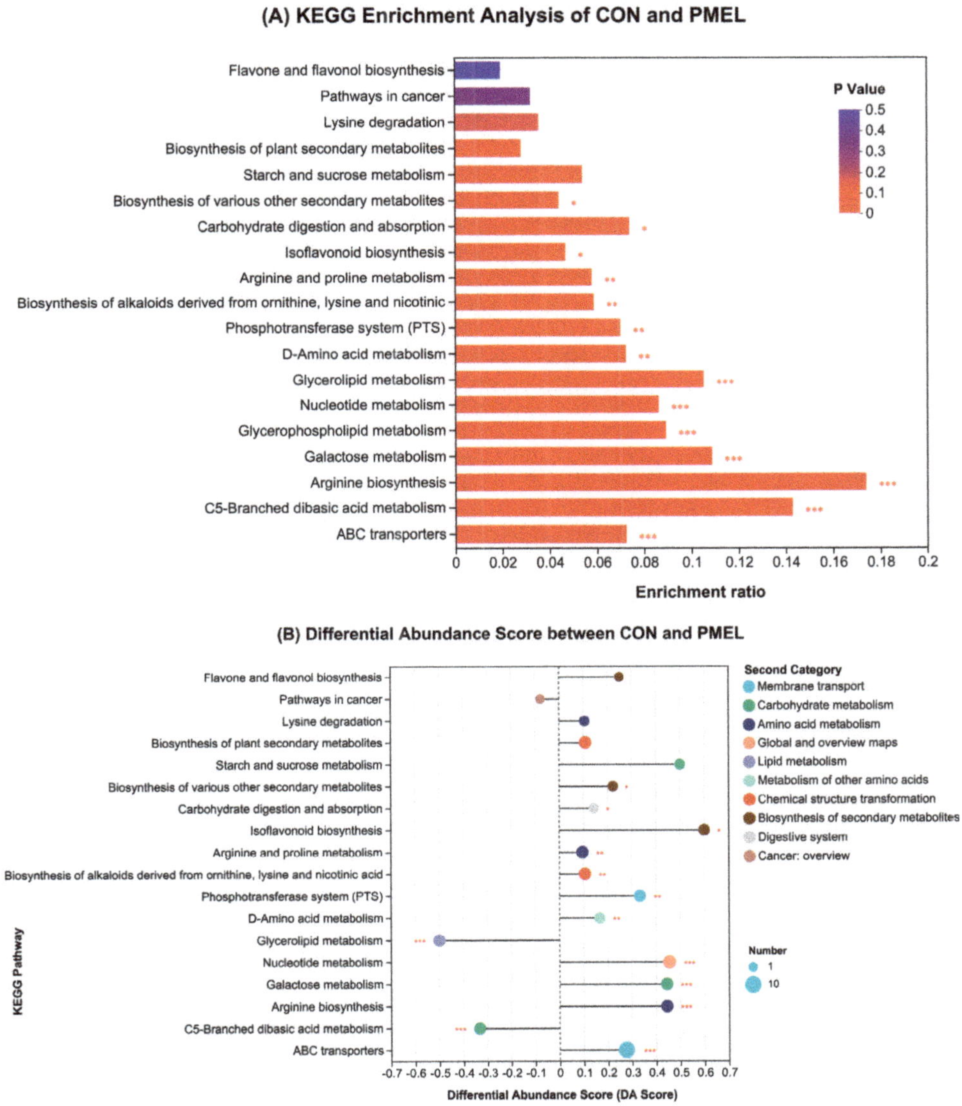

Figure 4. *Cont.*

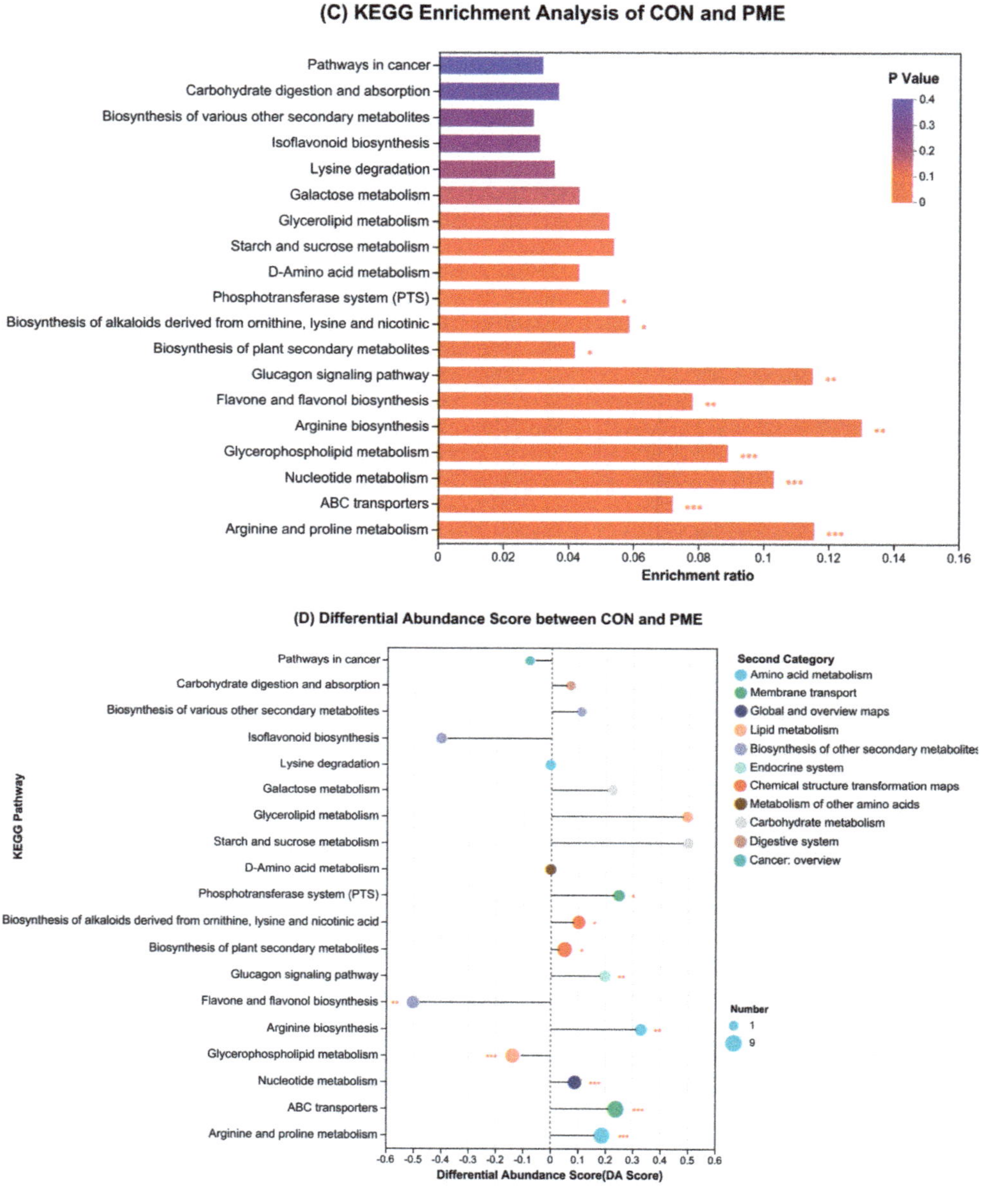

Figure 4. *Cont.*

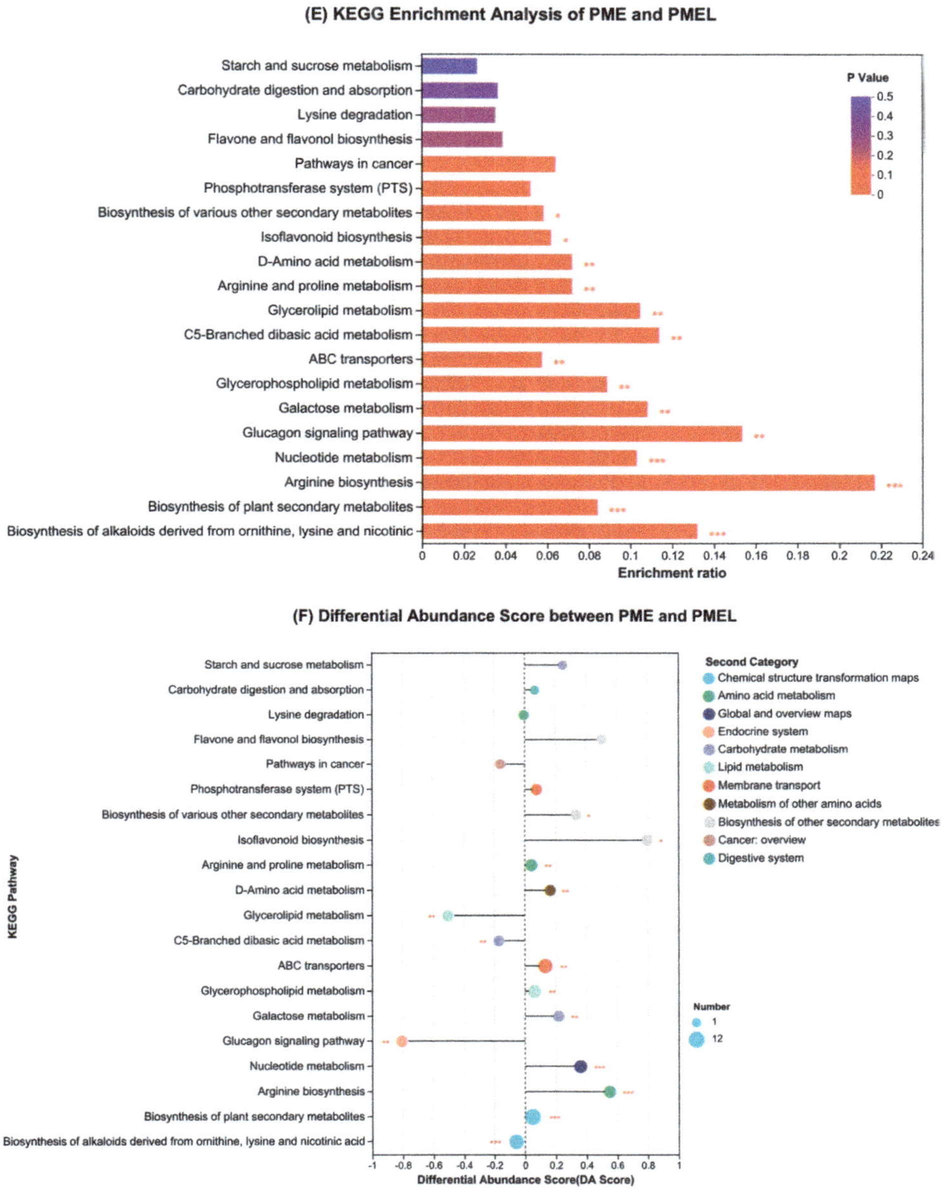

Figure 4. Enrichment analysis of KEGG pathway and Differential Abundance Score: (**A**) KEGG enrichment analysis of CON and PMEL; (**B**) Differential Abundance Score between CON and PMEL; (**C**) KEGG enrichment analysis of CON and PME; (**D**) Differential Abundance Score between CON and PME; (**E**) KEGG enrichment analysis of PME and PMEL, and (**F**) Differential Abundance Score between PME and PMEL. *** $p < 0.001$; ** $p < 0.01$; * $p < 0.05$.

The KEGG enrichment assessment of CON vs. PME showed that metabolic pathways linked to the ABC transporters, nucleotide metabolism, metabolism of arginine and proline, and glycerophospholipid metabolism had the highest augmented ratio in PME ($p < 0.001$). Similarly, the flavone and flavonol biosynthesis and arginine biogenesis pathways were also substantially improved pathways in PME ($p < 0.01$) (Figure 4C). The DAS evaluation of CON versus PME revealed that the chemical pathways showing increasing trends ($p < 0.001$) were related to ABC transporters, nucleotide metabolism, and arginine and proline metabolism in PME. The pathway indicating the downregulation ($p < 0.05$) of metabolites was relevant to flavone and flavonol biosynthesis; however, the isoflavonoid biosynthesis pathway tended to be downregulated in PME (Figure 4D).

The KEGG enrichment examination of PME versus PMEL discovered that metabolic processes linked to the biosynthesis of alkaloids, arginine biosynthesis, and biosynthesis of plant secondary metabolites possessed an enrichment ratio that was noticeably higher in PMEL ($p < 0.001$). The pathways related to the biosynthesis of isoflavonoids and other various secondary metabolites were also substantially augmented ($p < 0.05$). However, flavone and flavonol biosynthesis pathways tended to be enriched pathways in PMEL (Figure 4E). The DAS of PME vs. PMEL revealed that the pathways having significantly upregulated ($p < 0.001$) biomolecules were linked to arginine biosynthesis and nucleotide metabolism. The isoflavonoid biosynthesis and biosynthesis of other various secondary metabolites were also significantly upregulated pathways ($p < 0.05$); however, flavone and flavonol biosynthesis pathways tended to be upregulated pathways in PMEL (Figure 4F). These findings suggested that the fermentation with *Lactobacillus* yielded an increased enrichment ratio and upregulation of isoflavonoid biosynthesis and biosynthesis of other various secondary metabolites pathways in PMEL, which, in turn, pre-eminent the antioxidant capability of PMEL.

3.5. Metabolites Possessing Antioxidant Potential and Their VIP Assessment, Total Flavonoids and Polyphenols Content, and Antioxidant Capability

To determine the substantial distinction in metabolites between the two groups, the compounds possessing antioxidant-related properties (flavonoids, isoflavonoids, lignan glycosides, cinnamic acids, dihydrocoumarins, indoles, and prenol lipids) were scrutinized, administered to the OPLS-DA approach, and the VIP score was assessed. The expression profile of CON and PME demonstrated that 12 metabolites were discovered, with diosbulbinoside, aucubin, isowertin 2″-rhamnoside, and 6‴-O-sinapoylsaponarin being considerably more prevalent ($p < 0.01$) in PME (Figure 5A).

A total of 25 metabolites were detected in the expression profile of CON and PMEL. The hydroxymelleolide, 4′,6-Dihydroxyflavone, and indole-3-acetaldehyde were notably increased ($p < 0.01$) compounds in the PMEL group whereas the extremely noteworthy ($p < 0.001$) metabolites in PMEL were methylflavanone, kaempferol derivatives, daidzein, geniposide, isoartocarpesin, dihydrocoumarin, occidentoside, maackiain, cycloalliin, aucubin, inuline, isowertin 2″-rhamnoside, tectorigenin, ononin, diferuloylputrescine, diosbulbinoside, indoleacetic acid 5-glucoside, and sinapoylsaponarin (Figure 5B).

The expression assessment of PME and PMEL confirmed that 29 metabolites were significantly different, and the notably highest ($p < 0.001$) metabolites were occidentoside, methylflavanone, kaempferol derivatives, daidzein, isoartocarpesin, dihydrocoumarin, maackiain, cycloalliin, aucubin, inuline, isowertin 2″-rhamnoside, tectorigenin, ononin, and indoleacetic acid 5-glucoside in PMEL. Similarly, ominously prominent ($p < 0.05$) metabolites detected in PMEL were malonylglycitin, ferulic acid, subaphylline, quercetin 3,7-dimethyl ether, and indole-3-acetaldehyde (Figure 5C). These findings verified that *Lactobacillus* fermentation considerably advanced the flavonoids, isoflavonoids, lignan glycosides, cinnamic acids, dihydrocoumarins, indoles, and prenol lipids in PMEL, which triggered a significant upsurge in the entire antioxidant ability of fermented PKM.

The anticipated biosynthesis pathways of antioxidant-related bioactive peptides are discussed in Figure 5D. Mainly, four biosynthesis pathways were identified, namely phenylpropanoid, isoflavonoid, flavonoid, and flavone and flavonol biosynthesis pathways. The p-coumaric acid is generated from tyrosine in the phenylpropanoid biosynthesis pathway and is converted to p-coumaroyl-CoA and caffeic acid. The caffeic acid is then converted to ferulic acid. After entering the flavonoid biosynthesis pathway, the p-coumaroyl-CoA is transformed into liquiritigenin and naringenin. The naringenin is converted to kaempferol and apigenin, and later, both enter the flavone and flavonol biosynthesis pathway. The kaempferol is then converted to kaempferol derivatives and quercetin. The quercetin is finally converted to quercetin 3,7-dimethyl ether. The liquiritigenin from the flavonoid biosynthesis pathway flows into the isoflavonoid biosynthesis pathway, where it is converted to 4′,6-dihydroxyflavone, 2,4′,7-trihydroxy-isoflavanone, and glycitein. The glycitein is then converted to 6″-O-malonylglycitin. The 2,4′,7-trihydroxy-isoflavanone is converted to daidzein and 2,7-dihydroxy-4′-methoxyisoflavanone. The latter one is then converted to formononetin, which is then converted to ononin and 2′,7-dihydroxy-4′,5′-methylenedioxyisoflavone. The latter one is finally converted to maackiain (Figure 5D).

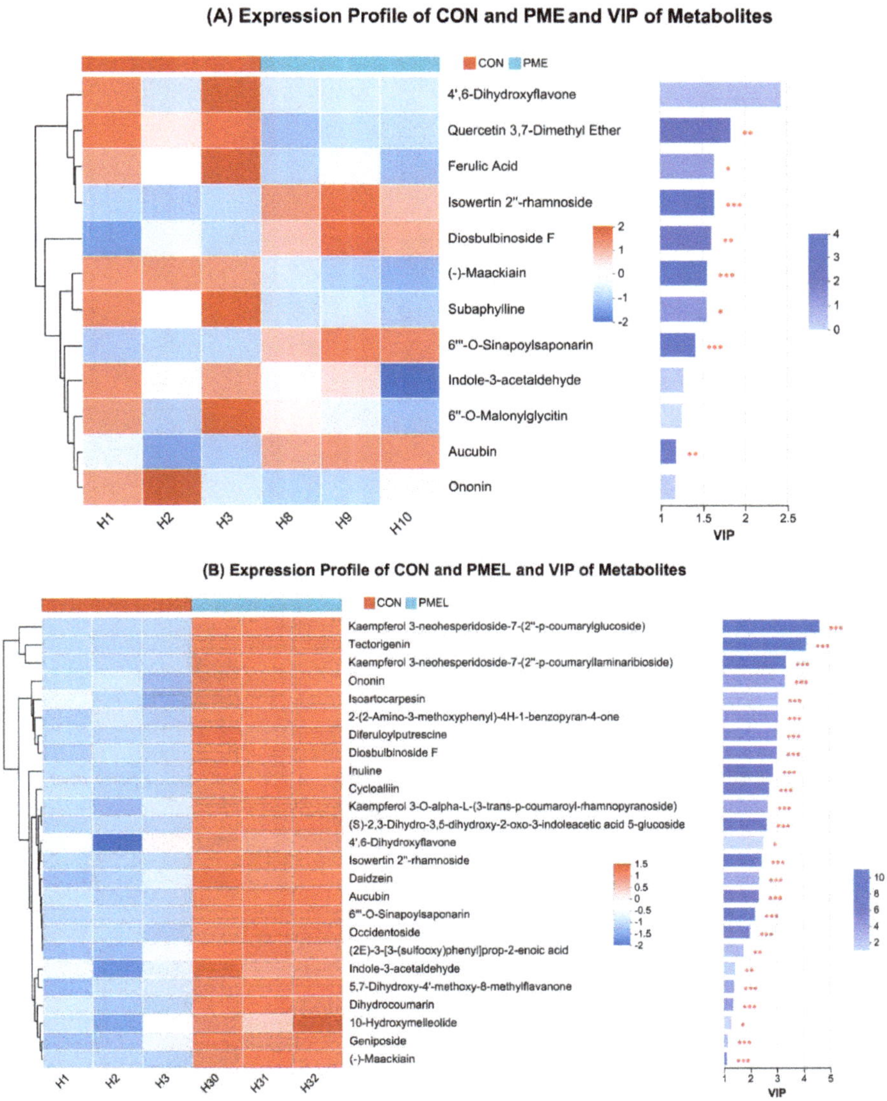

Figure 5. *Cont.*

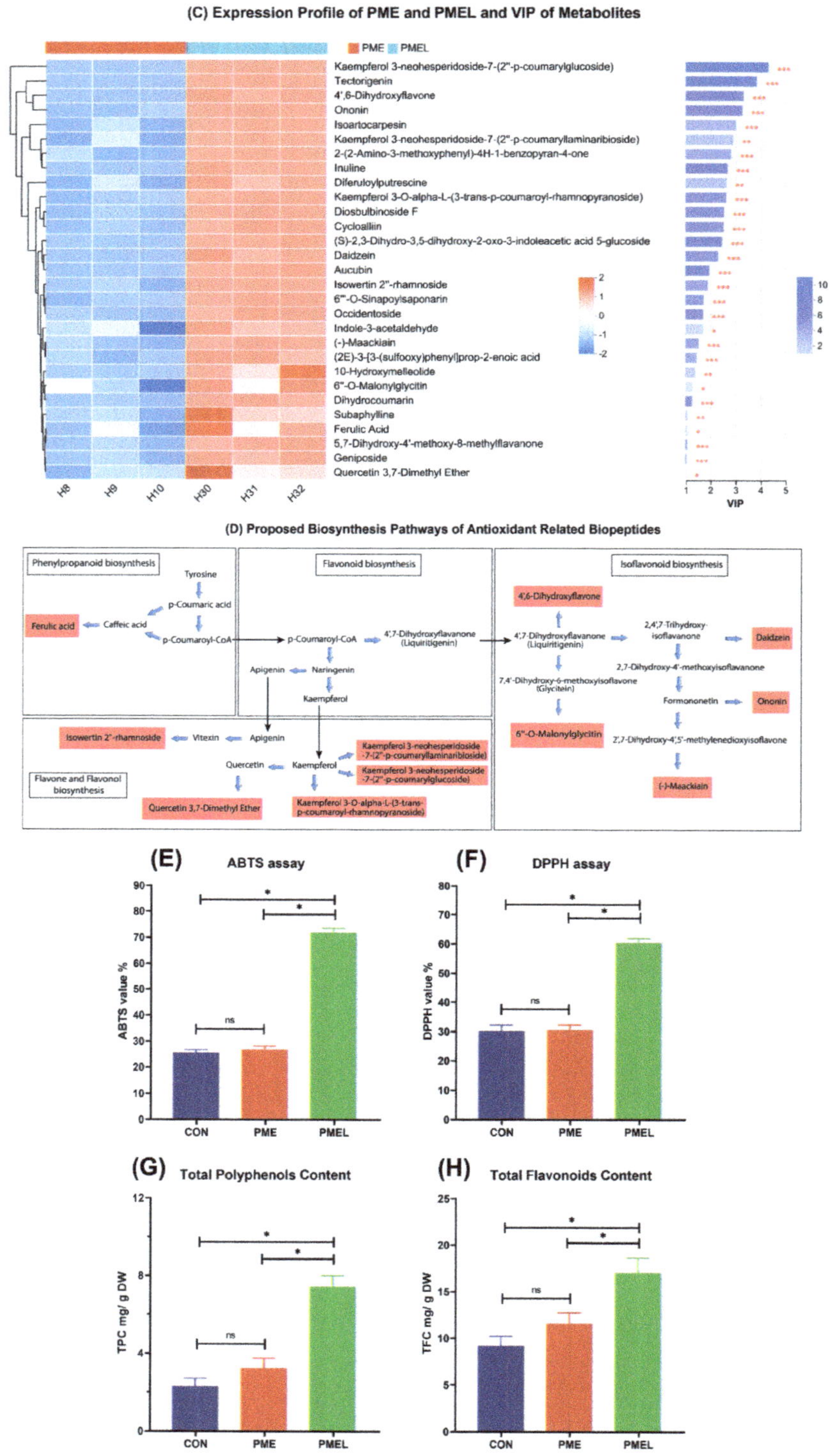

Figure 5. Heatmap Analysis and VIP of metabolites related to antioxidant activity, scavenging values, total flavonoids, and total polyphenols content. (**A**) Expression profile of CON and PME and VIP

of metabolites. The significantly upregulated metabolites in PME were diosbulbinoside, aucubin, isowertin 2″-rhamnoside, and 6‴-O-sinapoylsaponarin; (**B**) Expression profile of CON and PMEL and VIP of metabolites. All the mentioned metabolites in PMEL were significantly upregulated; (**C**) Expression profile of PME and PMEL and VIP of metabolites. All the mentioned metabolites in PMEL were significantly upregulated; (**D**) Proposed biosynthesis pathway of antioxidant-related peptides. Mainly, 4 biosynthesis pathways were identified, namely phenylpropanoid, isoflavonoid, flavonoid, and flavone and flavonol biosynthesis pathways. The p-coumaroyl-CoA generation is crucial because its successors are liquiritigenin, kaempferol, and apigenin. The kaempferol is then converted to kaempferol derivatives and quercetin. The quercetin is finally converted to quercetin 3,7-dimethyl ether. The liquiritigenin conversion to its successors like daidzein, ononin, and maackiain in the isoflavonoid biosynthesis pathway is very crucial because it may regulate the antioxidant status of fermented PMEL; (**E**) ABTS analysis to check the scavenging value; (**F**) DPPH analysis to check the scavenging value; (**G**) Total polyphenols content and (**H**) total flavonoids content. *** $p < 0.001$; ** $p < 0.01$; * $p < 0.05$; ns, non-significant.

The overall antioxidant potential of cultured PKM was assessed by employing a DPPH and ABTS test. The DPPH and ABTS scavenging efficiencies of CON and PME appeared equivalent. Interestingly, the DPPH and ABTS scavenging rates of PMEL were substantially greater ($p < 0.05$) as opposed to CON and PME (Figure 5E,F). The overall flavonoid and polyphenol levels of PMEL were considerably stronger ($p < 0.05$) compared to that of the remaining groups (Figure 5G,H). There existed no difference in the overall flavonoid or polyphenol values of CON and PME. The increased levels of polyphenols and flavonoids in the PMEL probably assisted in its elevated overall antioxidant efficacy.

3.6. Metagenome of Bacterial Microbial Community

To assess the bacterial microbial community of PME and PMEL, 16sRNA sequencing was performed. Generally, the most dominant microbial community in PME was related to the order *Enterobacterales*, whereas in PMEL, it was related to the order *Lactobacillales*. At the species level, the PME mainly possessed *s__unclassified_g__Enterobacter* (72.16%) and *Enterobacter roggenkampii* (6.34%), whereas PMEL consisted of *Limosilactobacillus reuteri* (32.49%), *Pediococcus acidilactici* (22.66%), and *Lacticaseibacillus paracasei* (12.14%) (Figure 6A). At the genus level, the PME mainly comprised the genus *Enterobacter* (82.95%) and the unclassified genus having the family *Enterobacteriaceae* (5.53%), whereas the PMEL mainly included the genus *Limosilactobacillus* (44.99%), *Pediococcus* (32.54%), *Lacticaseibacillus* (12.76%), and *Lactiplantibacillus* (3.03%) (Figure 6B).

The heatmap analysis was performed to determine the species abundance and composition in PME and PMEL. The dominant species in PME were *Enterobacter roggenkampii, Enterobacter asburiae, Enterobacter hormaechei*, and *Escherichia coli*, whereas, in PMEL, the dominant species were *Limosilactobacillus reuteri, Pediococcus acidilactici, Lacticaseibacillus paracasei*, and *Lactiplantibacillus plantarum* (Figure 6C).

The species relative abundance analysis revealed that *Limosilactobacillus reuteri, Pediococcus acidilactici, Lacticaseibacillus paracasei, Pediococcus pentosaceus*, and *Lactiplantibacillus plantarum* had significantly higher ($p < 0.05$) proportion in PMEL as compared to PME, whereas in PME, *s__unclassified_g__Enterobacter, Enterobacter roggenkampii*, and *Enterobacter cloacae* showed significantly higher ($p < 0.001$) proportion as compared to PMEL (Figure 6D). These results demonstrated that PMEL harbored more *Lactobacillus* species and had a diverse bacterial community.

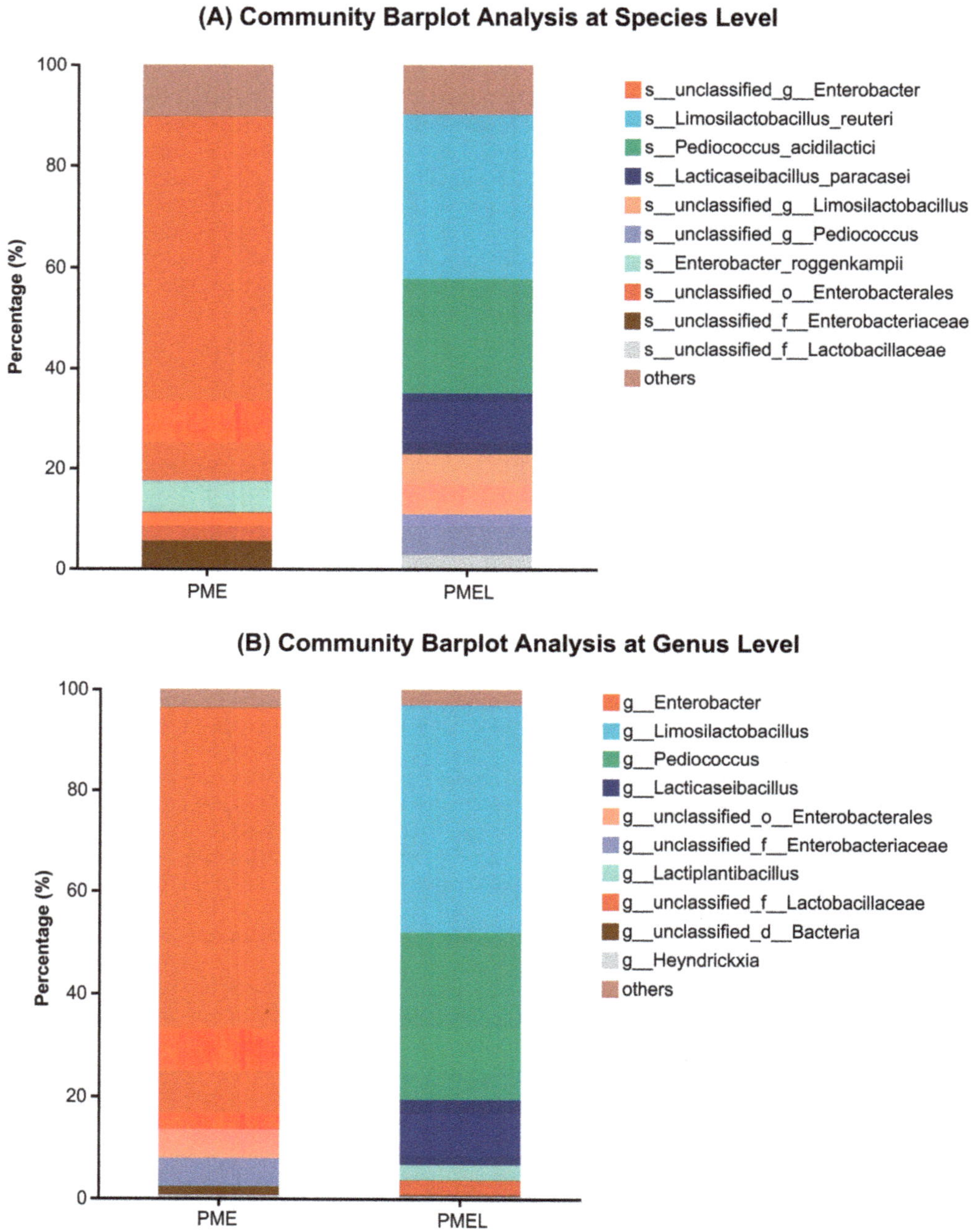

Figure 6. *Cont.*

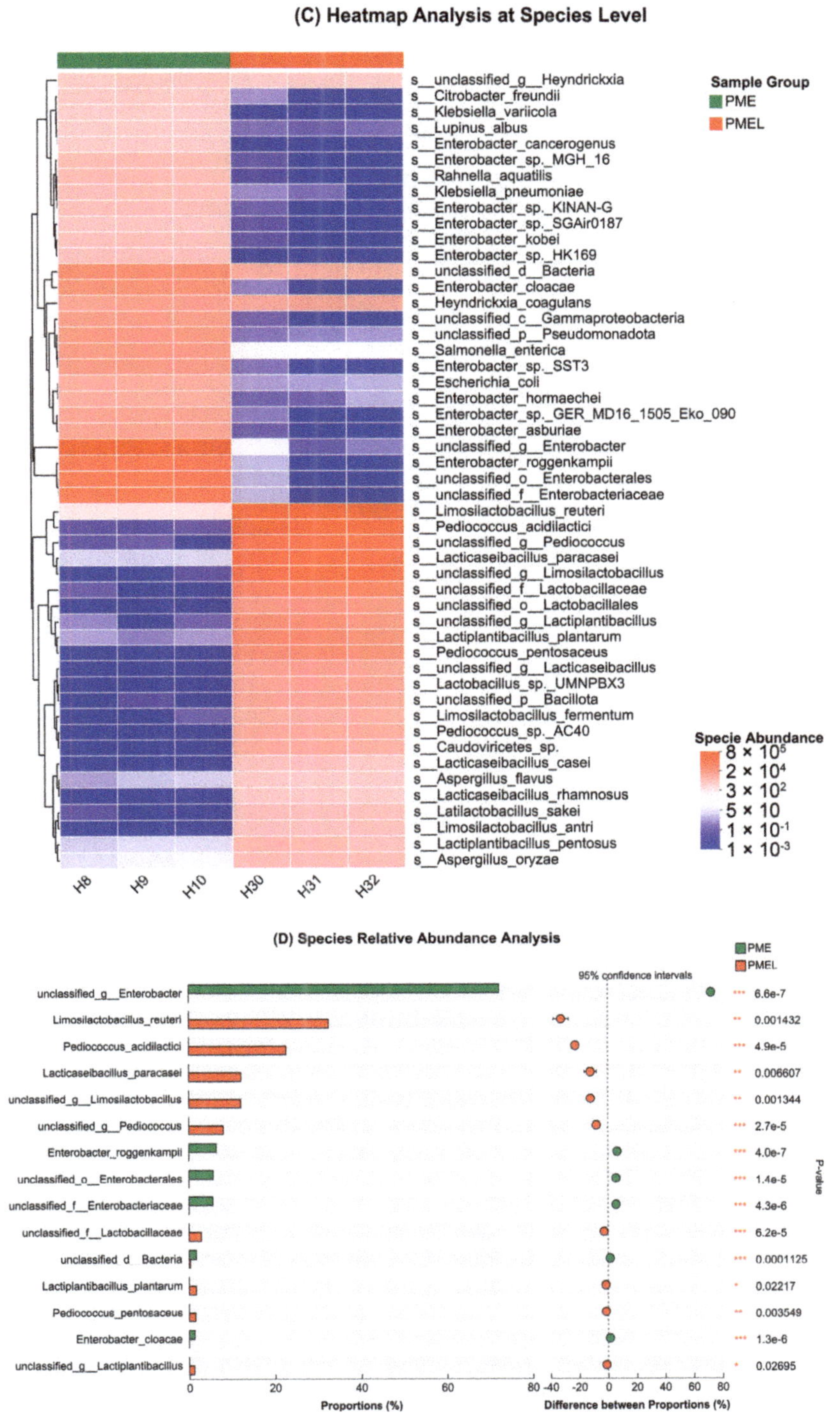

Figure 6. Metagenomic analysis of bacterial microbial community. (**A**) Community barplot analysis of PME and PMEL at the species level; (**B**) Community barplot analysis of PME and PMEL at genus

level; (**C**) Species heatmap analysis of PME and PMEL, and (**D**) species relative abundance analysis of PME and PMEL. The PME mainly possessed *s__unclassified_g__Enterobacter* (72.16%) species, which might be responsible for CAZy in PME. The PMEL consisted of *Limosilactobacillus reuteri* (32.49%) and *Pediococcus acidilactici* (22.66%), which may have contributed to its CAZy and antioxidant production. *** $p < 0.001$; ** $p < 0.01$; * $p < 0.05$.

3.7. Carbohydrates Active Enzymes (CAZy) Analysis

The circos diagram illustrated the CAZy in the PME and PMEL groups. The PME had six classes of enzymes, namely glycoside hydrolases (GH) (42%), glycosyl transferases (GT) (33%), carbohydrate esterases (CE) (18%), auxiliary activities (AA) (7%), carbohydrate-binding modules (0.005%), and polysaccharide lyases (PL) (0.0001%). The PMEL also had six classes of enzymes, namely glycoside hydrolases (34%), glycosyl transferases (39%), carbohydrate esterases (20%), auxiliary activities (6%), carbohydrate-binding modules (0.008%), and polysaccharide lyases (0.03%) (Figure 7A).

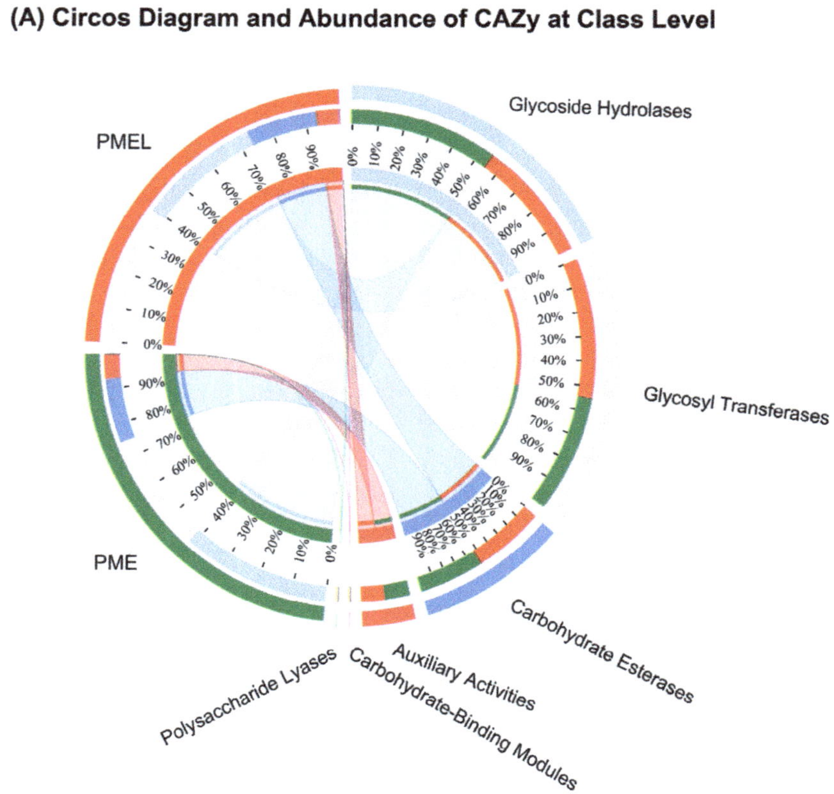

Figure 7. Cont.

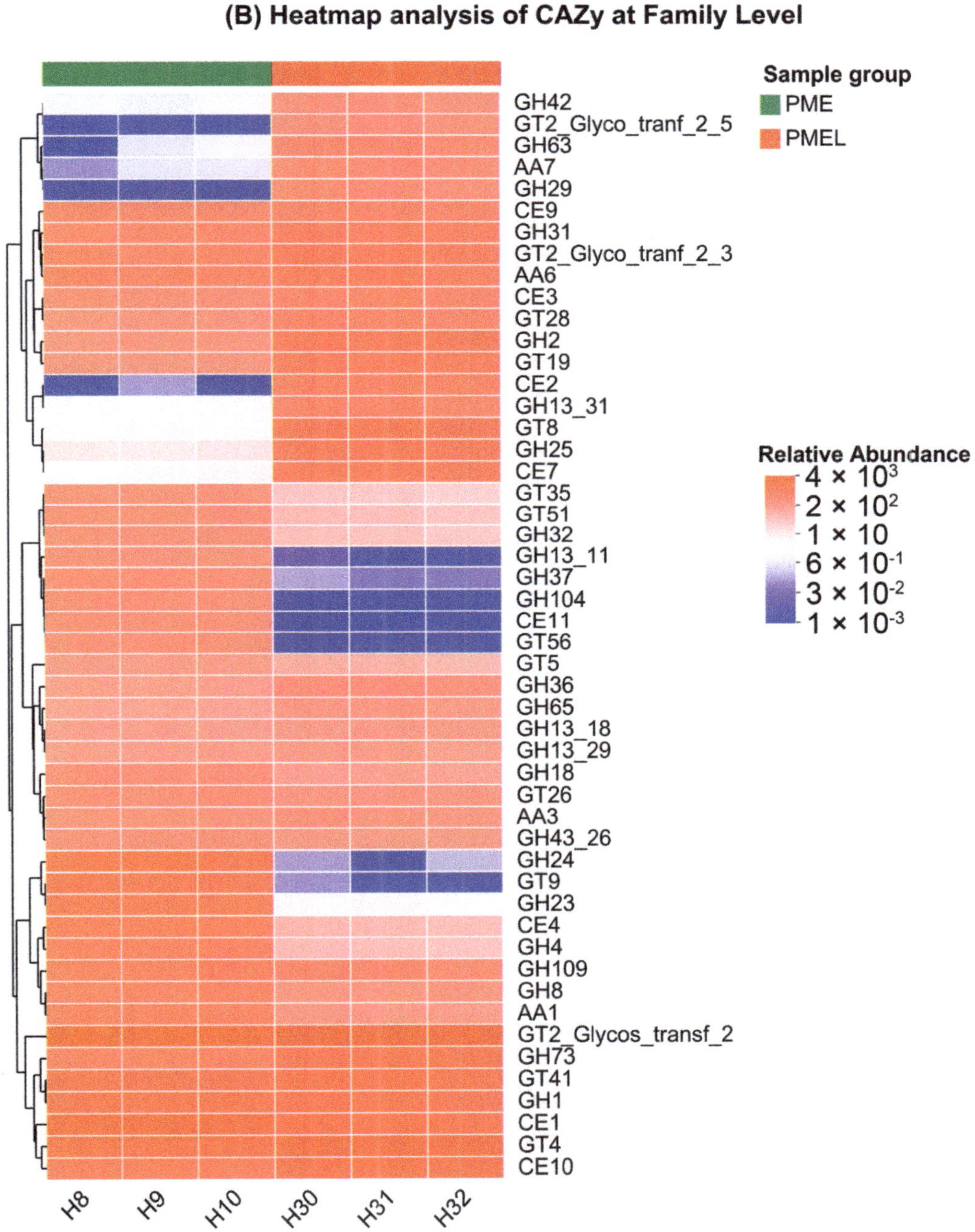

Figure 7. Cont.

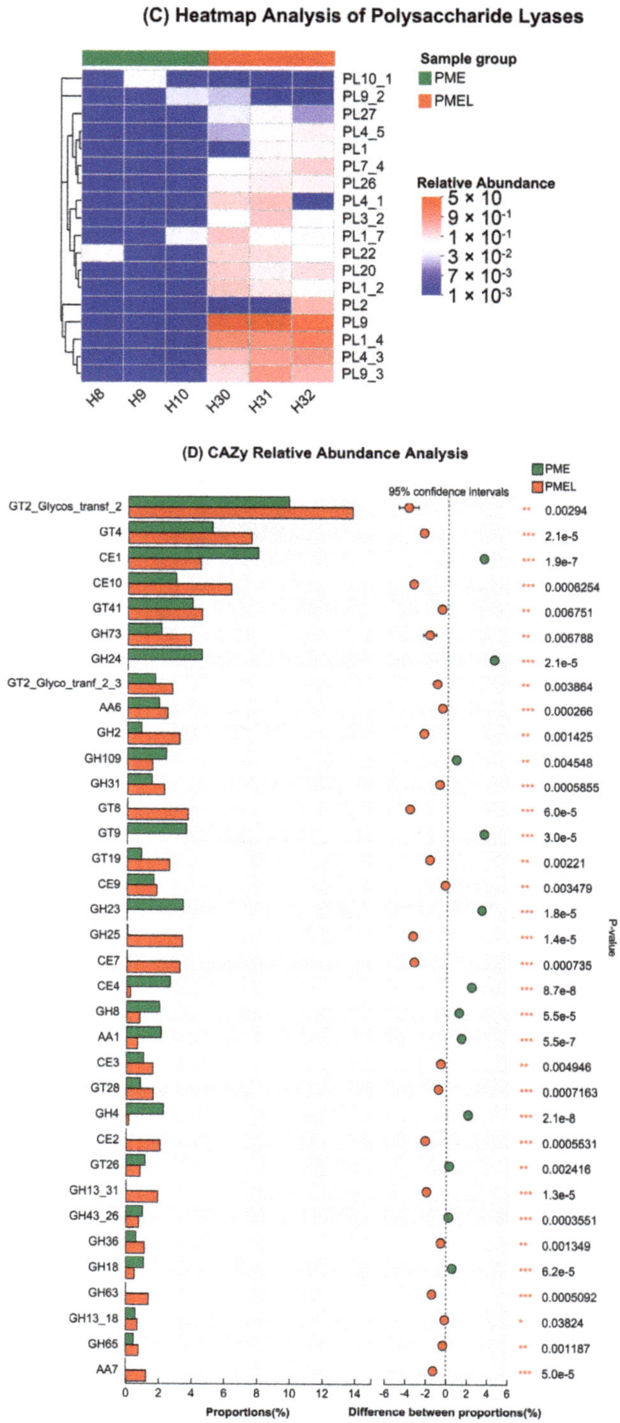

Figure 7. Cont.

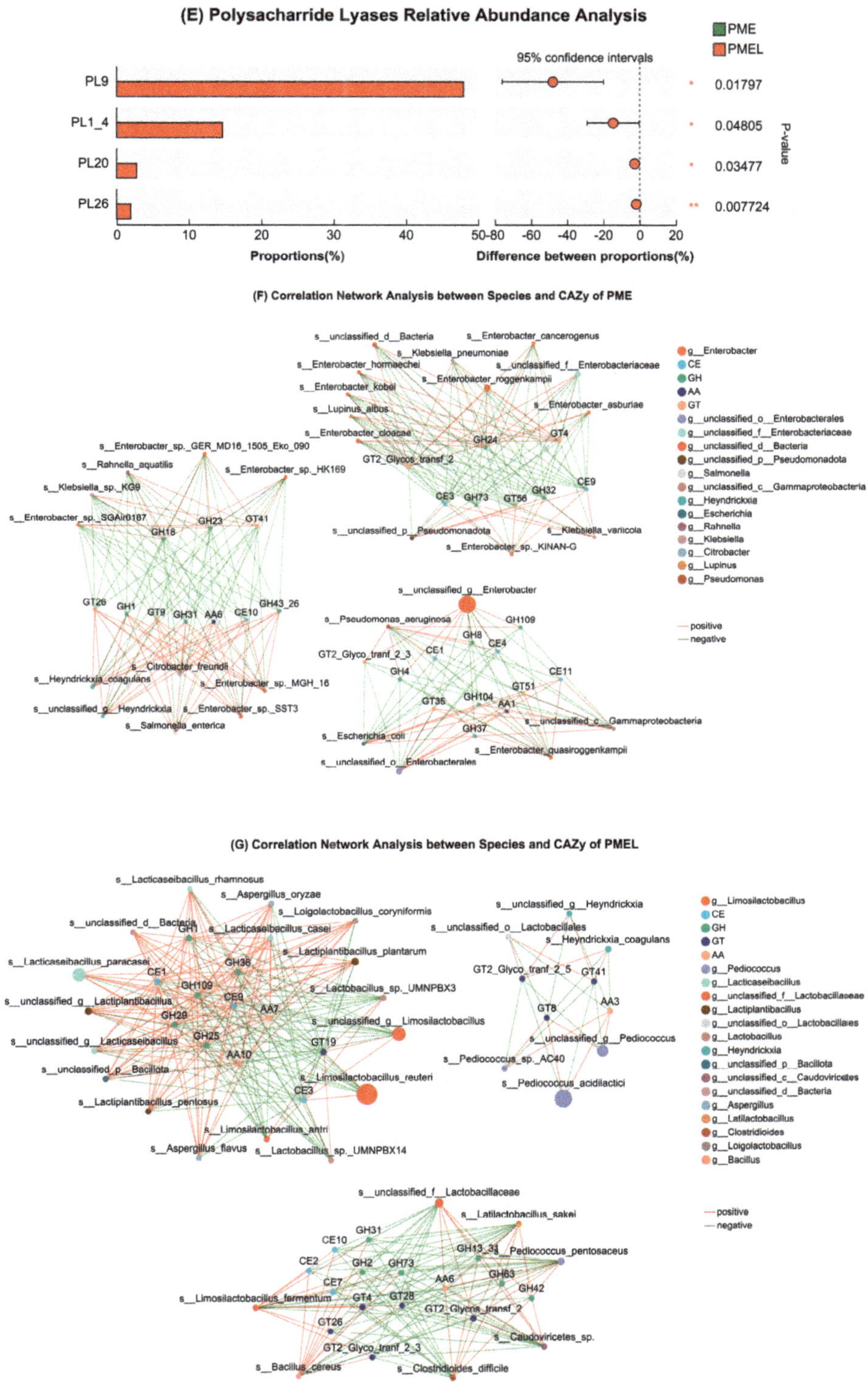

Figure 7. CAZy analysis. (**A**) Circos diagram and abundance of CAZy of PME and PMEL, (**B**) Heatmap analysis of CAZy of PME and PMEL at the family level, (**C**) Heatmap analysis of

polysaccharide lyases of PME and PMEL, (**D**) CAZy relative abundance analysis between PME and PMEL, (**E**) Polysaccharide lyases relative abundance analysis between PME and PMEL, (**F**) Correlation network analysis among species and CAZy of PME, and (**G**) Correlation network analysis among species and CAZy of PMEL. The PMEL had significantly higher ($p < 0.01$) abundance levels of GT4, GT2_Glycos_transf_2, GH73, GT41, CE10, GH2, AA6, GT8, GH31, and GT2_Glyco_tranf_2_3, which possess the cellulose synthase, xylanase, chitin synthase, endoglucanase, and beta-galactosidase-like activity. The PMEL also possessed a significantly higher ($p < 0.05$) relative abundance of PL26, PL20, PL1_4, and PL9, which may be responsible for the degradation of the cell wall, specifically pectin complex polysaccharides. *** $p < 0.001$; ** $p < 0.01$; * $p < 0.05$.

The heatmap analysis of CAZy at the family level revealed that PME had a higher relative abundance of GH13_11, GH37, GH104, CE11, GT56, GT9, and GH24 families, whereas PMEL comprised a higher relative abundance of GH42, GT2_Glyco_tranf_2_5, GH63, AA7, GH29, CE2, GH13_31, GT8, GH25, and CE7 families (Figure 7B). The relative abundance of PL families in PME and PMEL varies greatly. The heatmap analysis showed that PMEL possessed a higher relative abundance of different PL families (PL9, PL1_4, PL4_3, PL9_3, PL22, PL20, and PL1_2) as compared to the PME (Figure 7C).

The relative abundance analysis of CAZy showed that PMEL had significantly higher ($p < 0.01$) abundance levels of GT4, GT2_Glycos_transf_2, GH73, GT41, CE10, GH2, AA6, GT8, GH31, and GT2_Glyco_tranf_2_3 as compared to PME (Figure 7D). Similarly, PMEL possessed a significantly higher ($p < 0.05$) relative abundance of PL26, PL20, PL1_4, and PL9 as compared to PME (Figure 7E).

The correlation network assessment was accomplished to comprehend the association among microbes and enzymes. The correlation network analysis between PME and CAZy exposed that *Pseudomonas aeruginosa*, *Escherichia coli*, *Enterobacter quasiroggenkampii*, *Salmonella enterica*, and *Heyndrickxia coagulans* were mainly responsible for CAZy occurrence in PME (Figure 7F). However, the correlation network analysis between PMEL and CAZy revealed that *Lactiplantibacillus plantarum*, *Limosilactobacillus reuteri*, *Latilactobacillus sakei*, *Pediococcus pentosaceus*, *Bacillus cereus*, *Limosilactobacillus fermentum*, *Aspergillus oryzae*, *Aspergillus flavus*, and *Lacticaseibacillus rhamnosus* were mainly accounted for the CAZy development in PMEL (Figure 7G). The above results demonstrated that *Lactobacillus* fermentation led to a diversified microbial community and was responsible for developing various CAZy in PMEL, ultimately boosting its nutritional and antioxidant value.

3.8. Species and Metabolites Correlation Analysis and WGCNA

Significant clusters were identified by the HCLUST analysis, suggesting strong relationships between certain microbial populations and metabolites. These clusters indicate possible functional connections or interactions among the components of the microbial community. The analysis revealed important compounds that are significantly associated with specific microbial groups, therefore emphasizing potential metabolic pathways. Clusters 2, 3, 7, and 8 were positively correlated with *Lactobacillus species*. Whereas clusters 1, 5, 6, 9, and 10 were positively correlated with *Enterobacter species* (Figure 8A).

The correlation heatmap analysis between bacterial species and antioxidant peptides revealed that most of the *Lactobacillus* and *Pediococcus species* were positively correlated with antioxidant peptides, whereas *Enterobacter species* were negatively correlated with antioxidant peptides. The *Enterobacter roggenkampii*, *Enterobacter cloacae*, *Enterobacter asburiae*, and *Enterobacter hormaechei* were considerably adversely ($p < 0.05$) correlated with antioxidant peptides, and these species have been reported in PME. The *Limosilactobacillus reuteri*, *Pediococcus acidilactici*, *Lacticaseibacillus paracasei*, *Lactiplantibacillus plantarum*, *Pediococcus pentosaceus*, *Limosilactobacillus fermentum*, and *Limosilactobacillus oris* were considerably positively ($p < 0.05$) interrelated with antioxidants peptides and these bacterial species have been found in PMEL (Figure 8B). The above findings implied that *Lactobacillus* fermentation enhances the *Lactobacillus* and *Pediococcus species* in PMEL, which were positively correlated with antioxidant peptides and hence increased its antioxidant potential.

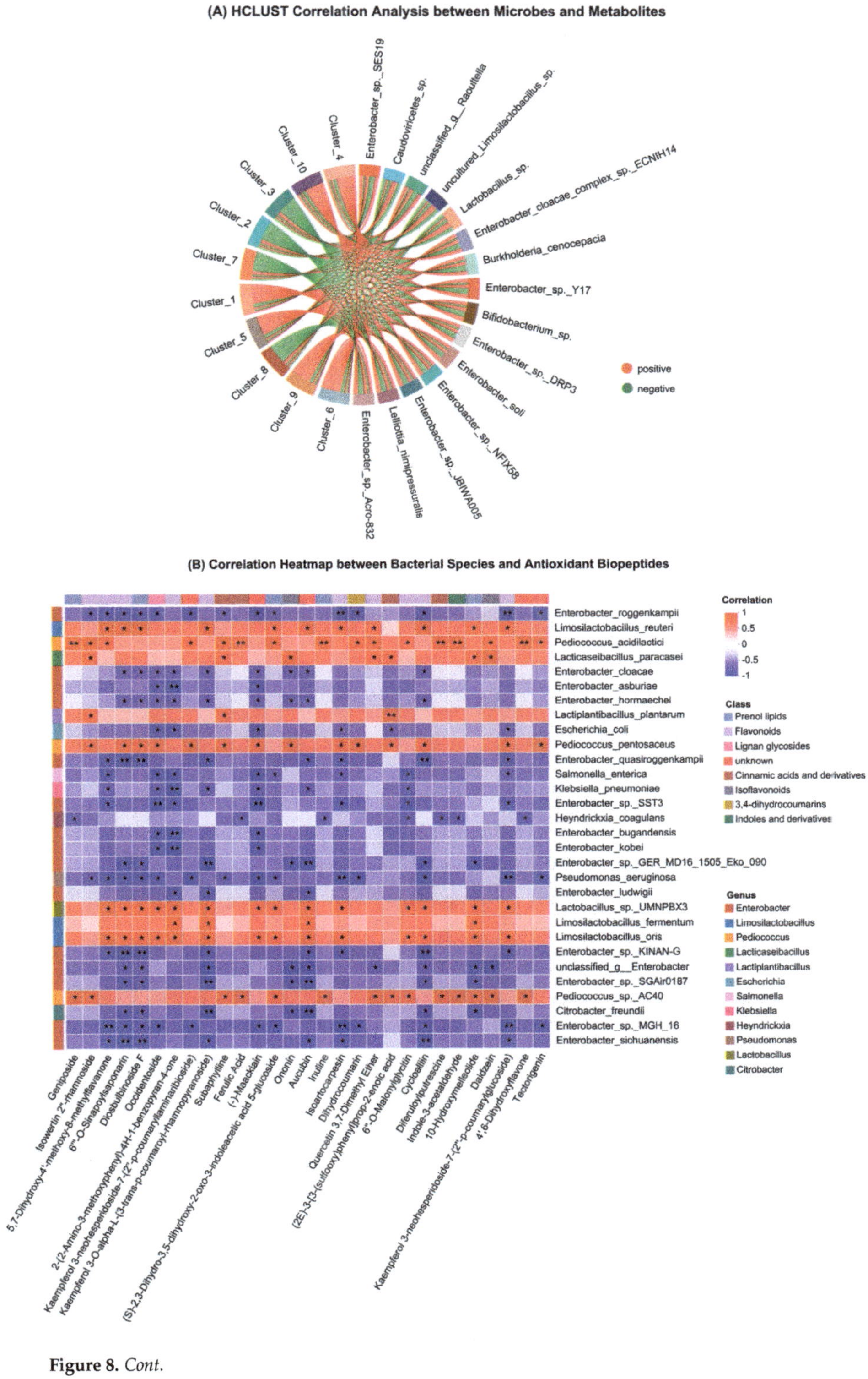

Figure 8. *Cont.*

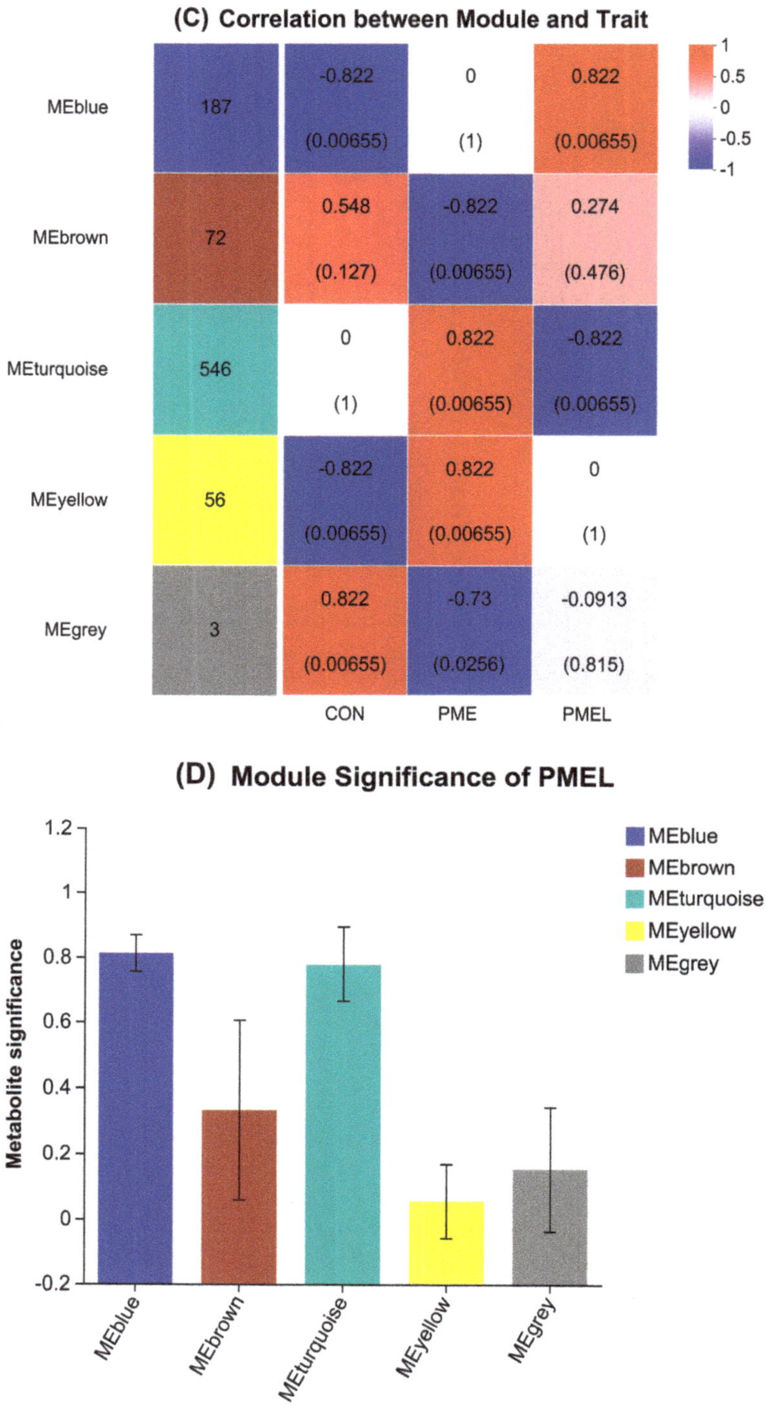

Figure 8. *Cont.*

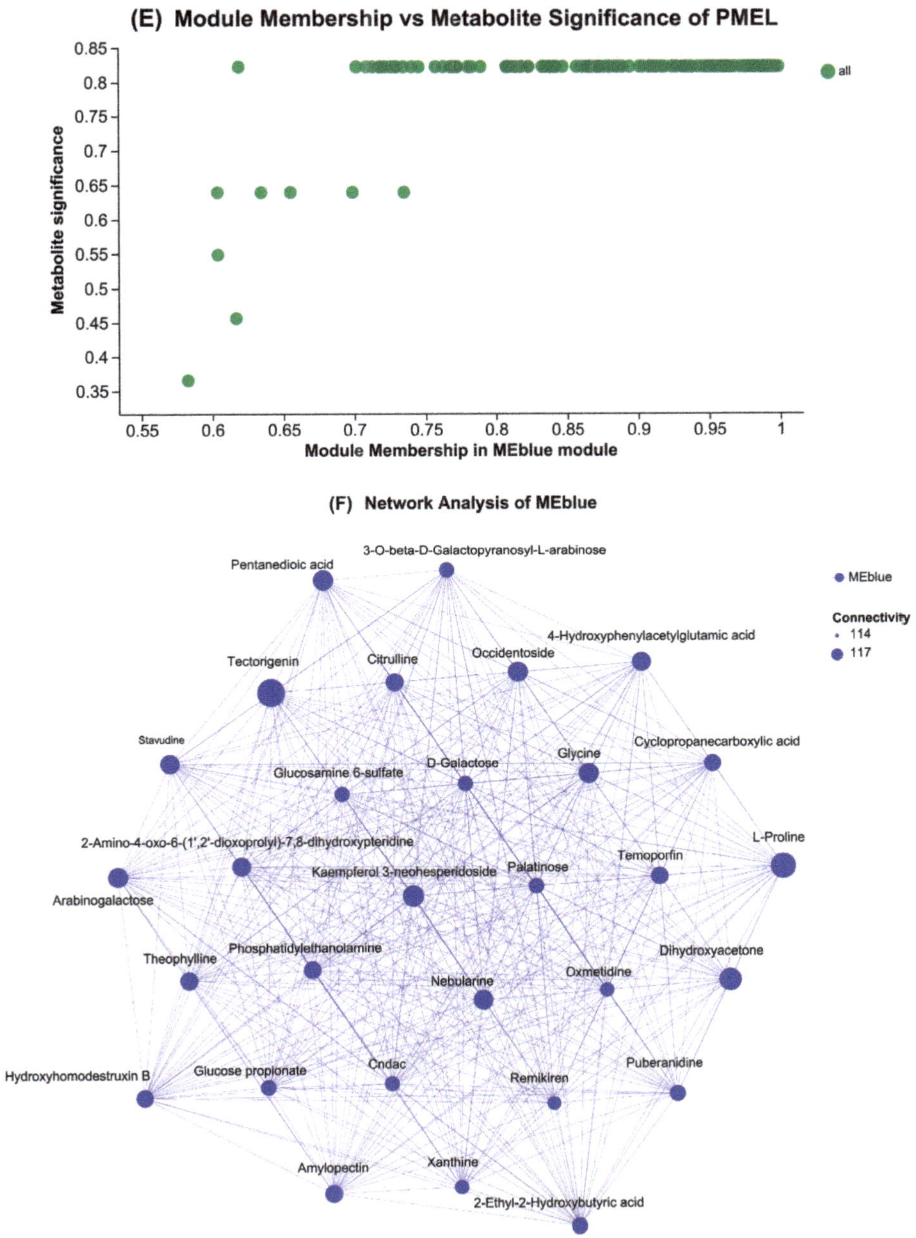

Figure 8. Species and Metabolites Correlation Analysis and WGCNA: (**A**) HCLUST correlation analysis between microbes and metabolites; (**B**) Correlation heatmap between bacterial species and antioxidant biopeptides; (**C**) Correlation between module and trait; (**D**) Module significance of PMEL; (**E**) Module membership vs. metabolite significance of PMEL; and (**F**) Network analysis of MEblue. ** $p < 0.01$; * $p < 0.05$.

In the context of metabolome data, Weighted Gene Co-expression Network Analysis (WGCNA) was employed to detect clusters (modules) of strongly correlated metabolites, evaluate the association between these modules and exterior traits, and identify significant metabolites that could potentially function as biomarkers of antioxidant activity. The sample clustering dendrogram, metabolites dendrogram, correlation between modules, and module clustering dendrogram can be found in Supplementary Material (Figure S1). The correlation between module and trait revealed that MEblue was significantly ($p < 0.05$) positively correlated, and MEturquoise was significantly ($p < 0.05$) negatively correlated with PMEL (*Lactobacillus* fermentation), whereas MEturquoise and MEyellow had a significant ($p < 0.05$) positive association with PME (Figure 8C). The module significance of PMEL revealed that MEblue and MEturquoise had more significance as compared to other modules (Figure 8D). To understand the MEblue significance, we performed (MM-GS analysis) module membership vs. gene/metabolite significance analysis, which revealed that most of the metabolites depicted higher MM and GS and were responsible for variation in PMEL (Figure 8E). The network analysis of MEblue revealed the hub metabolites, which included different antioxidant peptides like tectorigenin and kaempferol (Figure 8F).

4. Discussion

Integrating metabolomics with metagenomics has been demonstrated to be an effective strategy for deciphering complicated biological processes along with finding novel biomarkers with significant applications across numerous fields. The current research used an integrated approach to investigate the antioxidant potential of fermented palm kernel meal. These products exhibit remarkable antioxidant activity and demonstrate the potential to serve as a functional food additive and bioactive supplement in animals' diets to mitigate toxicity owing to their exceptional nutritional composition and bioactive substances [46,47].

Lactobacillus is a strain of Gram-positive bacteria capable of flourishing in both anaerobic and aerobic conditions. Through carbohydrate fermentation, it is capable of generating two distinct isomers of lactic acid [48]. Due to its exceptional probiotic, antioxidant, and immune-enhancing properties, it has been widely used as a probiotic [49,50]. The current research project utilized integrative metabolomic and metagenomic approaches to acquire a deep understanding of the distinguishing aspects of the metabolome and metagenome, along with the antioxidant-related substances in PKM fermented with *L. plantarum* and *L. reuteri*.

4.1. Fermented Palm Kernel Meal, Metabolomic, and Metagenomic Profile

The metabolomics assessment presented a thorough understanding of the biochemical features of fermented PKM. Employing high-throughput approaches, i.e., high-performance liquid chromatography-mass spectrometry (HPLC-MS), we discovered a wide range of metabolites comprising phenolic substances, isoflavonoids, flavonoids, lignan glycosides, cinnamic acids, prenol lipids, and indoles derivatives. These biomolecules are well-recognized for their antioxidant capabilities and perform significant responsibilities in eliminating free radicals, decreasing oxidative damage, and imparting positive health effects [51,52].

The heatmap suggested a substantial upregulation of flavonoids and isoflavonoids in the PMEL group. The prominent flavonoids such as dihydroxyflavone, isowertin, methylflavanone, saponarin, benzopyran, quercetin, isoartocarpesin, cycoalliin, and kaempferol derivatives were substantially upregulated in PMEL. The flavonoids have been extensively reported for their involvement in decreasing the risk of chronic illnesses via their antioxidant and anti-inflammatory activities [53–55]. Dihydroxyflavone (flavones) illustrates antioxidant capabilities and neuroprotective benefits towards glutamate-induced toxic effects [56]. The antioxidant action of methylflavanone (O-methylated flavonoids) could include processes such as effective quenching of free radicals and stimulation of cellular detoxifying enzymes [57]. The benzopyrans (O-methylated flavonoids) possess antioxidative properties, which could be linked to their possible medicinal implementations in oxidative stress-related disorders [58]. The isoartocarpesin is a flavone that functions as a powerful antioxidant by regulating surplus reactive oxygen species (ROS) generation [59].

The kaempferol derivatives products are flavonoid glycosides that exhibit diversified biological properties comprising antioxidant, anticancer, and anti-inflammatory capabilities [60]. This demonstrated that *Lactobacillus* fermentation boosted flavonoid concentration and, consequently, the antioxidant properties of PMEL.

The isoflavonoids, including tectorigenin, maackiain, daidzein, malonylglycitin, and ononin, exhibited a considerable increase in PMEL. These isoflavonoids are renowned for their antioxidant capabilities, which help to lower the stress caused by oxidative damage [61–63]. Tectorigenin is an organic chemical categorized as an isoflavone that has many biological actions, which include anti-inflammatory, antioxidant, cancer-fighting, and estrogen-like properties [64]. Daidzein is an isoflavone that performs its antioxidant action by preventing the peroxidation of lipids in the membranes of cells, preventing renal impairment [65]. Ononin is an organic isoflavone glycoside typically found in many plants and displays antioxidant capabilities, which are connected to minimizing ROS formation and the amount of inflammatory mediators [66]. The elevated amounts of these isoflavonoids revealed *Lactobacillus* fermentation boosted the antioxidant ability of PMEL.

The impact of the *Lactobacillus* fermentation process on the isoflavonoid biosynthesis pathway in PKM has been thoroughly researched. Our work found that the isoflavonoids production pathway exhibited significant enrichment ratios in PMEL, which led to improved antioxidant capability. Similar findings have been published earlier. An investigation by Merenkova et al. [67] highlighted the synthesis of isoflavone aglycones in fermented grains, suggesting the prospect for enhanced isoflavonoid yield via fermentation techniques. Likewise, Verni et al. [68] examined the modification of the formation mechanism of flavonoids through fermentation, demonstrating an increase in isoflavone percentage post-fermentation. Yang et al. [69] further corroborated these findings, revealing an enormous improvement in isoflavone levels in polyphenol-rich foodstuffs after the process of fermentation. In other research, Gao et al. [70] evaluated the flavonoid production during alfalfa silage microbial fermentation and reported a surge in the antioxidant capacity post-fermentation. These findings collectively indicate the intricate connection between fermentation, isoflavonoid synthesis, and antioxidant properties, explaining the potential of fermentation approaches to boost the nutritional value and biological activity of food products.

Investigations revealed that *Lactobacillus* fermentation strengthens the activation of pathways that contribute to carbohydrate decomposition, including glycolysis and the pathway of pentose phosphate [71]. Additionally, the fermentation process carried out by *Lactobacillus* strains results in the synthesis of short-chain fatty acids (SCFAs) via the disintegration of complex carbohydrates. This eventually impacts the pathways that regulate the metabolism of SCFA and, thus, energy production [72]. Furthermore, the fermentation of PKM by *Lactobacillus* influences the pathways that govern the metabolism of amino acids, involving either their synthesis or breakdown, which is in agreement with the findings of our experiment. This process could possibly boost the nutritional value of the fermented item by augmenting the accessibility of key amino acids and enhancing overall protein quality [73].

Lactobacillus fermentation of PKM contributes to pathways associated with the synthesis of bioactive substances and vitamins. In particular, this fermentation strengthens the enrichment of pathways that regulate the synthesis of B vitamins, such as riboflavin and folate, which are vital for multiple processes in cells involving the utilization of energy and nucleic acid production [74]. Furthermore, bioactive molecules developed throughout *Lactobacillus* fermentation, which include antioxidants and peptides that inhibit microbes, could impact pathways related to oxidative stress reactions and immune regulation. This might improve the functional properties of the fermented item, as demonstrated by our present study [75,76].

We developed a metagenomics assessment to evaluate the microbial communities that occur in the process of fermentation of PKM. The findings of our research demonstrated the presence of an extensive number of microbial classes, which comprise lactic acid-producing bacteria, that are known for their potential to create antioxidant peptides via distinct metabolic pathways. Soriano et al. [77] and Mamuad et al. [78] studied the influence of *Lactobacillus* fermentation on microbial biodiversity in grains. This research demonstrated microbial community modifications, including increased populations in LAB such as *Lactobacillales*. These results correspond with our microbial community barplot and the network evaluation of PMEL, indicating that most of the bacteria fall under *Lactobacillales* and contribute about 94% of the overall bacterial community. Multiple factors lead to the multiplication of LAB throughout *Lactobacillus* fermentation. First of all, *Lactobacillus* species make lactic acid as a consequence of their metabolic processes, which provides an acidic atmosphere by decreasing the pH value that favors their growth [79]. More importantly, these bacteria typically go beyond other species of bacteria in concurrence by successfully utilizing fermentable sugars that exist in the substrate [80]. In the end, the mechanism of *Lactobacillus* fermentation might end up in the development of antibiotic compounds that impede the growth of other microorganisms, thereby promoting the predominance of LAB [26,81].

Lactobacillus fermentation serves an important part in modifying the metabolite profile and microbiome of PKM, predominantly through enzymatic reactions that break down intricate polysaccharides, protein molecules, and fatty substances. During fermentation, *Lactobacillus* species contribute to multiple enzymes, such as glycoside hydrolases, proteases, and lipases, which deteriorate the fiber, protein, and fat constituents of PKM into less complicated molecules like monosaccharides, peptides, and fatty acids. This enzyme digestion not only enhances the nutritional importance of PKM by improving the bioavailability of nutrients but also boosts the release of bioactive components, including phenolics and flavonoids, which have been recognized for their antioxidant characteristics [82,83]. Furthermore, fermentation by *Lactobacillus* may promote the synthesis of organic acids such as lactic acid, which reduces the pH and hinders the proliferation of spoilage microbes, thereby enhancing the shelf life and safety of the food item [84]. This modification in pH assists in deliberately strengthening the populations of fermentative microbes, thereby enhancing the whole microbiome composition [85]. The metabolic switches resulting from *Lactobacillus* fermentation led to an altered metabolite profile that increases the functional and nutritional attributes of PKM, resulting in a more advantageous component in animal feed as well as applications involving biotechnology.

4.2. Correlation Evaluation of Microbial Population with Antioxidant Biopeptides and CAZy

We revealed the complicated interaction between the metabolism of microbes and the synthesis of bioactive compounds in fermented PKM by incorporating metabolomics and metagenomics results. Our investigation discovered different metabolite–microbe relationships, emphasizing critical metabolic processes and enzyme expression necessary for the generation of antioxidant compounds. Moreover, Spearman correlation studies uncovered the probable synergistic relationship between metabolites and microbial taxa, offering a mechanistic understanding of the antioxidant potential of the fermenting technique.

During fermentation, *L. plantarum* could promote the excretion and modification of flavonoids through enzymatic reactions, catalysing the disintegration of complex polyphenols into less complicated forms that are easier to absorb by the body. This enzymatic activity improves the bioavailability of flavonoids, possibly enhancing their physiological advantages [86]. Previous studies have illustrated that *L. plantarum* has a close connection with certain flavonoids like daidzein and kaempferol and their by-products, vanillic acid and genistin. The fermentation procedure involving LAB elevated the overall quantity of flavonoids and transformed the structure of certain ones, leading to improved antioxidant properties and enhanced endurance against tyrosinase [87]. The fermentation of kiwifruit by employing *L. plantarum* demonstrated a positive effect on its phenolic characteristics by increasing the concentrations of compounds, including dihydroxy coumarin, particularly protocatechuic acid. This enhancement inevitably strengthened the antioxidant and scavenging capacities of the processed kiwifruit [88]. These modifications in the phenolic substances may be attributed to the glucosidase enzyme generated by *L. plantarum*, which leads to alterations in the chemical structure of these molecules [89]. Feruloyl esterase enzymes assist in dispensing ferulic acid or p-coumarate from embedded phenols. *L. plantarum* yields a type of phenolic acid decarboxylase peptide, which influences these phenolic molecules and contributes to the overall antioxidant effectiveness [90]. Our correlation study findings revealed significant associations between the occurrence of LAB and higher levels of specific flavonoids, demonstrating an obvious connection between microbial development and the elevation of flavonoids.

Lactobacillus fermentation influences the profile of carbohydrate-active enzymes (CAZy), strengthening the decomposition of complex carbohydrates and boosting the absorption and digestion of essential nutrients [91]. Investigations on the association between species and CAZy enzymes offered a comprehensive understanding of the wide range of enzymes utilized by *Lactobacillus* during the process of fermentation, including glycoside hydrolases, polysaccharide lyases, carbohydrate esterases, and glycosyl transferases. These enzymes are essential in breaking down complicated carbohydrates, such as plant cell wall polymers, into simpler digestible sugars like glucose and galactose. The treatment leads to the creation of several different fermentation products, including organic acids and bioactive molecules, which strengthen the sensory qualities and dietary benefits of fermented items [92–94]. The most recent studies illustrate that specific CAZy communities exhibit a strong connection to the metabolic activities of *Lactobacillus* species, amending their abilities to break down intricate polysaccharides into simpler carbohydrates [95]. The elevated level of glycoside hydrolases appears to be linked to the higher concentration of *L. plantarum* during the process of fermenting, implying a beneficial association that enhances the productivity of the fermentation procedure [96]. In the present research, multiple CAZy enzymes have been identified to interact strongly with bacterial species, involving carbohydrate esterases, glycoside hydrolases, polysaccharide lyases, and glycosyl transferases. These enzymes may have accounted for the increased nutritional content and antioxidant potency of PMEL.

4.3. Polyphenolic and Flavonoid Content and Antioxidant Capability

The LAB, primarily *Lactobacillus* spp., produces metabolites that consist of exopolysaccharides, phenolic molecules, and organic acids. These substances offer the cornerstone for the fermented item's antioxidant benefits [97]. The findings of our experiment demonstrate that these compounds, which contain antioxidant properties, considerably boost the fermented item's antioxidant activity by effectively counteracting harmful unbound radicals and minimizing oxidative-related stress.

Flavonoids and polyphenols are metabolically active compounds that are renowned for their antioxidant and beneficial health capabilities [98]. Investigations have demonstrated that *Lactobacillus* fermentation could lead to a considerable increase in both flavonoid and polyphenol contents of fermented food [99]. Meng et al. [86] performed a study in which they revealed that a procedure of fermentation employing *Lactobacillus* contributed to a massive increase in the flavonoid composition of loquat juice that finally resulted in enhanced antioxidant capabilities. Verni et al. [100] discovered a high elevation in the polyphenol and flavonoid content and antioxidant capacity of grains after it was fermented using *Lactobacillus plantarum*.

The scavenging proportion of PMEL was enhanced following *Lactobacillus* fermentation. *Lactobacillus* bacteria exhibited enzymatic attributes that may degrade complex compounds present in PKM into simpler molecules, thereby boosting the availability of bioactive compounds [68,101]. Moreover, *Lactobacillus* fermentation might end up in the production and dispersion of antioxidant compounds through metabolic mechanisms involving the alteration of phenolic intermediates into bioactive substances [102]. Furthermore, the method of fermentation provides an acidic environment that promotes the retrieval and accessibility of bioactive compounds, therefore enhancing their level of concentration in the fermented food [68]. These could potentially be the foundations of PMEL's improving scavenging capabilities.

Modifications in the composition of unprocessed palm kernel meal, especially fiber composition and nutritional value, might influence fermentation accuracy and the production of antioxidant peptides, possibly restricting the applicability of the study's findings. These variations were not fully accounted for, which might impact the coherence of metabolomic and metagenomic outcomes. The variation in nutritional and fiber content can affect microbial growth, which in turn can affect the whole microbiome composition and metabolite production. The high-fat level in certain PKM batches may restrict microbial growth, as *Lactobacillus* strains opt for carbohydrate-rich materials. Additionally, the excessive lignocellulosic fiber percentage of PKM exhibits a further challenge, as these fibers are not digested readily by the microbes. Variation in fiber levels among various PKM batches lessens the accessibility of fermentable sugars, which impedes fermentation efficacy [15,25,103]. This study concentrated primarily on specific strains of *Lactobacillus*, probably overlooking contributions from other communities of microbes. Fermentation is a complicated procedure usually driven by a consortium of microorganisms, each incorporating various enzymes and metabolic processes. By emphasizing primarily *Lactobacillus*, the study likely neglected other microbial populations (such as yeasts, fungi, or other bacteria) that might improve antioxidant peptide biosynthesis [104]. Some low-abundance metabolites might have been neglected due to detection sensitivity. These metabolites, although present in smaller quantities, may contribute to the overall bioactivity and functionality of the fermentation process. The research might depend predominantly on in vitro or computational approaches, and without verification in vivo, it restricts the knowledge of the real-world performance and bioavailability of the newly identified antioxidant peptides. The variety of microbial strains employed in this research could be limited, which could fail to adequately represent the potential diversity of peptide production pathways or wider ecological relationships that could take place in more diverse microbial environments.

5. Conclusions

In conclusion, the combined application of metabolome and metagenome analyses has effectively unveiled the biosynthesis pathways of novel peptides with antioxidant properties in the fermented PKM. This study revealed the potential of *Lactobacillus* fermentation to improve the bioactive functions of PKM by facilitating the synthesis and release of antioxidant peptides via the enzymatic breakdown of complicated carbohydrates and lignocellulosic structures. The outcomes offer valuable insights into the underlying metabolic pathways and provide a foundation to enhance the nutritional and functional benefits of PKM. The results demonstrate the enormous potential of microbial fermentation in valorizing PKM, transforming it into an efficient food component with enriched antioxidant properties. The research not only broadens the understanding of microbial transformation mechanisms on lignocellulosic substrates but also unlocks new avenues toward creating bioactive substances through sustainable biotechnological methods. However, further in vivo testing and study of different bacterial strains and palm kernel meal sources is essential for fully comprehending the feasible applications of these bioactive substances in the animal nutrition and food industry. Furthermore, extending the diversity of bacterial strains employed during the fermentation process could assist in exploring a broader range of metabolic processes for novel antioxidant peptide production. Scaling up the fermentation technique for industrial application while improving conditions for optimal yield is also crucial. Future studies ought to explore the practical uses of these peptides in the food and pharmaceutical industry to evaluate their efficacy as health-promoting ingredients. Conclusively, carrying out long-term stability research studies might guarantee the antioxidant peptides preserve their efficacy during commercial manufacturing and delivery.

Supplementary Materials: The following supporting information can be downloaded at: https://www.mdpi.com/article/10.3390/antiox13101253/s1, Figure S1. (A) Sample clustering dendrogram; (B) Metabolites dendrogram; (C) Correlation between modules; and (D) Module clustering dendrogram.

Author Contributions: Conceptualization, H.Q.; Data curation, H.Q. and R.H.; Formal analysis, H.Q. and R.H.; Funding acquisition, M.S. and X.M.; Investigation, H.Q., Y.L. and M.Y.; Methodology, H.Q. and X.M.; Project administration, D.D., M.Y. and X.M.; Resources, H.Q., Y.C. and M.Y.; Software, H.Q.; Supervision, M.Y. and X.M.; Validation, H.Q., M.Y. and X.M.; Visualization, H.Q.; Writing—original draft, H.Q.; Writing—review and editing, H.Q. and X.M. All authors have read and agreed to the published version of the manuscript.

Funding: This work was supported by Science and Technology Program of Guangdong Province (2023B0202010020), the Modern Agricultural Industrial Park Project of Qingcheng District, Qingyuan City, Guangdong Province, State Key Laboratory of Swine and Poultry Breeding Industry (2023QZ-NK09, ZQQZ-03), Guangzhou Science and Technology Program (2023B03J1368), Special Fund for Rural Revitalization Strategy of Guangdong (2023KJ465, 2023TS-3), the Low Carbon Agriculture and Carbon Neutralization Research Center, GDAAS (XTXM202204), and the Opening Project of Guangdong Provincial Key Laboratory of Animal Breeding and Nutrition (2022SZ04).

Institutional Review Board Statement: Not applicable.

Informed Consent Statement: Not applicable.

Data Availability Statement: Data will be available upon request.

Acknowledgments: The enzymes (pectinase and cellulase) were courteously delivered by VTR Bio-Tech, located in Zhuhai, China.

Conflicts of Interest: The writers state no conflicts of interest.

References

1. Gustavsson, J.; Cederberg, C.; Sonesson, U.; Van Otterdijk, R.; Meybeck, A. *Global Food Losses and Food Waste*; Save Food Congress. The Swedish Institute for Food and Biotechnology: Düsseldorf, Germany, 2011.
2. Teixeira, G.L.; Ibañez, E.; Block, J.M. Emerging Lipids from Arecaceae Palm Fruits in Brazil. *Molecules* **2022**, *27*, 4188. [CrossRef]
3. USDA. Global Market Analysis, Foreign Agricultural Service, United States Department of Agriculture. Oilseeds: World Markets and Trade. Available online: https://apps.fas.usda.gov/psdonline/circulars/oilseeds.pdf (accessed on 4 September 2024).
4. Safi, C.; Humblet, N.-P.; Geerdink, P.; Theunissen, M.; Beelen, B.; Voogt, J.; Mulder, W. Valorisation of proteins from palm kernel meal. *Bioresour. Technol. Rep.* **2022**, *18*, 101050. [CrossRef]
5. Galanakis, C.M. *Proteins: Sustainable Source, Processing and Applications*; Elsevier Science: Amsterdam, The Netherlands, 2019.
6. Amaral-Júnior, J.M.d.; Morais, E.d.; Lima, A.C.S.; Martorano, L.G.; Nahúm, B.d.S.; Sousa, L.F.; Lourenço-Júnior, J.d.B.; Rodrigues, T.C.G.d.C.; Silva, J.A.R.d.; Silva, A.L.d.C.; et al. Effect of Palm Kernel Cake Supplementation on Voluntary Feed Intake, In Situ Rumen Degradability and Performance in Buffaloes in the Eastern Amazon. *Animals* **2023**, *13*, 934. [CrossRef]
7. Mohamad Asri, N.; Muhialdin, B.J.; Zarei, M.; Saari, N. Low molecular weight peptides generated from palm kernel cake via solid state lacto-fermentation extend the shelf life of bread. *LWT* **2020**, *134*, 110206. [CrossRef]
8. Sulaiman, F.; Abdullah, N.; Gerhauser, H.; Shariff, A. An outlook of Malaysian energy, oil palm industry and its utilization of wastes as useful resources. *Biomass Bioenergy* **2011**, *35*, 3775–3786. [CrossRef]
9. Pelaez, R.D.R.; Oliveira, M.E.C.; Miller, R.N.G.; de Almeida, J.R.M.; de Siqueira, F.G. Biotechnological valorization of lignocellulosic residues from the oil palm industry: Status and perspectives. *Biomass Convers. Biorefinery* **2024**, *14*, 3077–3099. [CrossRef]
10. Vakili, M.; Rafatullah, M.; Ibrahim, M.H.; Salamatinia, B.; Gholami, Z.; Zwain, H.M. A review on composting of oil palm biomass. *Environ. Dev. Sustainability* **2015**, *17*, 691–709. [CrossRef]
11. Sundu, B.; Kumar, A.; Dingle, J. Palm kernel meal in broiler diets: Effect on chicken performance and health. *Worlds Poult. Sci. J.* **2006**, *62*, 316–325. [CrossRef]
12. Cheah, W.Y.; Siti-Dina, R.P.; Leng, S.T.K.; Er, A.C.; Show, P.L. Circular bioeconomy in palm oil industry: Current practices and future perspectives. *Environ. Technol. Innov.* **2023**, *30*, 103050. [CrossRef]
13. Eggleston, H.; Buendia, L.; Miwa, K.; Ngara, T.; Tanabe, K. *2006 IPCC Guidelines for National Greenhouse Gas Inventories*; Institute for Global Environmental Strategies (IGES): Hayama, Japan, 2006.
14. Sathitkowitchai, W.; Ayimbila, F.; Nitisinprasert, S.; Keawsompong, S. Selection of pretreatment method and mannanase enzyme to improve the functionality of palm kernel cake. *J. Biosci. Bioeng.* **2022**, *134*, 301–306. [CrossRef]
15. Stein, H.H.; Casas, G.A.; Abelilla, J.J.; Liu, Y.; Sulabo, R.C. Nutritional value of high fiber co-products from the copra, palm kernel, and rice industries in diets fed to pigs. *J. Anim. Sci. Biotechnol.* **2015**, *6*, 56. [CrossRef]
16. Sindhu, R.; Binod, P.; Pandey, A. Biological pretreatment of lignocellulosic biomass–An overview. *Bioresour. Technol.* **2016**, *199*, 76–82. [CrossRef]
17. Hou, Y.; Wu, Z.; Dai, Z.; Wang, G.; Wu, G. Protein hydrolysates in animal nutrition: Industrial production, bioactive peptides, and functional significance. *Bioact. Pept. Food* **2022**, 209–232. [CrossRef]

18. Shahidi, F.; Kamil, Y.J. Enzymes from fish and aquatic invertebrates and their application in the food industry. *Trends Food Sci. Technol.* **2001**, *12*, 435–464. [CrossRef]
19. Khubber, S.; Marti-Quijal, F.J.; Tomasevic, I.; Remize, F.; Barba, F.J. Lactic acid fermentation as a useful strategy to recover antimicrobial and antioxidant compounds from food and by-products. *Curr. Opin. Food Sci.* **2022**, *43*, 189–198. [CrossRef]
20. Feng, T.; Wang, J. Oxidative stress tolerance and antioxidant capacity of lactic acid bacteria as probiotic: A systematic review. *Gut Microbes* **2020**, *12*, 1801944. [CrossRef]
21. Olukomaiya, O.O.; Fernando, W.C.; Mereddy, R.; Li, X.; Sultanbawa, Y. Solid-state fermentation of canola meal with Aspergillus sojae, Aspergillus ficuum and their co-cultures: Effects on physicochemical, microbiological and functional properties. *LWT* **2020**, *127*, 109362. [CrossRef]
22. Gänzle, M.G. Lactic metabolism revisited: Metabolism of lactic acid bacteria in food fermentations and food spoilage. *Curr. Opin. Food Sci.* **2015**, *2*, 106–117. [CrossRef]
23. Hakim, A.H.; Zulkifli, I.; Farjam, A.S.; Awad, E.A. Feeding fermented palm kernel cake with higher levels of dietary fat improved gut bacterial population and blood lipid concentration but not the growth performance in broiler chickens. *Ital. J. Anim. Sci.* **2021**, *20*, 1671–1680. [CrossRef]
24. Hammes, W.P.; Hertel, C. The genera lactobacillus and carnobacterium. *Prokaryotes* **2006**, *4*, 320–403.
25. Kumar, V.; Ahluwalia, V.; Saran, S.; Kumar, J.; Patel, A.K.; Singhania, R.R. Recent developments on solid-state fermentation for production of microbial secondary metabolites: Challenges and solutions. *Bioresour. Technol.* **2021**, *323*, 124566. [CrossRef]
26. Xu, D.; Ding, W.; Ke, W.; Li, F.; Zhang, P.; Guo, X. Modulation of metabolome and bacterial community in whole crop corn silage by inoculating homofermentative Lactobacillus plantarum and heterofermentative Lactobacillus buchneri. *Front. Microbiol.* **2019**, *9*, 3299. [CrossRef]
27. Yu, M.; Li, Z.; Rong, T.; Tian, Z.; Deng, D.; Lu, H.; Zhang, R.; Ma, X. Integrated metagenomics-metabolomics analysis reveals the cecal microbial composition, function, and metabolites of pigs fed diets with different starch sources. *Food Res. Int.* **2022**, *154*, 110951. [CrossRef]
28. Cui, Y.; Wang, X.; Yue, Y.; Du, G.; Chen, H.; Ning, M.; Yuan, Y.; Yue, T. Metagenomic features of Tibetan kefir grains and its metabolomics analysis during fermentation. *LWT* **2023**, *175*, 114502. [CrossRef]
29. Tavares, E.Q.; De Souza, A.P.; Buckeridge, M.S. How endogenous plant cell-wall degradation mechanisms can help achieve higher efficiency in saccharification of biomass. *J. Exp. Bot.* **2015**, *66*, 4133–4143. [CrossRef]
30. Kuhad, R.C.; Gupta, R.; Singh, A. Microbial cellulases and their industrial applications. *Enzym. Res.* **2011**, *2011*, 280696. [CrossRef]
31. Sawatdeenarunat, C.; Surendra, K.C.; Takara, D.; Oechsner, H.; Khanal, S.K. Anaerobic digestion of lignocellulosic biomass: Challenges and opportunities. *Bioresour. Technol.* **2015**, *178*, 178–186. [CrossRef]
32. Singh, S.B. Enzyme Catalysis and Its Role in Food Processing Industries. In *Enzymes in Food Technology: Improvements and Innovations*; Springer: Berlin/Heidelberg, Germany, 2018; pp. 143–165.
33. Agrawal, R.; Kumari, P.; Sivagurunathan, P.; Satlewal, A.; Kumar, R.; Gupta, R.P.; Puri, S.K. Pretreatment process and its effect on enzymatic hydrolysis of biomass. In *Current Status and Future Scope of Microbial Cellulases*; Elsevier: Amsterdam, The Netherlands, 2021; pp. 145–169.
34. Xue, Y.; Guo, C.; Hu, F.; Liu, J.; Mao, S. Hepatic metabolic profile reveals the adaptive mechanisms of ewes to severe undernutrition during late gestation. *Metabolites* **2018**, *8*, 85. [CrossRef]
35. Li, M.; Liu, M.; Wang, B.; Shi, L. Metabonomics Analysis of Stem Extracts from Dalbergia sissoo. *Molecules* **2022**, *27*, 1982. [CrossRef]
36. Jia, X.; Chen, H.; Wang, X.; Nie, X.; Xiang, L.; Liu, D.; Zhao, Z. Effects of Fermentation Period on the Non-Volatile Metabolites of Chinese Ultra-Long-Term Solid Fermented Kohlrabi Based on Non-Targeted Metabolomic Analysis. *Fermentation* **2023**, *9*, 753. [CrossRef]
37. Qamar, H.; Li, Y.; He, R.; Waqas, M.; Song, M.; Deng, D.; Cui, Y.; Yang, P.; Liu, Z.; Qammar, B.; et al. Integrated Metabolomics and Metagenomics Unveiled Biomarkers of Antioxidant Potential in Fermented Brewer's Grains. *Antioxidants* **2024**, *13*, 872. [CrossRef]
38. Chen, S.; Zhou, Y.; Chen, Y.; Gu, J. fastp: An ultra-fast all-in-one FASTQ preprocessor. *Bioinformatics* **2018**, *34*, i884–i890. [CrossRef]
39. Li, D.; Liu, C.-M.; Luo, R.; Sadakane, K.; Lam, T.-W. MEGAHIT: An ultra-fast single-node solution for large and complex metagenomics assembly via succinct de Bruijn graph. *Bioinformatics* **2015**, *31*, 1674–1676. [CrossRef]
40. Hyatt, D.; Chen, G.-L.; LoCascio, P.F.; Land, M.L.; Larimer, F.W.; Hauser, L.J. Prodigal: Prokaryotic gene recognition and translation initiation site identification. *BMC Bioinform.* **2010**, *11*, 119. [CrossRef]
41. Fu, L.; Niu, B.; Zhu, Z.; Wu, S.; Li, W. CD-HIT: Accelerated for clustering the next-generation sequencing data. *Bioinformatics* **2012**, *28*, 3150–3152. [CrossRef]
42. Li, R.; Li, Y.; Kristiansen, K.; Wang, J. SOAP: Short oligonucleotide alignment program. *Bioinformatics* **2008**, *24*, 713–714. [CrossRef]
43. Buchfink, B.; Xie, C.; Huson, D.H. Fast and sensitive protein alignment using DIAMOND. *Nat. Methods* **2015**, *12*, 59–60. [CrossRef]
44. Zhu, J.; Wang, R.; Zhang, Y.; Lu, Y.; Cai, S.; Xiong, Q. Metabolomics Reveals Antioxidant Metabolites in Colored Rice Grains. *Metabolites* **2024**, *14*, 120. [CrossRef]
45. Xiong, Q.; Zhang, J.; Sun, C.; Wang, R.; Wei, H.; He, H.; Zhou, D.; Zhang, H.; Zhu, J. Metabolomics revealed metabolite biomarkers of antioxidant properties and flavonoid metabolite accumulation in purple rice after grain filling. *Food Chem. X* **2023**, *18*, 100720. [CrossRef]

46. Qamar, H.; Waqas, M.; Li, A.; Iqbal, M.; Mehmood, K.; Li, J. Plastrum Testudinis Extract Mitigates Thiram Toxicity in Broilers via Regulating PI3K/AKT Signaling. *Biomolecules* **2019**, *9*, 784. [CrossRef]
47. Waqas, M.; Qamar, H.; Zhang, J.; Yao, W.; Li, A.; Wang, Y.; Iqbal, M.; Mehmood, K.; Jiang, X.; Li, J. Puerarin enhance vascular proliferation and halt apoptosis in thiram-induced avian tibial dyschondroplasia by regulating HIF-1α, TIMP-3 and BCL-2 expressions. *Ecotoxicol. Environ. Saf.* **2020**, *190*, 110126. [CrossRef]
48. Seddik, H.A.; Bendali, F.; Gancel, F.; Fliss, I.; Spano, G.; Drider, D. Lactobacillus plantarum and its probiotic and food potentialities. *Probiotics Antimicrob. Proteins* **2017**, *9*, 111–122. [CrossRef]
49. Liu, D.-M.; Huang, Y.-Y.; Liang, M.-H. Analysis of the probiotic characteristics and adaptability of Lactiplantibacillus plantarum DMDL 9010 to gastrointestinal environment by complete genome sequencing and corresponding phenotypes. *LWT* **2022**, *158*, 113129. [CrossRef]
50. Yang, S.-J.; Lee, J.-E.; Lim, S.-M.; Kim, Y.-J.; Lee, N.-K.; Paik, H.-D. Antioxidant and immune-enhancing effects of probiotic Lactobacillus plantarum 200655 isolated from kimchi. *Food Sci. Biotechnol.* **2019**, *28*, 491–499. [CrossRef]
51. Pizzino, G.; Irrera, N.; Cucinotta, M.; Pallio, G.; Mannino, F.; Arcoraci, V.; Squadrito, F.; Altavilla, D.; Bitto, A. Oxidative stress: Harms and benefits for human health. *Oxid. Med. Cell Longev.* **2017**, *2017*, 8416763. [CrossRef]
52. Gulcin, İ. Antioxidants and antioxidant methods: An updated overview. *Arch. Toxicol.* **2020**, *94*, 651–715. [CrossRef]
53. Panche, A.N.; Diwan, A.D.; Chandra, S.R. Flavonoids: An overview. *J. Nutr. Sci.* **2016**, *5*, e47. [CrossRef]
54. Karak, P. Biological activities of flavonoids: An overview. *Int. J. Pharm. Sci. Res* **2019**, *10*, 1567–1574.
55. Calderon-Montano, J.M.; Burgos-Morón, E.; Pérez-Guerrero, C.; López-Lázaro, M. A review on the dietary flavonoid kaempferol. *Mini Rev. Med. Chem.* **2011**, *11*, 298–344. [CrossRef]
56. Chen, J.; Chua, K.-W.; Chua, C.C.; Yu, H.; Pei, A.; Chua, B.H.; Hamdy, R.C.; Xu, X.; Liu, C.-F. Antioxidant activity of 7, 8-dihydroxyflavone provides neuroprotection against glutamate-induced toxicity. *Neurosci. Lett.* **2011**, *499*, 181–185. [CrossRef]
57. Tsuji, P.A.; Stephenson, K.K.; Wade, K.L.; Liu, H.; Fahey, J.W. Structure-activity analysis of flavonoids: Direct and indirect antioxidant, and antiinflammatory potencies and toxicities. *Nutr. Cancer* **2013**, *65*, 1014–1025. [CrossRef]
58. Costantino, L.; Rastelli, G.; Gamberini, M.C.; Vinson, J.A.; Bose, P.; Iannone, A.; Staffieri, M.; Antolini, L.; Del Corso, A.; Mura, U. 1-Benzopyran-4-one antioxidants as aldose reductase inhibitors. *J. Med. Chem.* **1999**, *42*, 1881–1893. [CrossRef]
59. Gupta, A.; Marquess, A.R.; Pandey, A.K.; Bishayee, A. Jackfruit (*Artocarpus heterophyllus* Lam.) in health and disease: A critical review. *Crit. Rev. Food Sci. Nutr.* **2023**, *63*, 6344–6378. [CrossRef]
60. Wang, J.; Fang, X.; Ge, L.; Cao, F.; Zhao, L.; Wang, Z.; Xiao, W. Antitumor, antioxidant and anti-inflammatory activities of kaempferol and its corresponding glycosides and the enzymatic preparation of kaempferol. *PLoS ONE* **2018**, *13*, e0197563. [CrossRef]
61. Rüfer, C.E.; Kulling, S.E. Antioxidant activity of isoflavones and their major metabolites using different in vitro assays. *J. Agric. Food Chem.* **2006**, *54*, 2926–2931. [CrossRef]
62. Heim, K.E.; Tagliaferro, A.R.; Bobilya, D.J. Flavonoid antioxidants: Chemistry, metabolism and structure-activity relationships. *J. Nutr. Biochem.* **2002**, *13*, 572–584. [CrossRef]
63. Miadoková, E. Isoflavonoids—An overview of their biological activities and potential health benefits. *Interdiscip. Toxicol.* **2009**, *2*, 211–218. [CrossRef]
64. Rong, J.; Fu, F.; Han, C.; Wu, Y.; Xia, Q.; Du, D. Tectorigenin: A review of its sources, pharmacology, toxicity, and pharmacokinetics. *Molecules* **2023**, *28*, 5904. [CrossRef]
65. Guru, A.; Sudhakaran, G.; Velayutham, M.; Murugan, R.; Pachaiappan, R.; Mothana, R.A.; Noman, O.M.; Juliet, A.; Arockiaraj, J. Daidzein normalized gentamicin-induced nephrotoxicity and associated pro-inflammatory cytokines in MDCK and zebrafish: Possible mechanism of nephroprotection. *Comp. Biochem. Physiol. Part C: Toxicol. Pharmacol.* **2022**, *258*, 109364. [CrossRef]
66. Bhuia, M.S.; Aktar, M.A.; Chowdhury, R.; Ferdous, J.; Rahman, M.A.; Al Hasan, M.S.; Islam, M.T. Therapeutic potentials of ononin with mechanistic insights: A comprehensive review. *Food Biosci.* **2023**, *56*, 103302. [CrossRef]
67. Merenkova, S.; Zinina, O.; Yu, K. Microbial Fermentation of Grain Raw Materials. Prospects for Food Technology: An Analytical Review. *Вестник Южно-Уральского государственного университета. Серия: Пищевые и биотехнологии* **2021**, *9*, 5–15.
68. Verni, M.; Verardo, V.; Rizzello, C.G. How fermentation affects the antioxidant properties of cereals and legumes. *Foods* **2019**, *8*, 362. [CrossRef]
69. Yang, F.; Chen, C.; Ni, D.; Yang, Y.; Tian, J.; Li, Y.; Chen, S.; Ye, X.; Wang, L. Effects of Fermentation on Bioactivity and the Composition of Polyphenols Contained in Polyphenol-Rich Foods: A Review. *Foods* **2023**, *12*, 3315. [CrossRef]
70. Gao, Y.; Zhou, H.; Wang, Y.; Nussio, L.G.; Yang, F.; Ni, K. Insights into fermentation with lactic acid bacteria on the flavonoids biotransformation of alfalfa silage. *Chem. Biol. Technol. Agric.* **2024**, *11*, 73. [CrossRef]
71. Salvetti, E.; Fondi, M.; Fani, R.; Torriani, S.; Felis, G.E. Evolution of lactic acid bacteria in the order Lactobacillales as depicted by analysis of glycolysis and pentose phosphate pathways. *Syst. Appl. Microbiol.* **2013**, *36*, 291–305. [CrossRef]
72. LeBlanc, J.G.; Chain, F.; Martín, R.; Bermúdez-Humarán, L.G.; Courau, S.; Langella, P. Beneficial effects on host energy metabolism of short-chain fatty acids and vitamins produced by commensal and probiotic bacteria. *Microb. Cell Factories* **2017**, *16*, 79. [CrossRef]
73. Outeiriño, D.; Costa-Trigo, I.; Ochogavias, A.; Pinheiro de Souza Oliveira, R.; Pérez Guerra, N.; Salgado, J.M.; Domínguez, J.M. Biorefinery of brewery spent grain to obtain bioproducts with high value-added in the market. *N. Biotechnol.* **2024**, *79*, 111–119. [CrossRef]

74. Linares, D.M.; Martín, M.; Ladero, V.; Alvarez, M.A.; Fernández, M. Biogenic Amines in Dairy Products. *Crit. Rev. Food Sci. Nutr.* **2011**, *51*, 691–703. [CrossRef]
75. Paul Ross, R.; Morgan, S.; Hill, C. Preservation and fermentation: Past, present and future. *Int. J. Food Microbiol.* **2002**, *79*, 3–16. [CrossRef]
76. Chai, K.F.; Voo, A.Y.H.; Chen, W.N. Bioactive peptides from food fermentation: A comprehensive review of their sources, bioactivities, applications, and future development. *Compr. Rev. Food Sci. Food Saf.* **2020**, *19*, 3825–3885. [CrossRef]
77. Soriano, A.P.; Mamuad, L.L.; Kim, S.-H.; Choi, Y.J.; Jeong, C.D.; Bae, G.S.; Chang, M.B.; Lee, S.S. Effect of Lactobacillus mucosae on In vitro Rumen Fermentation Characteristics of Dried Brewers Grain, Methane Production and Bacterial Diversity. *Asian-Australas J. Anim. Sci.* **2014**, *27*, 1562–1570. [CrossRef]
78. Mamuad, L.; Kim, S.; Choi, Y.; Soriano, A.; Cho, K.; Lee, K.; Bae, G.; Lee, S. Increased propionate concentration in Lactobacillus mucosae–fermented wet brewers grains and during in vitro rumen fermentation. *J. Appl. Microbiol.* **2017**, *123*, 29–40. [CrossRef]
79. Hugenholtz, J. Citrate metabolism in lactic acid bacteria. *FEMS Microbiol. Rev.* **1993**, *12*, 165–178. [CrossRef]
80. Makarova, K.; Slesarev, A.; Wolf, Y.; Sorokin, A.; Mirkin, B.; Koonin, E.; Pavlov, A.; Pavlova, N.; Karamychev, V.; Polouchine, N. Comparative genomics of the lactic acid bacteria. *Proc. Natl. Acad. Sci.* **2006**, *103*, 15611–15616. [CrossRef]
81. Šušković, J.; Kos, B.; Beganović, J.; Leboš Pavunc, A.; Habjanič, K.; Matošić, S. Antimicrobial activity–the most important property of probiotic and starter lactic acid bacteria. *Food Technol. Biotechnol.* **2010**, *48*, 296–307.
82. Bairagi, A.; Ghosh, K.; Sen, S.; Ray, A. Evaluation of the nutritive value of Leucaena leucocephala leaf meal, inoculated with fish intestinal bacteria Bacillus subtilis and Bacillus circulans in formulated diets for rohu, Labeo rohita (Hamilton) fingerlings. *Aquac. Res.* **2004**, *35*, 436–446. [CrossRef]
83. Abdeshahian, P.; Samat, N.; Hamid, A.A.; Yusoff, W.M. Utilization of palm kernel cake for production of beta-mannanase by Aspergillus niger FTCC 5003 in solid substrate fermentation using an aerated column bioreactor. *J. Ind. Microbiol. Biotechnol.* **2010**, *37*, 103–109. [CrossRef]
84. Wang, L.; Zhou, H.; He, R.; Xu, W.; He, G. Effects of soybean meal fermentation by Lactobacillus plantarum P8 on growth, immune responses, and intestinal morphology in juvenile turbot (*Scophthalmus maximus* L.). *Aquaculture* **2016**, *464*, 87–94. [CrossRef]
85. Gupta, S.; Abu-Ghannam, N. Probiotic fermentation of plant based products: Possibilities and opportunities. *Crit. Rev. Food Sci. Nutr.* **2012**, *52*, 183–199. [CrossRef]
86. Meng, F.-B.; Lei, Y.-T.; Li, Q.-Z.; Li, Y.-C.; Deng, Y.; Liu, D.-Y. Effect of Lactobacillus plantarum and Lactobacillus acidophilus fermentation on antioxidant activity and metabolomic profiles of loquat juice. *LWT* **2022**, *171*, 114104. [CrossRef]
87. Xu, H.; Sun, J.; Zhao, Z.; Ma, X.; Li, C.; Liu, L.; Zhang, G. Lactobacillus plantarum ZLC-18 fermentation improve tyrosinase inhibition activity and antioxidant capacity in soybean hulls. *Int. J. Food Sci. Technol.* **2022**, *57*, 4518–4527. [CrossRef]
88. Zhou, Y.; Wang, R.; Zhang, Y.; Yang, Y.; Sun, X.; Zhang, Q.; Yang, N. Biotransformation of phenolics and metabolites and the change in antioxidant activity in kiwifruit induced by Lactobacillus plantarum fermentation. *J. Sci. Food Agric.* **2020**, *100*, 3283–3290. [CrossRef]
89. Scalbert, A.; Manach, C.; Morand, C.; Rémésy, C.; Jiménez, L. Dietary polyphenols and the prevention of diseases. *Crit. Rev. Food Sci. Nutr.* **2005**, *45*, 287–306. [CrossRef]
90. Rodríguez, H.; Landete, J.M.; de las Rivas, B.; Muñoz, R. Metabolism of food phenolic acids by Lactobacillus plantarum CECT 748T. *Food Chem.* **2008**, *107*, 1393–1398. [CrossRef]
91. Michlmayr, H.; Kneifel, W. β-Glucosidase activities of lactic acid bacteria: Mechanisms, impact on fermented food and human health. *FEMS Microbiol. Lett.* **2014**, *352*, 1–10. [CrossRef]
92. Liang, T.; Jiang, T.; Liang, Z.; Zhang, N.; Dong, B.; Wu, Q.; Gu, B. Carbohydrate-active enzyme profiles of Lactiplantibacillus plantarum strain 84-3 contribute to flavor formation in fermented dairy and vegetable products. *Food Chem. X* **2023**, *20*, 101036. [CrossRef]
93. Liu, X.; Wu, Z.; Hu, T.; Lin, X.; Liang, H.; Li, W.; Jin, X.; Xiao, L.; Fang, X.; Zou, Y. Comparative genomic analysis reveals niche adaption of Lactobacillus acidophilus. *J. Appl. Microbiol.* **2023**, *134*, lxad287. [CrossRef]
94. Li, Z.; Li, Y.; Xiao, C.; Yan, Z.; Pan, R.; Gao, Y.; Li, B.; Wei, J.; Qiu, Y.; Liu, K. Genomic and metabolic features of the Lactobacillus sakei JD10 revealed potential probiotic traits. *Microbiol. Res.* **2022**, *256*, 126954. [CrossRef]
95. Kostrzynska, M.; Bachand, A. Use of microbial antagonism to reduce pathogen levels on produce and meat products: A review. *Can. J. Microbiol.* **2006**, *52*, 1017–1026. [CrossRef]
96. Kaoutari, A.E.; Armougom, F.; Gordon, J.I.; Raoult, D.; Henrissat, B. The abundance and variety of carbohydrate-active enzymes in the human gut microbiota. *Nat. Rev. Microbiol.* **2013**, *11*, 497–504. [CrossRef]
97. Rodríguez, H.; Curiel, J.A.; Landete, J.M.; de las Rivas, B.; de Felipe, F.L.; Gómez-Cordovés, C.; Mancheño, J.M.; Muñoz, R. Food phenolics and lactic acid bacteria. *Int. J. Food Microbiol.* **2009**, *132*, 79–90. [CrossRef]
98. Akbari, B.; Baghaei-Yazdi, N.; Bahmaie, M.; Mahdavi Abhari, F. The role of plant-derived natural antioxidants in reduction of oxidative stress. *Biofactors* **2022**, *48*, 611–633. [CrossRef]
99. Ryu, J.Y.; Kang, H.R.; Cho, S.K. Changes over the fermentation period in phenolic compounds and antioxidant and anticancer activities of blueberries fermented by Lactobacillus plantarum. *J. Food Sci.* **2019**, *84*, 2347–2356. [CrossRef]
100. Verni, M.; Pontonio, E.; Krona, A.; Jacob, S.; Pinto, D.; Rinaldi, F.; Verardo, V.; Díaz-de-Cerio, E.; Coda, R.; Rizzello, C.G. Bioprocessing of brewers' spent grain enhances its antioxidant activity: Characterization of phenolic compounds and bioactive peptides. *Front. Microbiol.* **2020**, *11*, 1831. [CrossRef]

101. Mahajan, G.; Sharma, V.; Ray, R.C.; Gupta, R. 10—Lactic acid bacteria as potential sources of enzymes: From genes to industrial applications. In *Lactic Acid Bacteria as Cell Factories*; Montet, D., Ray, R.C., De Carvalho Azevedo, V.A., Paramithiotis, S., Eds.; Woodhead Publishing: Amsterdam, The Netherlands, 2023; pp. 183–197.
102. Filannino, P.; Di Cagno, R.; Gobbetti, M. Metabolic and functional paths of lactic acid bacteria in plant foods: Get out of the labyrinth. *Curr. Opin. Biotechnol.* **2018**, *49*, 64–72. [CrossRef]
103. Ezejiofor, T.I.N.; Enebaku, U.E.; Ogueke, C. Waste to wealth-value recovery from agro-food processing wastes using biotechnology: A review. *Br. Biotechnol. J.* **2014**, *4*, 418–481. [CrossRef]
104. Hammes, W.P.; Brandt, M.J.; Francis, K.L.; Rosenheim, J.; Seitter, M.F.; Vogelmann, S.A. Microbial ecology of cereal fermentations. *Trends Food Sci. Technol.* **2005**, *16*, 4–11. [CrossRef]

antioxidants

MDPI

Article

Synergistic Antioxidant Effects of Cysteine Derivative and Sm-Cluster for Food Applications

Lingxia Chen [1], Lijun Wang [1], Lifu Ma [2], Chao Wang [1], Xinshu Qin [1], Minlong Wang [1], Xiaohe Zhang [1], Ruoyan Yang [1], Bing Fang [1,*] and Jie An [1,*]

[1] Department of Nutrition and Health, China Agricultural University, Beijing 100083, China; lxchen@cau.edu.cn (L.C.); lijunwang@cau.edu.cn (L.W.); b20233311213@cau.edu.cn (C.W.); b20233311203@cau.edu.cn (X.Q.); mlwang@cau.edu.cn (M.W.); b20223311102@cau.edu.cn (X.Z.); 2019310060114@cau.edu.cn (R.Y.)

[2] Tianjin Rianlon Corporation Research Institute Analytic Center, Tianjin 300457, China; malifu@rianlon.com

* Correspondence: bingfang@cau.edu.cn (B.F.); jie_an@cau.edu.cn (J.A.)

Abstract: The incorporation of antioxidants in food products is essential to prevent or delay deterioration, thereby addressing food spoilage. Thiol compounds, recognized for their natural antioxidant properties, are widely used in various foods; however, their antioxidant capacity is often limited. This study investigates the potential enhancement of thiol antioxidant capacity through the addition of a soluble, low-toxic inorganic Sm-cluster. Our findings demonstrate that the Sm-cluster significantly bolsters the antioxidant efficacy of thiol compounds. We explored, for the first time, the in vitro antioxidant activities of an Sm-oxo/hydroxy cluster combined with a cysteine derivative for potential food applications. The composition exhibited a robust inhibition of aromatic aldehyde flavor compound oxidation and displayed strong, dose-dependent DPPH (2,2-diphenyl-1-picrylhydrazine) radical scavenging activity. Notably, the antioxidant activity of the Sm-cluster/cysteine derivative was further enhanced under strong visible light conditions, which typically increased the likelihood of oxidation. These results suggest that the combination of inorganic cluster and thiol compounds presents a promising natural alternative to traditional antioxidants in the food industry.

Keywords: oxidation; antioxidant; cysteine derivative; Sm-cluster; thiol; DPPH

1. Introduction

Food spoilage has long been a major concern, particularly in countries with relatively warm climates [1]. The process of food spoilage is closely linked to oxidative reactions and the decomposition of oxidative products [2]. To prevent or delay food spoilage, antioxidants are widely used as additives in food processing [3]. They play a crucial role in the food industry and are one of the most common methods of food preservation. Antioxidants used in the food industry can be classified into natural antioxidants and synthetic organic antioxidants.

Biological thiols, or biologically derived thiols, are among the most important natural antioxidants [4]. Examples include glutathione, cysteine, N-acetylcysteine, homocysteine, and γ-glutamylcysteine [5,6]. However, despite their potential, thiol antioxidants often suffer from limited antioxidant potency. Although synthetic antioxidants, such as butylated hydroxytoluene and tert-butylhydroquinone [7], are more effective compared to natural antioxidants, they have been shown to pose potential health hazards, including toxicological and carcinogenic effects [8].

In recent years, many less toxic inorganic nanoparticles with antioxidant properties (average size < 100 nanometers) have been discovered to exhibit excellent antioxidant performance [9,10]. Examples include CeO_3, TiO_2, and Fe_3O_4 [11–13]. In addition to their standalone use, inorganic nanoparticles can enhance the antioxidant activity of organic substances through synergistic effects [14,15]. Nanoparticles have been incorporated into various biodegradable polymer matrices to create enhanced nanocomposites with high performance and improved properties [16–18]. For instance, samarium (Sm) and its oxides are widely used in many products due to their low toxicity [19]. Although inorganic nanoparticles are generally considered less toxic compared to synthetic organic antioxidants, they can potentially induce oxidative stress and lead to adverse health issues [20,21]. Moreover, the agglomeration of nanoparticles can significantly impact their practical application [22].

Herein, we report the discovery of the first soluble inorganic metal oxide cluster that exhibits synergistic antioxidant effects with thiols. Soluble metal oxide clusters can avoid issues such as toxicity and agglomeration associated with metal oxide nanoparticles. In our previous work, we developed a novel Sm-oxo/hydroxy cluster for applications as a catalyst in photoredox and thermal redox reactions, which prompted us to explore its potential as an antioxidant to scavenge oxidative radicals [23]. Our findings demonstrate that the in vitro antioxidant activities of a composition comprising Sm-oxo/hydroxy clusters and a cysteine derivative exhibited a robust inhibition of the oxidation process and displayed strong radical scavenging activity. Notably, due to its excellent photoredox activity, the antioxidant efficacy of the Sm-cluster/cysteine derivative was further enhanced under strong visible light conditions. While oxidation processes were accelerated under these conditions, the enhanced antioxidant activity under strong light suggests broader potential applications for this antioxidant strategy.

In this study, we specifically selected a series of aromatic aldehyde flavor compounds and DPPH (2,2-diphenyl-1-picrylhydrazine) to evaluate the antioxidant activity of the Sm-cluster/cysteine derivative composition. Aromatic aldehydes, such as benzaldehyde, benzyl aldehyde, vanillin, and cinnamaldehyde, naturally occur in many foods and food ingredients and are commonly used as flavorings and fragrances (Figure 1) [24]. These compounds are prone to free radical-initiated auto-oxidation when exposed to air, which can alter their sensory properties, produce toxic benzoic acid derivatives, cause food spoilage and color changes, and alterations in protein functionality, thereby reducing food quality [25]. Therefore, controlling their oxidative degradation is of significant concern in the food industry. Additionally, the DPPH assay is a well-established method for measuring radical scavenging activity [26], providing a reliable and quantifiable means to assess the antioxidant efficacy of the Sm-cluster/cysteine derivative composition. The combination of these tests allows for a comprehensive evaluation of the antioxidant potential and practical applications of our novel composition in improving food quality and stability.

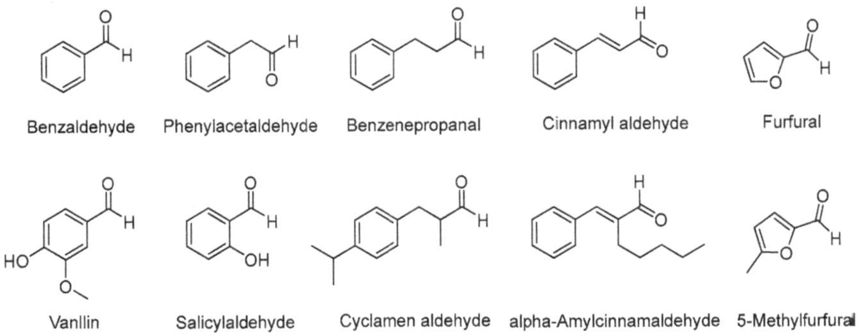

Figure 1. Some common aromatic aldehyde spices.

2. Materials and Methods

2.1. Materials and Reagents

Sm metal (25 mesh) was purchased from Hebei Zhongyue Metal Materials Technology Co., Ltd. (Shijiazhuang, China) 1,2-Diiodoethane (98%) was purchased from Macklin. Benzaldehyde (98%), vanillin (98%), Methyl (tert-butoxycarbonyl)-L-cysteinate (*N*-Boc-

L-Cysteine methyl ester, 90%), 6-hydroxy-2,5,7,8-tetramethylchroman-2-carboxylic acid (Trolox, 99%), Deuterated chloroform ($CDCl_3$, 0.03 vol% tetramethylsilane (TMS), CAS no. 865–49-6, >99.8% D), Dimethyl sulfoxide-d6 (DMSO-d_6, 0.03 vol% tetramethylsilane (TMS), CAS no. 2206-27-1, >99.8% D), and 1,1,2,2-Tetrachloroethane (99%) were purchased from Energy Chemical (Anqing, China). Cinnamaldehyde (98%) was purchased from Heowns (Tianjin, China). 1,1-Diphenyl-2-picrylhydrazyl radical 2,2-Diphenyl-1-(2,4,6-trinitrophenyl)hydrazyl (DPPH, 97%) was purchased from Aladdin (Shanghai, China). Deionized water was utilized in all experimental procedures, and oxygen was obtained from Jinghui Gas (Beijing, China). All solvents (analytical grade) are obtained from commercial sources and used as received, except as otherwise noted.

2.2. Instruments

Multifunctional microplate detector (BioTek Synergy HTX, BioTek Instruments, Inc., Winooski, VT, USA) and Bruker Avance NOE 500 (Bruker Corporation, Billerica, MA, USA). The SEM (scanning electron microscope) image of $[Sm_6O(OH)_8(H_2O)_{24}]I_8(H_2O)_8$ cluster (Sm-cluster) was obtained on a Hitachi SU3500 (Hitachi High-Technologies Corporation, Tokyo, Japan). UV–vis diffuse reflectance spectroscopy was performed on a Hitachi U4150 (Hitachi High-Technologies Corporation, Tokyo, Japan). A 300 W Xe lamp (Zhongjiaojinyuan Co., Ltd., Beijing, China) was used as the artificial solar light source.

NMR data were measured with a Bruker Avance NOE 500 and manipulated directly from the spectrometer or via a networked PC with appropriate software. Reference values for residual solvent were taken as $\delta = 7.26$ ($CDCl_3$), $\delta = 2.50$ (DMSO-d_6) for ^1H NMR. Multiplicities for coupled signals were designated using the following abbreviations: s = singlet, d = doublet, t = triplet, q = quartet, quin = quintet, br = broad signal, and are given in Hz.

2.3. Preparation of Sm-Cluster

Sm-cluster was prepared following the reported procedure in our previous work [23]. An excess amount of Sm metal (1.80 g, 12.0 mmol) was reacted with purified 1,2-diiodoethane (1.69 g, 6.00 mmol) in a 250 mL round-bottom flask. Carefully, argon gas was introduced into the flask for approximately 20 min to ensure an air-free atmosphere. Then, 60.0 mL of extra dry THF was added to the round-bottom flask containing Sm metal and 1,2-diiodoethane using syringes. The reaction mixture was stirred under argon atmosphere at room temperature. After 18 h, a dark blue SmI_2 solution was formed. The reaction mixture was allowed to settle for 30 min to precipitate the unreacted Sm metal. Then, 24.0 mL of the SmI_2 solution was transferred to a flask sealed with a rubber stopper and filled with Ar. Oxygen was carefully introduced to oxidize the SmI_2 solution. After the solution turned yellow, deionized water (0.274 g, 15.2 mmol) was added under argon, and the mixture was stirred at room temperature for 5 h to obtain an orange-red solution. The solution was concentrated by rotary evaporation to remove most of the liquid. The remaining trace solvent was exposed overnight in a vacuum oil pump to yield a yellow solid Sm-cluster weighing 1.267 g. The structure of Sm-cluster was characterized through scanning electron microscopy (SEM) and ultraviolet–visible (UV–vis) diffuse reflectance spectroscopy.

2.4. Scanning Electron Microscope (SEM) of Synthesized Sm-Cluster

A crystal of the Sm-cluster was acquired using the layer diffusion method. A single crystal of the Sm-cluster was used to obtain the SEM image. Sm-cluster samples were mounted on aluminum stubs with sticky double-sided carbon tape and sputtered with a gold layer using an ion coater under vacuum. The samples were examined with a Hitachi SU3500 scanning electron microscope (Hitachi Ltd., Tokyo, Japan) operating at 30.0 keV.

2.5. Ultraviolet–Visible (UV–Vis) Diffuse Reflectance Spectroscopy of Sm-Cluster

In order to study the optical characteristics of synthesized Sm-cluster and verify its structure, the absorption spectrum was obtained on a Hitachi U4150 with an integrating sphere (Hitachi, Tokyo, Japan) in between a wavelength scan of 200–800 nm, using $BaSO_4$ as the reflectance standard.

2.6. Evaluation of the Antioxidation Effect on Aromatic Aldehydes

To a quartz flask, aromatic aldehyde (0.200 mmol) and thiol (0.004–0.020 mmol, 2.00–10.0 mol%) were added, and the flask was sealed with rubber plug. Carefully, oxygen was introduced into the quartz flask for approximately 10 min to ensure an oxygen atmosphere. Then, Sm-cluster (1.3–6.6 mg, 0.10–1.0 mol%) dissolved in extra dry EtOAc was added to the system and then supplemented with EtOAc until a total volume of 6.0 mL. The mixture was stirred and illuminated with a 300 W Xe lamp under an oxygen atmosphere at room temperature for a period of 16 h. The mixture was concentrated under vacuum, and the recovery rate of aromatic aldehyde was determined through ^{1}H NMR by an internal standard.

2.6.1. ^{1}H NMR Acquisition Method and Parameters

The concentrated mixture was dissolved in 0.5 mL of $CDCl_3$ or DMSO-d_6. The internal standard, 1,1,2,2-tetrachloroethane, was weighted and added to the solution. The solution was transferred into 5 mm NMR tubes for ^{1}H NMR analysis. The one-dimensional ^{1}H NMR spectra acquisition were recorded on a Bruker Avance NOE 500 MHz Digital FT-NMR Spectrometer (Peabody, MA, USA), equipped with a high-resolution, 5 mm wideband Z gradient probe. The ^{1}H-NMR spectra were obtained using number of scans (NS) = 16, spectral width (SWH) of 10,000.0 Hz, size of fid (TD) = 32 K, relaxation delay = 1.0 s, and pulse width = 8.0 µs. And the spectra were processed using MestReNova 9.0.1 software (Santiago de Compostela, Spain, EU), where all spectra were Fourier transformed, zero filled to 64 K data points and the 0.2 Hz line broadening function was applied. Phase and baseline corrections were automatically performed.

2.6.2. Calculation of the Recovery Rate by Internal Standard Using ^{1}H NMR

First, 1,1,2,2-tetrachloroethane was used as the internal standard for yield calculations employing the ^{1}H NMR internal standard approach. The procedure entailed the integration of ^{1}H NMR signals corresponding to the target product and an internal standard of known quantity introduced into the reaction mixture. The ratio of the integrated areas of these signals furnished a quantitative assessment of product formation. Through comparison of this ratio to the known quantity of the internal standard, the yield of the desired product could be precisely ascertained. Recovery rates were determined via ^{1}H NMR using the following equation:

$$\text{Yield} = \left(\frac{\text{Area}_{\text{reagent}}}{\text{Area}_{\text{internal standard}}}\right)\left(\frac{n_{\text{internal standard}}}{n_{\text{theoretical reagent}}}\right) \times 100\% \tag{1}$$

$\text{Area}_{\text{reagent}}$ represents the integration of the reagent peak, $\text{Area}_{\text{internal standard}}$ represents the integration of the internal standard peak, $n_{\text{internal standard}}$ represents the quantity of moles of the internal standard, $n_{\text{theoretical reagent}}$ represents the theoretical quantity of moles of the reagent.

2.7. Evaluation of DPPH Free Radical Scavenging Activity

2.7.1. General Method

The radical scavenging activity was evaluated through the DPPH radical scavenging assay, employing Trolox as the standard antioxidant compound. Experiments were conducted under dark conditions or xenon lamp irradiation. Utilizing the correlation between antioxidant activity and the standard, the obtained values were transformed from the equivalence of one specific antioxidant standard to another.

For the DPPH radical scavenging assay, sample **a** solution (0.1 mM *N*-Boc-L-Cysteine methyl ester and 0.2 mM Sm-cluster), sample **b** solution (0.1 mM *N*-Boc-L-Cysteine methyl ester), standard antioxidant **c** solution (0.1 mM Trolox), and the DPPH radical solution (50 mg/L) were all prepared with EtOAc. The sample or Trolox solutions with varying volumes (20, 40, 60, 80, and 100 µL) were added to a 2 mL DPPH radical solution, and a corresponding volume of EtOAc (980, 960, 940, 920, and 900 µL) was supplemented. The mixtures were then shaken vigorously and left to incubate for 30 min at room temperature under ambient indoor light or Xe lamp irradiation. For the blank group, the DPPH radical solution was replaced by EtOAc. For the control group, the sample or Trolox was replaced by EtOAc.

In a 96-well microplate, 200 µL of each mixture was added to each well. The absorbance was measured at 517 nm using a BioTek Synergy HTX microplate reader (BioTek Instruments, Inc., Winooski, VT, USA). The inhibition percentage of DPPH radical dot discoloration was calculated using the following Equation (2):

$$\% \text{ Inhibition} = \frac{\left[1 - \left(A_{sample} - A_{blank}\right)\right]}{A_{control}} \times 100\% \quad (2)$$

A_{sample} represents the absorbance of the samples, A_{blank} represents the absorbance of the blank (contained solvent instead of DPPH solution), $A_{control}$ represents t the absorbance of the control (contained solvent instead of samples).

Trolox was used as the standard compound, and all the experiments were conducted in triplicate.

2.7.2. Statistical Analysis

The results, presented as the means ± standard deviation (SD) of three replicate assays, were analyzed using SPSS 26.0 software (SPSS Inc., Chicago, IL, USA). Significant differences were determined by two-way analysis of variance (ANOVA) completed by Ducan's test. Pearson's correlation test was used to compare the different values of assays obtained in our extracts. Significant levels were defined at $p < 0.05$, $p < 0.01$, and $p < 0.001$. All these figures were performed by Origin 2024 (OriginLab Corporation, Northampton, MA, USA).

3. Results

3.1. Synthesis and Structural Characterization of Sm-Cluster

The Sm-cluster was synthesized using an auxiliary ligand-free oxidative hydrolysis method, and the product was obtained as a yellow-brown powder, as shown in Figure 2A. The SEM image (Figure 2B) of the Sm-cluster crystal revealed clear layered and hollow structures, with edge lengths of approximately 5–10 µm, consistent with previously reported data [23]. Further structural characterization was performed by UV–Vis diffuse reflectance spectroscopy (Figure S1). The results were also consistent with the reported data [23]. Those structural characterizations confirmed the successful synthesis of Sm-clusters with the expected structure (Figure 2C).

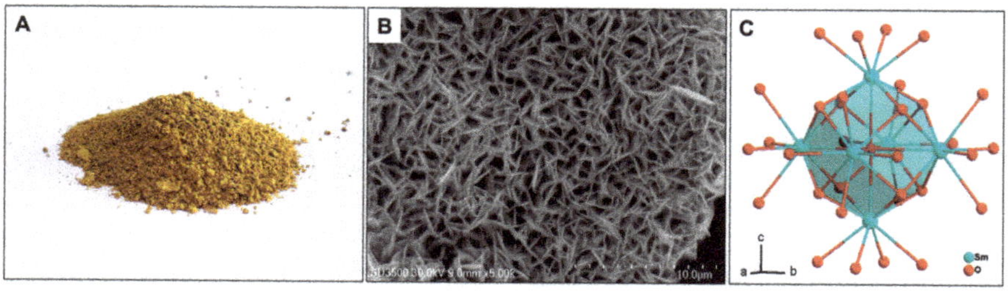

Figure 2. The Structure of Sm-cluster. (**A**) Image of Sm-cluster power; (**B**) SEM image of Sm-cluster single crystal; (**C**) Schematic of Sm-cluster Unit.

3.2. Evaluation of the Antioxidation Efficiency of Sm-Cluster/Cysteine Derivative

To evaluate the antioxidation efficiency of the Sm-cluster/cysteine derivative, we selected three aromatic aldehydes—benzaldehyde, vanillin, and cinnamaldehyde—as model compounds due to their significance as flavor compounds. When aldehydes are exposed to oxygen, they can undergo autoxidation, which is a radical chain process (Figure 3A). This process typically begins with the formation of an aldehyde radical through hydrogen abstraction (Figure 3A, (1)), followed by the reaction with molecular oxygen to form a peroxy radical (Figure 3A, (2)). The peroxy radical can then react with another aldehyde molecule, generating a hydroperoxide and propagating the chain reaction (Figure 3A, (3)). This series of reactions can ultimately lead to the formation of peroxy acids or carboxylic acids, significantly affecting the stability and sensory properties of the aromatic aldehydes (Figure 3B).

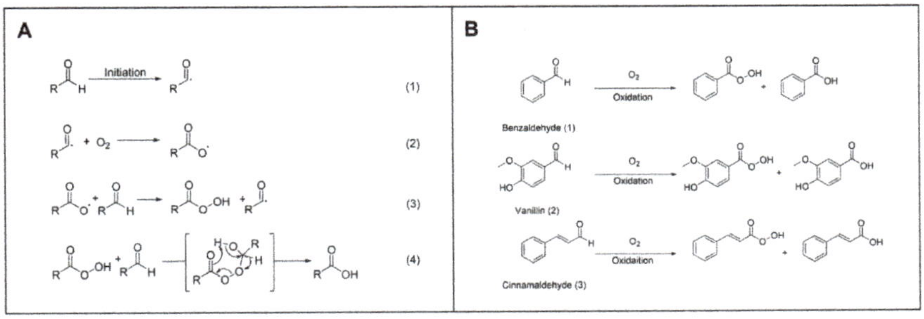

Figure 3. The autoxidation of aldehydes. (**A**) Mechanism of aldehydes autoxidation. (**B**) Reactions in the progress of selected aromatic aldehydes autooxidation.

The antioxidation efficiency of the Sm-cluster/cysteine derivative was rigorously evaluated under accelerated conditions. This involved simulating the autoxidation of aromatic aldehydes by exposing them to pure oxygen and/or irradiation with a 300 W Xe lamp for a duration of 16 h. The following four experimental groups were established: a blank control group, processing group **I** (with the addition of 1 mol% Sm-cluster), processing group **II** (with the addition of 2 mol% *N*-Boc-L-cysteine methyl ester), and processing group **III** (with the addition of 2 mol% *N*-Boc-L-cysteine methyl ester and 1 mol% Sm-cluster).

The recovery rates of aldehydes were determined by ^{1}H NMR using 1,1,2,2-tetrachloroethane as the internal standard (Figure 4). The ^{1}H NMR spectra of all the tested aromatic aldehydes listed below were used for the identification and quantification of the recovered aldehydes in the antioxidation tests (Figures S2–S4).

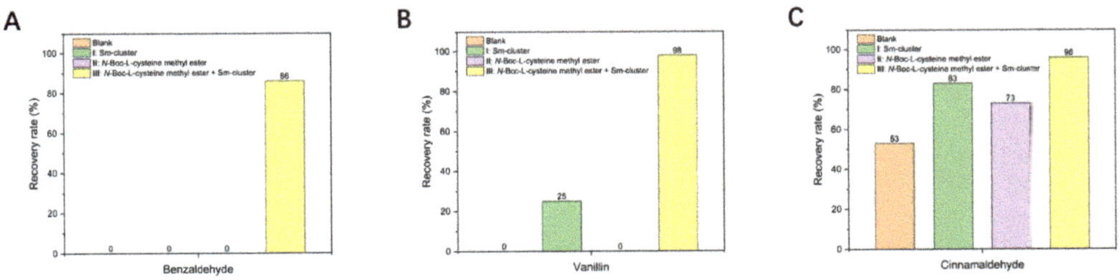

Figure 4. The recovery rates of three selected aromatic aldehydes with different processing groups. (**A**) Benzaldehyde. (**B**) Vanillin. (**C**) Cinnamaldehyde.

Benzaldehyde (**1**) [27]: ^{1}H NMR (500 MHz, $CDCl_3$) δ 10.00 (s, 1H), 7.86 (d, *J* = 8.0 Hz, 2H), 7.60 (t, *J* = 8.0 Hz, 1H), 7.51 (d, *J* = 8.0 Hz, 2H).

Vanillin (**2**) [28]: ^{1}H NMR (500 MHz, DMSO-d_6) δ 10.24 (s, 1H), 9.77 (s, 1H), 7.43–7.37 (m, 2H), 6.96 (d, *J* = 8.0 Hz, 1H), 3.84 (s, 3H).

Cinnamaldehyde (**3**) [29]: ^{1}H NMR (500 MHz, $CDCl_3$) δ 9.66 (d, *J* = 7.7 Hz, 1H), 7.54–7.50 (m, 2H), 7.42 (s, 1H), 7.41–7.36 (m, 3H), 6.73–6.62 (m, 1H).

Taking the benzaldehyde experimental group as an example, the ^{1}H NMR spectra of pure benzaldehyde, the blank control group, and the processing groups are presented in Figure 5. The singlet observed at 5.96 ppm corresponds to the internal standard, 1,1,2,2-tetrachloroethane. The peak at 7.45 ppm is attributed to the meta-position hydrogen of the benzene ring in the aldehyde group, while the chemical shift at 7.55 ppm corresponds to the para-position hydrogen of the benzene ring. The peak at 7.80 ppm is assigned to the ortho-position hydrogen of the benzene ring. In the blank, **I**, and **II** groups, the peaks at 8.12 ppm and 8.00 ppm are assigned to the aromatic hydrogen of benzoic acid and benzohydroperoxide, respectively. The ^{1}H NMR spectra clearly demonstrate that both benzoic acid and benzohydroperoxide were formed in a ratio of 1:1 in the blank group, with no benzaldehyde recovered. In groups **I** and **II**, although no benzaldehyde was recovered, benzoic acid was formed as the major oxidation product. In group **III**, no oxidation products were observed in the spectra.

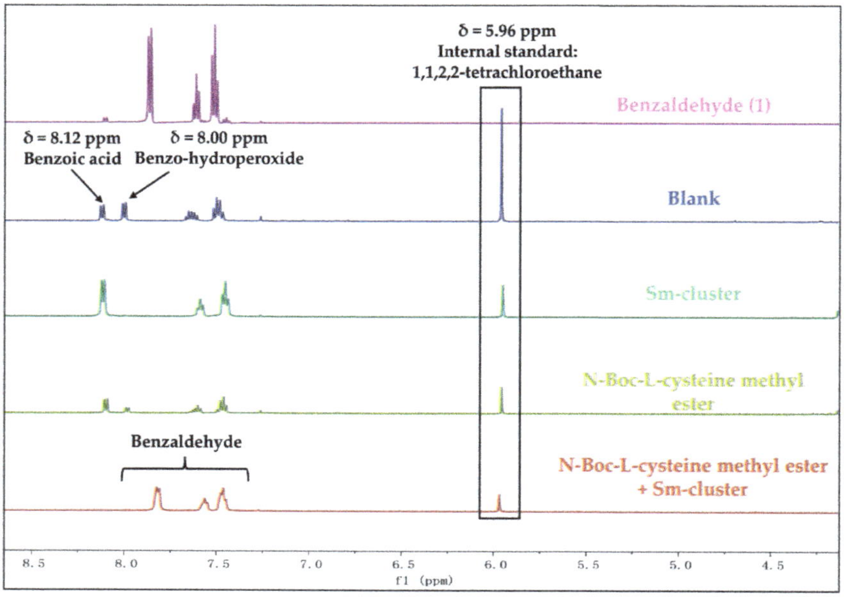

Figure 5. ^{1}H NMR spectra ($CDCl_3$, 500 MHz) of experiments for benzaldehyde at room temperature (298 K).

The accurate recovery rates of benzaldehyde were calculated based on the peak area of the ortho-position hydrogen at 7.80 ppm relative to the peak area of the internal standard at 5.96 ppm. A similar method was employed to calculate the recovery rates of vanillin and cinnamaldehyde (Figures S5 and S6). The calculated recovery rates of the three aldehydes across the four experimental groups (Blank, **I**, **II**, and **III**) are illustrated in Figure 4.

According to the calculated recovery rates of aldehydes, the recovery rates in processing group **III** were significantly higher than those in the other groups. Specifically, even after exposure to pure oxygen under a 300 W Xe lamp for 16 h, the recovery rates of benzaldehyde, vanillin, and cinnamaldehyde were 86%, 98%, and 96%, respectively. In contrast, while cysteine derivatives are also commonly used antioxidants, the effectiveness was limited in this study. Only 0% of benzaldehyde and 25% of vanillin were recovered in processing group **II**. However, cinnamaldehyde, which is relatively more resistant to oxidation, showed an acceptable recovery rate of 83%. On the other hand, when the Sm-cluster was used alone (processing group **I**), it demonstrated minimal antioxidant effects. These findings clearly indicate the synergistic effect between the Sm-cluster and *N*-Boc-L-cysteine methyl ester, resulting in enhanced antioxidant efficacy. The combined use of Sm-cluster and cysteine derivative in processing group **III** significantly improved the stabilization and recovery rates of the aromatic aldehydes, even under harsh oxidative conditions. This synergistic interaction suggests a promising approach for improving the antioxidant properties of food preservatives.

3.3. Determination of Radical Scavenging Ability of Sm-Cluster/Cysteine Derivative

Free radical chain reactions are widely recognized as a common mechanism of lipid peroxidation (a major manifestation of food oxidation), including steps of initiation, reproduction, and chain termination [30]. The autoxidation of aldehydes is also a free radical chain reaction, following the same mechanism as free radical oxidation.

Therefore, the ability of Sm-cluster/cysteine derivative to scavenge DPPH free radicals was measured to quantitatively represent its antioxidant capacity under ambient indoor light and 300 W Xenon lamp irradiation (Figure 6), where the Sm-cluster/*N*-Boc-L-cysteine methyl ester, *N*-Boc-L-cysteine methyl ester, Sm-cluster and Trolox were represented by **a**, **b**, **c** and **d**, respectively. Pearson's test was used to find the correlation between concentration and antioxidant activity (Table 1). For both ambient indoor light and 300 W Xenon lamp irradiation, the DPPH free radical scavenging curves of Trolox exhibited better linearity (R^2 = 0.998, 0.995, respectively), demonstrating the high reliability of this experiment, and their antioxidant activities were indexed in 50% inhibitory concentration (IC_{50}) values (Table 2). A two-way ANOVA model was conducted using the IC_{50} value of different groups' individual assays under two light conditions.

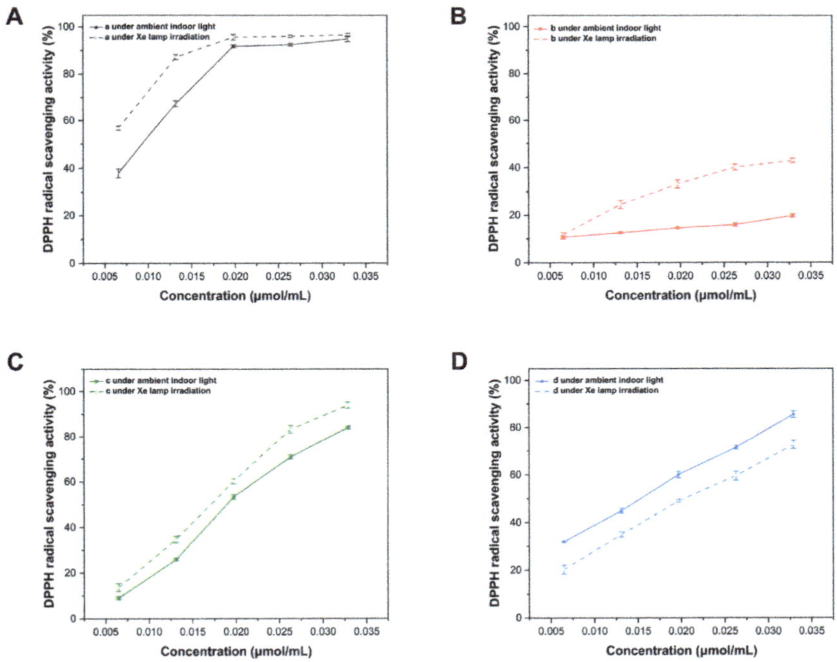

Figure 6. Determination of DPPH radical scavenging ability (%). (**A**) *N*-Boc-L-cysteine methyl ester – Sm-cluster. (**B**) *N*-Boc-L-cysteine methyl ester. (**C**) Sm-cluster. (**D**) Trolox.

Table 1. Pearson's correlation analysis of DPPH scavenging activity of different groups. ** $p < 0.01$.

Groups	Minimal Light Exposure	Xenon Lamp Irradiation
a	0.897 **	0.820 **
b	0.973 **	0.968 **
c	0.993 **	0.992 **
d	0.998 **	0.995 **

Table 2. The IC_{50} values of DPPH scavenging activity of different groups. Different letters in the same column indicate significant differences ($p < 0.01$); ** $p < 0.01$.

Light Condition	Sample	IC_{50} (1×10^{-3} μmol/mL)
ambient indoor light	a	8.49 ± 0.16 ^{b}
	b	1049 ± 43 ^{a}
	c	18.48 ± 0.26 ^{b}
	d	14.4 ± 2.4 ^{b}
Xe lamp irradiation	a	5.70 ± 0.17 ^{d}
	b	40.0 ± 1.2 ^{a}
	c	15.35 ± 0.46 ^{c}
	d	18.96 ± 0.37 ^{b}
sample		**
light condition		**
sample × light condition		**

All samples quenched the DPPH free radicals in a dose-dependent manner. As the concentration of the samples increased, the DPPH scavenging activities also increased (Figure 6). Under ambient indoor light, the DPPH scavenging activities for the three samples, **a**, **b**, **c** and **d** were found to be in the following order: $A_a > A_d > A_c > A_b$, where A_a, A_b, A_c and A_d were the DPPH scavenging activities for the sample **a**, **b**, **c** and **d**, respectively. Furthermore, the IC_{50} values of the samples were $(8.49 \pm 0.16) \times 10^{-3}$, $(1048.67 \pm 43.25) \times 10^{-3}$, $(18.48 \pm 0.26) \times 10^{-3}$, and $(14.41 \pm 2.44) \times 10^{-3}$ μmol/mL for **a**, **b**, **c**, and **d**, respectively. The DPPH radical scavenging ability of the *N*-Boc-L-cysteine methyl ester or Sm-cluster was much weaker than that of Trolox; however, the addition of Sm-cluster to the cysteine derivative resulted in a significant enhancement of DPPH free radical quenching activity. In addition, the lower IC_{50} values of **d** under Xe lamp irradiation ($(5.70 \pm 0.17) \times 10^{-3}$ μmol/mL) showed that the DPPH free radical quenching activity of the Sm-clusters/cysteine derivative was further enhanced under Xe lamp irradiation, while the DPPH free radical quenching activity of positive control Trolox under Xe lamp irradiation ($IC_{50} = (14.41 \pm 2.44) \times 10^{-3}$ μmol/mL) was lower than that under ambient indoor light ($IC_{50} = (18.96 \pm 0.37) \times 10^{-3}$ μmol/mL).

All these results suggested that the addition of Sm-cluster was beneficial for elevating the antioxidant activity of *N*-Boc-L-cysteine methyl ester, and that the high illumination could promote the antioxidant activity of *N*-Boc-L-cysteine methyl ester, especially the combination of Sm-cluster and *N*-Boc-L-cysteine methyl ester. Given that Sm-cluster was known as a catalyst in photoredox and thermal redox reactions, it was suggested that the synergistic antioxidant effects of the cysteine derivative and Sm-cluster may also arise from Sm-cluster-catalyzed hydrogen atom transfer processes.

4. Conclusions

This study demonstrated the synergistic antioxidant effects of cysteine derivatives and Sm-cluster for food applications. The addition of soluble, low-toxic inorganic Sm-clusters significantly enhanced the antioxidant capacity of natural thiol compounds. The composition of Sm-oxo/hydroxy cluster and a cysteine derivative exhibited robust antioxidative effects, as evidenced by the inhibition of aromatic aldehyde oxidation and strong, dose-dependent DPPH radical scavenging activity. Notably, the antioxidant activity was further enhanced under strong visible light conditions. These findings suggest that the combination of inorganic clusters and thiol compounds presents a promising natural alternative to traditional synthetic antioxidants in the food industry.

Supplementary Materials: The following supporting information can be downloaded at: https://www.mdpi.com/article/10.3390/antiox13080910/s1, Figure S1: UV–Vis diffuse reflectance spectroscopy of Sm-cluster; Figure S2: ^{1}H NMR spectrum of benzaldehyde at room temperature (298 K); Figure S3: ^{1}H NMR spectrum of vanillin at room temperature (298 K); Figure S4: ^{1}H NMR spectrum of cinnamaldehyde at room temperature (298 K); Figure S5: ^{1}H NMR spectra of experiments for vanillin at room temperature (298 K); Figure S6: ^{1}H NMR spectra of experiments for cinnamaldehyde at room temperature (298 K).

Author Contributions: Conceptualization, J.A.; synthesis, L.C. and L.W.; NMR (nuclear magnetic resonance), L.C., X.Q. and C.W.; methodology, L.C., X.Z. and R.Y.; software, L.C., C.W. and X.Q.; investigation, L.M. and L.C.; data curation, L.C. and L.M., writing—original draft preparation, L.C.; writing—review and editing, M.W., B.F. and J.A., funding acquisition, J.A. All authors have read and agreed to the published version of the manuscript.

Funding: This research was funded by the National Key Research and Development Program of China (2022YFF0710402) and Qidian Shineng (Jiangxi) Technology Co., Ltd. (202105423140109).

Institutional Review Board Statement: Not applicable.

Informed Consent Statement: Not applicable.

Data Availability Statement: The data presented in this study are contained within the article or in Supplementary Materials, or are available on request from the corresponding author Jie An.

Conflicts of Interest: Author Lifu Ma is from Rianlon Corporation; Rianlon Corporation had no role in the design of the study; in the collection, analyses, or interpretation of data; in the writing of the manuscript, or in the decision to publish the results. The remaining authors declare that the research was conducted in the absence of any commercial or financial relationships that could be construed as a potential conflict of interest.

References

1. Zeb, A. Concept, Mechanism, and Applications of Phenolic Antioxidants in Foods. *J. Food Biochem.* **2020**, *44*, e13394. [CrossRef] [PubMed]
2. Sun, X.; Wang, J.; Dong, M.; Zhang, H.; Li, L.; Wang, L. Food Spoilage, Bioactive Food Fresh-Keeping Films and Functional Edible Coatings: Research Status, Existing Problems and Development Trend. *Trends Food Sci. Technol.* **2022**, *119*, 122–132. [CrossRef]
3. Bayram, I.; Decker, E.A. Underlying Mechanisms of Synergistic Antioxidant Interactions during Lipid Oxidation. *Trends Food Sci. Technol.* **2023**, *133*, 219–230. [CrossRef]
4. Demirkol, O.; Adams, C.; Ercal, N. Biologically Important Thiols in Various Vegetables and Fruits. *J. Agric. Food Chem.* **2004**, *52*, 8151–8154. [CrossRef] [PubMed]
5. Neal, R.; Matthews, R.H.; Lutz, P.; Ercal, N. Antioxidant Role of N-Acetyl Cysteine Isomers Following High Dose Irradiation. *Free Radic. Biol. Med.* **2003**, *34*, 689–695. [CrossRef] [PubMed]
6. Meister, A.; Anderson, M.E. Glutathione. *Annu. Rev. Biochem.* **1983**, *52*, 711–760. [CrossRef] [PubMed]
7. Yehye, W.A.; Rahman, N.A.; Ariffin, A.; Abd Hamid, S.B.; Alhadi, A.A.; Kadir, F.A.; Yaeghoobi, M. Understanding the Chemistry behind the Antioxidant Activities of Butylated Hydroxytoluene (BHT): A Review. *Eur. J. Med. Chem.* **2015**, *101*, 295–312. [CrossRef] [PubMed]
8. Xu, X.; Liu, A.; Hu, S.; Ares, I.; Martínez-Larrañaga, M.-R.; Wang, X.; Martínez, M.; Anadón, A.; Martínez, M.-A. Synthetic Phenolic Antioxidants: Metabolism, Hazards and Mechanism of Action. *Food Chem.* **2021**, *353*, 129488. [CrossRef] [PubMed]
9. Hosseini, S.N.; Pirsa, S.; Farzi, J. Biodegradable Nano Composite Film Based on Modified Starch-Albumin/MgO; Antibacterial, Antioxidant and Structural Properties. *Polym. Test* **2021**, *97*, 107182. [CrossRef]
10. Nunes, M.R.; de Souza Maguerroski Castilho, M.; de Lima Veeck, A.P.; da Rosa, C.G.; Noronha, C.M.; Maciel, M.V.O.B.; Barreto, P.M. Antioxidant and Antimicrobial Methylcellulose Films Containing *Lippia Alba* Extract and Silver Nanoparticles. *Carbohydr. Polym.* **2018**, *192*, 37–43. [CrossRef]
11. Clark, A.; Zhu, A.; Sun, K.; Petty, H.R. Cerium Oxide and Platinum Nanoparticles Protect Cells from Oxidant-Mediated Apoptosis. *J. Nanopart. Res.* **2011**, *13*, 5547–5555. [CrossRef]
12. Singh, K.; Chopra, D.S.; Singh, D.; Singh, N. Optimization and Ecofriendly Synthesis of Iron Oxide Nanoparticles as Potential Antioxidant. *Arab. J. Chem.* **2020**, *13*, 9034–9046. [CrossRef]

13. Tee, J.K.; Ong, C.N.; Bay, B.H.; Ho, H.K.; Leong, D.T. Oxidative Stress by Inorganic Nanoparticles. *WIREs Nanomed Nanobi* **2016**, *8*, 414–438. [CrossRef]
14. Gil, D.; Rodriguez, J.; Ward, B.; Vertegel, A.; Ivanov, V.; Reukov, V. Antioxidant Activity of SOD and Catalase Conjugated with Nanocrystalline Ceria. *Bioengineering* **2017**, *4*, 18. [CrossRef]
15. Vieira, I.R.S.; de Carvalho, A.P.A.d.; Conte-Junior, C.A. Recent Advances in Biobased and Biodegradable Polymer Nanocomposites, Nanoparticles, and Natural Antioxidants for Antibacterial and Antioxidant Food Packaging Applications. *Compr. Rev. Food. Sci. Food Saf.* **2022**, *21*, 3673–3716. [CrossRef]
16. Bi, F.; Zhang, X.; Liu, J.; Yong, H.; Gao, L.; Liu, J. Development of Antioxidant and Antimicrobial Packaging Films Based on Chitosan, D-α-Tocopheryl Polyethylene Glycol 1000 Succinate and Silicon Dioxide Nanoparticles. *Food Packag. Shelf Life* **2020**, *24*, 100503. [CrossRef]
17. Bumbudsanpharoke, N.; Ko, S. The Green Fabrication, Characterization and Evaluation of Catalytic Antioxidation of Gold Nanoparticle-Lignocellulose Composite Papers for Active Packaging. *Int. J. Biol. Macromol.* **2018**, *107*, 1782–1791. [CrossRef]
18. Pachaiappan, R.; Rajendran, S.; Show, P.L.; Manavalan, K.; Naushad, M. Metal/Metal Oxide Nanocomposites for Bactericidal Effect: A Review. *Chemosphere* **2021**, *272*, 128607. [CrossRef] [PubMed]
19. Malhotra, N.; Hsu, H.-S.; Liang, S.-T.; Roldan, M.J.M.; Lee, J.-S.; Ger, T.-R.; Hsiao, C.-D. An Updated Review of Toxicity Effect of the Rare Earth Elements (REEs) on Aquatic Organisms. *Animals* **2020**, *10*, 1663. [CrossRef]
20. Chen, J.; Guo, Y.; Zhang, X.; Liu, J.; Gong, P.; Su, Z.; Fan, L.; Li, G. Emerging Nanoparticles in Food: Sources, Application, and Safety. *J. Agric. Food Chem.* **2023**, *71*, 3564–3582. [CrossRef]
21. Abbasi, R.; Shineh, G.; Mobaraki, M.; Doughty, S.; Tayebi, L. Structural Parameters of Nanoparticles Affecting Their Toxicity for Biomedical Applications: A Review. *J. Nanopart. Res.* **2023**, *25*, 43. [CrossRef] [PubMed]
22. Shrestha, S.; Wang, B.; Dutta, P. Nanoparticle Processing: Understanding and Controlling Aggregation. *Adv. Colloid Interface Sci.* **2020**, *279*, 102162. [CrossRef] [PubMed]
23. Wang, L.; Chen, L.; Qin, Z.; Zhao, B.; Ni, K.; Li, H.; Li, J.; Duan, H.; Ren, F.; An, J. Samarium-Oxo/Hydroxy Cluster: A Solar Photocatalyst for Chemoselective Aerobic Oxidation of Thiols for Disulfide Synthesis. *J. Org. Chem.* **2024**, *89*, 8357–8362. [CrossRef] [PubMed]
24. Arias-Pérez, I.; Sáenz-Navajas, M.P.; de-la-Fuente-Blanco, A.; Ferreira, V.; Escudero, A. Insights on the Role of Acetaldehyde and Other Aldehydes in the Odour and Tactile Nasal Perception of Red Wine. *Food Chem.* **2021**, *361*, 130081. [CrossRef] [PubMed]
25. Alcolado, C.I.; Garcia-Rio, L.; Mejuto, J.C.; Moreno, I.; Poblete, F.J.; Tejeda, J. Oxidation of Aldehydes Used as Food Additives by Peroxynitrite. *Antioxidants* **2023**, *12*, 743. [CrossRef] [PubMed]
26. Baliyan, S.; Mukherjee, R.; Priyadarshini, A.; Vibhuti, A.; Gupta, A.; Pandey, R.P.; Chang, C.-M. Determination of Antioxidants by DPPH Radical Scavenging Activity and Quantitative Phytochemical Analysis of Ficus Religiosa. *Molecules* **2022**, *27*, 1326. [CrossRef]
27. Sun, G.; Lv, X.; Zhang, Y.; Lei, M.; Hu, L. Palladium-Catalyzed Formylation of Aryl Iodides with HCOOH as CO Source. *Org. Lett.* **2017**, *19*, 4235–4238. [CrossRef] [PubMed]
28. Fu, X.; Pu, W.; Zhang, G.; Wang, C. Synthesis of Salicylaldehydes from Phenols via Copper-Mediated Duff Reaction. *Res. Chem. Intermed.* **2015**, *41*, 8147–8158. [CrossRef]
29. Sultane, P.R.; Bielawski, C.W. Burgess Reagent Facilitated Alcohol Oxidations in DMSO. *J. Org. Chem.* **2017**, *82*, 1046–1052. [CrossRef]
30. Yin, H.; Xu, L.; Porter, N.A. Free Radical Lipid Peroxidation: Mechanisms and Analysis. *Chem. Rev.* **2011**, *111*, 5944–5972. [CrossRef]

antioxidants

MDPI

Article

Essential Oil Emulsion from Caper (*Capparis spinosa* L.) Leaves: Exploration of Its Antibacterial and Antioxidant Properties for Possible Application as a Natural Food Preservative

Maria Merlino, Concetta Condurso *, Fabrizio Cincotta, Luca Nalbone, Graziella Ziino and Antonella Verzera

Department of Veterinary Sciences, University of Messina, Polo Universitario dell'Annunziata, Viale G. Palatucci, 98168 Messina, Italy; maria.merlino@unime.it (M.M.); fabrizio.cincotta@unime.it (F.C.); luca.nalbone@unime.it (L.N.); graziella.ziino@unime.it (G.Z.); antonella.verzera@unime.it (A.V.)

* Correspondence: concetta.condurso@unime.it

Abstract: This study explored, for the first time, the chemical composition and in vitro antioxidant and antibacterial activities of a caper leaf essential oil (EO) emulsion for possible food applications as a natural preservative. The EO was extracted by hydrodistillation from the leaves of *Capparis spinosa* growing wild in the Aeolian Archipelago (Sicily, Italy) and exhibited a pungent, sulphurous odour. The volatile fraction of the emulsion, analysed by SPME-GC-MS, consisted of over 100 compounds and was dominated by compounds with recognised antibacterial and antioxidant properties, namely dimethyl tetrasulfide (18.41%), dimethyl trisulfide (12.58%), methyl isothiocyanate (7.97%), and terpinen-4-ol (6.76%). The emulsion was effective against all bacterial strains tested (*Listeria monocytogenes*, *Staphylococcus aureus*, *Escherichia coli*, *Salmonella enterica* subsp. enterica serovar Enteritidis, *Pseudomonas fluorescens*), with *L. monocytogenes* exhibiting the lowest minimum inhibitory concentration (MIC = 0.02 mg/mL) while *E. coli* had the highest (MIC = 0.06 mg/mL). The emulsion had a good DPPH (2,2-diphenyl-1-picrylhydrazine) radical scavenging activity that was dose-dependent and equal to 42.98% at the 0.08 mg/mL level with an IC_{50} value of 0.099 mg/mL. Based on the results, the caper leaf EO emulsion has the potential to be proposed as a natural alternative to chemical preservatives in the food industry.

Keywords: leaf essential oil; *Capparis spinosa* L.; essential oil emulsion; volatile composition; pathogenic and spoilage foodborne bacteria; radical scavenging activity

1. Introduction

Chemical additives are successfully used in food processing and preservation to prevent and control pathogenic and spoilage microorganisms and limit oxidation; however their safety and impact on human health are under discussion. The increased awareness of the health risks connected with synthetic additives is driving consumer preferences towards more "green foods" and prompting the food industry to use alternative preservation methods to ensure food quality and safety [1].

In this context, essential oils extracted from various parts of plants and herbs are being extensively investigated for their potential application in the food industry [1–5]. Indeed, essential oils exhibit interesting biological activities, including antimicrobial, antioxidant, and anticancer properties, and are generally recognised as safe (GRAS) substances for human consumption by the Food and Drug Administration [6].

Caper (*Capparis spinosa* L.) is a perennial suffruticose shrub belonging to the Capparidaceae (or Capparaceae) family and the *Capparis* genus; the latter includes 142 species of tropical or subtropical origin [7]. The plant is widespread mainly in the Mediterranean basin, the Arabian Peninsula, and the Middle East, as well as in India, Malaysia, and Australia [8]. It grows both wild and as cultivated plants, preferring sunny stony soils in semi-arid coastal zones, the slopes of calcareous rocks, and the cracks of old walls. In addition, *C. spinosa* is salt tolerant and resistant to drought. Thanks to its ability to grow in harsh environments, *C. spinosa* is recommended for preventing land degradation, controlling soil erosion, and maintaining and promoting agriculture in regions subject to extreme climate changes [9].

This shrubby plant has a woody base and herbaceous branches with alternate simple leaves that are thick and shiny, an intense green colour, and oval-shaped. The flowers are solitary, sweetly fragrant, and showy, with white to pinkish-white petals and numerous long violet stamens. The fruits are ovate-oblong, dark green berries [8,10].

C. spinosa is one of the most common aromatic plants in Mediterranean cuisine. It is mainly known for its edible flower buds, named capers. The flowers, harvested in spring before they hatched, are usually processed in brine, pickled in vinegar, or preserved in granular salt, and used as a seasoning in salads, pasta, meat, sauces, and garnishes to add a tangy and briny flavour and aroma to food [8,11]. However, the whole plant is edible: the fruits, leaves, and younger shoots are also consumed salted or pickled in vinegar, and as fresh or cooked vegetables [12].

C. spinosa is also an important medicinal plant traditionally used to treat several health problems. Different parts of the plant (roots, bark, leaves, buds, and fruits) have been used, since ancient times, as a remedy for joint disease, haemorrhoids, rheumatism, rheumatoid arthritis, gout, fever, coughs, asthma, and inflammation [8]. In addition, anti-hypertensive, anti-hepatotoxic, anti-diabetic, anti-obesity, broncho-relaxant, anti-allergic and anti-histaminic, antibacterial, antioxidant, and anticancer effects have been recognised for caper extracts [8,13].

On the contrary, the composition and the biological activities of *C. spinosa* essential oil (EO) have been poorly investigated. Literature data are limited to the composition of leaf EOs from *C. spinosa* harvested in the Middle East [14,15], the composition and antioxidant activity of EOs obtained from leaves and flower buds and flower buds from Croatia [16] and Iran [17], respectively, and the composition and antimicrobial activity of the EO extracted from the aerial parts of *C. spinosa* harvested in Algeria [18] and Jordan [19].

To the best of our knowledge, the antioxidant and antimicrobial activity of caper leaf EOs from *C. spinosa* leaves has never been investigated. Therefore, considering the numerous phytochemical and pharmacological properties of *C. spinosa* and its appreciated aroma and flavour notes, especially in Mediterranean cuisine, the present study aimed to explore the potential of a caper leaf EO emulsion as a natural food preservative, through the assessment of its volatile composition, in vitro antioxidant activity, and in vitro antimicrobial effects against common food-born pathogenic and spoilage bacteria. The essential oil was extracted through the hydrodistillation of leaves of *C. spinosa* harvested in the Aeolian Archipelago, one of the main production areas in Italy, and formulated as an oil-in-water emulsion for preventing volatilisation and loss and increasing the oil dispersion capacity.

2. Materials and Methods

2.1. Plant Material

Leaves of *C. spinosa* were harvested from wild-grown caper plants in the northwest region of the Island of Lipari (Aeolian Archipelago, Sicily, Italy) in the first half of June 2023.

The collected leaves were quickly transported to the laboratory, washed under cold running water, and dried with a cloth. Finally, the leaves were dried with hot air at 50 °C at a constant air velocity (1.5 m/s) until a moisture loss of 95–96% was reached (~6 h) and ground into powder.

The taxonomic identification of the plant was carried out by Dr. Fabio Mondello at the Department of Chemical, Biological, Pharmaceutical, and Environmental Sciences of Messina University, and a voucher specimen was deposited in the same department.

2.2. Essential Oil Extraction

Approximately 200 g of powdered dried caper leaves was mixed with 1.0 L of distilled water and subjected to hydrodistillation for 3 h. The oil was dried overnight over anhydrous sodium sulphate; packaged in sealed, amber-coloured glass vials; and stored at room temperature in a dark, cool, and dry place until use. The extraction procedure was performed in triplicate.

2.3. Emulsion Preparation

An oil-in-water emulsion was prepared by mixing 40 mg of caper leaf EO and 500 mL of distilled water (conc. 0.08 mg/mL). The mixture was homogenised at 12,000 rpm for 2 min using an Ultra-Turrax T18 homogeniser (Janke & Kunkel, IKA Instruments, Staufen, Germany). The prepared emulsion was stable until use. All the analyses on the oil-in-water emulsion were performed in triplicate.

2.4. SPME-GC-MS Analysis

The volatile compounds of the caper leaf EO emulsion were extracted using the solid-phase microextraction technique (SPME) using a DVB/CAR/PDMS fibre with a 50/30 μm film thickness (Supelco, Bellefonte, PA, USA). For the extraction, an aliquot of 18 mL of the EO emulsion plus 6 g of NaCl (Sigma-Aldrich, Milan, Italy) were placed in a 40 mL vial sealed with a 'mininert' valve (Supelco, Bellefonte, PA, USA). The sample was thermally equilibrated at 35 °C for 20 min in a water bath and, subsequently, the SPME fibre was exposed to the sample headspace for 20 min under magnetic stirring.

The extracted volatiles were analysed by a Shimadzu GC 2010 Plus gas chromatograph coupled with a TQMS 8040 triple-quadrupole mass spectrometer (Shimadzu, Milan, Italy). Two capillary columns with different polarities were used: (1) a VF-WAXms, 60 m, 0.25 mm i.d., 0.25 μm film thickness polar column (Agilent Technologies Italia S.p.A.; Milan, Italy) and (2) a DB-5 ms, 30 m, 0.25 mm i.d., 0.25 μm film thickness apolar column (Agilent Technologies Italia S.p.A.; Milan, Italy). The conditions were as follows: injection mode, splitless; oven temperature, (1) 50 °C for 5 min, up to 190 °C at 3 °C/min and to 240 °C at 6 °C/min for the polar column, (2) 45 °C for 5 min, increased to 110 °C at 4 °C/min, to 240 °C at 2 °C/min, and to 260 °C at 20 °C/min, held at 260 °C for 2 min for the apolar column; carrier gas, helium at a constant flow of 1 mL/min; transfer line temperature, 250 °C; acquisition range, 40 to 400 m/z; and scan speed, 1250. The identification of the volatiles was performed using the mass spectral data, the NIST' 14 (NIST/EPA/NIH Mass Spectra Library, version 2.0, Gaithersburg, MD, USA), and FFNSC 3.0 database (Shimadzu, Kyoto, Japan), linear retention indices (LRIs) calculated according to the Van Den Dool and Kratz equation, and injection of available standards [20]. Quantitative results were expressed as a percentage of the total peak area.

2.5. Antibacterial Activity

The antimicrobial activity of the caper leaf EO emulsion was tested against different ATCC bacterial strains, namely *Listeria monocytogenes* 7644, *Staphylococcus aureus* 25923, *Escherichia coli* 25922, *Salmonella enterica* subsp. *enterica* serovar Enteritidis 13076, and *Pseudomonas fluorescens* 13525. All the strains were stored at −80 °C at the microbial collection of the "Food Microbiology Laboratory" of the Department of Veterinary Sciences, University of Messina (Messina, Italy). The tested strains were prepared by plating a loopful of the frozen stock onto nutrient agar plates (Biolife, Milan, Italy) and incubating overnight at 37 °C before each analysis.

The minimum inhibitory concentration (MIC) and minimum bactericidal concentration (MBC) were assessed for each of the strains. The MIC was evaluated using the broth microdilution method according to Kowalska-Krochmal and Dudek-Wicher [21] using the o/w emulsion prepared as described above. The o/w emulsion was diluted using brain heart infusion broth (BHI) (Biolife, Milan, Italy) to final concentrations of 0.06, 0.04, 0.02, and 0.01 mg/mL. Several U Eppendorf tubes (Biosigma, Cona, Italy; 2 mL volume) were arranged in rows on racks and filled with 1 mL of caper leaf EO emulsion at different concentrations. A fresh inoculum of each bacterium was grown in BHI at 37 °C for 24 h. Subsequently, a microbial suspension from the broth cultures was inoculated in each U Eppendorf tube obtaining a final concentration of ~10^4 CFU/mL. Then, the U Eppendorf tubes were incubated at 37 °C for 24 h. The positive control consisted of broth medium inoculated with microbial suspensions while the uninoculated broth medium with the highest concentration of each compound served as the negative control. The lowest concentration of each compound in which there was no visible growth (turbidity of the broth medium) was determined to be the MIC.

The MBC was determined by inoculating the suspensions from each U Eppendorf on nutrient agar plates (Biolife, Milan, Italy) and incubating them at 37 °C for 24 h. The lowest concentration in which there was no microbial growth (no colonies in the culture medium) was determined to be the MBC.

The MIC and MBC analyses were performed in triplicate.

2.6. Antioxidant Activity

The antioxidant activity of the caper leaf EO emulsion was assessed by measuring the free radical scavenging ability towards the DPPH (2,2-diphenyl-1-picrylhydrazine) radical (Merk Life Science S.r.l., Milan, Italy).

For the DPPH assay, 2000 μL of a DPPH methanolic solution (0.149 mM) was added to 500 μL of the EO emulsion and incubated at room temperature for 1 h. Similarly, a reference sample was prepared using 500 μL of solvent (distilled water) as a substitution for the EO emulsion. The absorbance of the mixtures was read at 517 nm using a double-beam UV–Vis spectrophotometer (UV-2550, Shimadzu, Milan, Italy) against a 4:1 v/v $MeOH/H_2O$ mixture used as a blank. The EO emulsion was tested in a concentration range of 0.016 to 0.080 mg/mL. Additionally, butylated hydroxytoluene (BHT) methanolic solutions (Merk Life Science S.r.l., Milan, Italy) were tested under the same experimental conditions and in the same concentration range, and they were utilised as positive controls.

The percentage radical scavenging activity [RSA(%)] of the caper leaf EO emulsion, as well as BHT, was calculated using the following equation:

$$RSA\ (\%) = \left(\frac{Ar - As}{Ar}\right) \times 100$$

where Ar is the absorbance at 517 nm of the reference sample, and As is the absorbance at 517 nm of the samples containing the EO emulsion or BHT solution.

The concentration required to scavenge 50% of the initial DPPH radical (IC_{50}) was determined by plotting the scavenging percentage against the test sample concentration (mg/mL); a linear correlation was found in both the EO emulsion ($y = 540.85\ x$, $R^2 = 0.999$) and BHT solution ($y = 1015.3\ x$, $R^2 = 0.995$).

2.7. Statistical Analysis

Excel 2010 software (Microsoft, Milan, Italy) was used to calculate the means and standard deviations of the three replicates. To determine significant differences, the means were analysed by one-way ANOVA and Duncan's multiple range test at a confidence level of 95% or more, using the Microsoft XLstat software 2014 (Addinsoft, Paris, France).

3. Results and Discussion

The hydrodistillation extraction of caper leaves allowed us to obtain a yellowish oil with a pungent sulphurous smell and a specific gravity of 1.1308 ± 0.0131 g/mL, consistent with the findings of Afsharypuor et al. [15]. The extraction yield was 0.021 ± 0.001%; this is a quite low yield but in line with the data reported in the literature when this extraction procedure was applied to caper leaves [15,18].

The antimicrobial and antioxidant activities of the caper leaf EO was tested using an oil-in-water emulsion. EO emulsification offers some advantages for food applications, such as a decrease in viscosity, an increase in dispersion capacity, and the prevention of EO volatilisation and loss [22,23]. Instead, the effect of emulsification on food microorganisms is still controversial; some studies verified that emulsification increases the antimicrobial effects of EOs against food microorganisms by increasing their ability to damage cell membranes [24], while others proved that the EO antimicrobial activity depends on its chemical composition rather than droplet size [22].

3.1. Volatile Fraction Composition

The oil-in-water emulsion was chemically characterised by HS-SPME-GC-MS: 34 terpenes, 13 C_{13}-norisoprenoids, 10 sulphur and 2 nitrogen compounds, 13 ketones, 15 alcohols, 7 aldehydes, 6 esters, and 4 furanic compounds were identified, for a total of over 100 compounds. The percentage composition of the volatile fraction of the caper leaf EO emulsion are reported in Table 1. The most quantitatively represented classes of compounds were terpenoids, sulphur compounds, and C_{13}-norisoprenoids, which together accounted for more than 86% of the total volatile fraction.

Table 1. Composition of volatile constituents and classes of substances of essential oil (EO) emulsion from caper (*Capparis spinosa* L.) leaves.

Compounds	LRI [1] on DB-5	LRI [1] on VF-WAX	Amount [2] (% ± SD)
Sulphur-containing compounds			
Methyl isothiocyanate	722	1243	7.97 ± 0.30
Dimethyl disulfide	725	1080	1.25 ± 0.08
Ethyl isothiocyanate	801	-	0.05 ± 0.00
Isopropyl isothiocyanate	827	1179	0.09 ± 0.00
Butyl isothiocyanate	928	1312	0.09 ± 0.00
Isobutyl isothiocyanate	951	1336	0.05 ± 0.00
Dimethyl trisulfide	969	1380	12.58 ± 0.93
Dimethyl tetrasulfide	1218	1750	18.41 ± 1.01
Dimethyl pentasulfide	1454	2185	4.28 ± 0.39
Cyclic octatonic sulphur	2014	-	0.57 ± 0.01
All			**45.33 ± 2.09**
Terpenenoids			
Eucalyptol	1032	1209	3.10 ± 0.10
Linalool	1102	1541	4.77 ± 0.18
Grandlure III	1111	1799	0.04 ± 0.00
Sabinol	1143	-	0.36 ± 0.02
(*E*)-*p*-Mentha-2,8-dien-1-ol	1123	1641	0.11 ± 0.01
Limona ketone	1132	-	0.05 ± 0.00
Dihydrolinalool	1137	1449	0.07 ± 0.00
(*Z*)-*p*-Mentha-2,8-dien-1-ol	1139	1656	0.07 ± 0.00
Camphor	1147	1498	0.27 ± 0.02
1,4-Dimethyl-4-acetylcyclohexene	1150	1504	0.13 ± 0.01
Karahanaenone	1153	-	0.06 ± 0.00
Menthone	1156	1472	1.63 ± 0.09
Pinocarvone	1163	1566	0.06 ± 0.00
Isomenthone	1166	1476	0.57 ± 0.02
Neomenthol	1171	1574	0.80 ± 0.03
Terpinen-4-ol	1182	1591	6.76 ± 0.32
p-Cymen-8-ol	1190	1850	0.55 ± 0.02
Methyl salicylate	1191	1745	0.12 ± 0.01
α-Terpineol	1198	1695	1.82 ± 0.05
Nerol	1225	1770	0.12 ± 0.01
(*Z*)-Carveol	1232	1846	0.09 ± 0.00
Pulegone	1237	1637	0.14 ± 0.01
Carvone	1243	1739	0.26 ± 0.01

Table 1. *Cont.*

Compounds	LRI [1] on DB-5	LRI [1] on VF-WAX	Amount [2] (% ± SD)
Geraniol	1251	1860	0.49 ± 0.02
β-Cyclohomocitral	1253	1598	0.15 ± 0.01
Perilla alcohol	1301	2021	0.62 ± 0.03
Thymol	1307	2154	2.03 ± 0.08
4-Vinyl-guaiacol	1310	2203	0.90 ± 0.03
1,5,9,9-Tetramethyl-2-methylene-spiro[3.5]non-5-ene	1329	-	1.64 ± 0.08
Eugenol	1350	2172	0.49 ± 0.02
Methyl-eugenol	1400	2011	0.12 ± 0.00
Caryophyllene oxide	1576	1990	0.14 ± 0.01
α-Bisabolol oxide B	1650	-	0.23 ± 0.02
Phytone	1840	2118	0.06 ± 0.00
All			**28.82 ± 0.72**
C13-Norisoprenoids			
2(1H)-Naphthalenone, 3,4,4a,5,6,7-hexahydro-1,1,4a-trimethyl	1277	-	0.25 ± 0.01
Dihydroedulan I	1287	1542	2.06 ± 0.09
(*Z*)-Theaspirane	1297	1561	0.24 ± 0.02
(*E*)-Theaspirane	1313	-	0.91 ± 0.04
Dihydroedulan II	1325	1513	0.60 ± 0.02
6-Methyl-5-(1-methylethylidene)-6,8-nonadien-2-one isomer I	1355	-	1.57 ± 0.01
(*E*)-β-Damascenone	1377	-	0.58 ± 0.01
6-Methyl-5-(1-methylethylidene)-6,8-nonadien-2-one isomer II	1389	-	1.02 ± 0.02
(*E*)-β-Damascone	1407	1782	0.10 ± 0.00
3,4-Didehydro-7,8-dihydro-β-ionone	1413	-	0.41 ± 0.00
(*E*)-α-Ionone	1421	1550	0.31 ± 0.00
7,8-Epoxy-α-ionone	1433	-	1.74 ± 0.00
(*E*)-β-Ionone	1478	1928	2.33 ± 0.12
All			**12.12 ± 0.15**
Nitrogen-containing compounds			
1H-Indole	1292	2435	0.96 ± 0.00
1H-Indole-3-methanol	1383	-	3.03 ± 0.18
All			**3.99 ± 0.10**
Alcohols			
(*Z*)-3-Hexen-1-ol	860	1384	0.04 ± 0.00
1-Hexanol	876	1357	0.03 ± 0.00
2,4-Dimethyl-3-heptanol	946	-	0.01 ± 0.00
1-Heptanol	979	1457	0.09 ± 0.00
1-Octen-3-ol	984	1448	0.22 ± 0.02
6-Methyl-5-hepten-2-ol	998	1468	0.06 ± 0.00
3-Ethyl-hexanol	1043	1483	0.13 ± 0.01
2-Octen-1-ol	1062	1637	0.30 ± 0.02
1-Nonen-4-ol	1109	-	0.76 ± 0.03
1-Octanol	1076	1557	1.05 ± 0.09
1-Nonanol	1176	1657	0.67 ± 0.03
1-Decanol	1273	1758	0.67 ± 0.03
1-Dodecanol	1475	1958	0.65 ± 0.03
1-Tetradecanol	1677	2159	0.17 ± 0.01
1-Hexadecanol	1880	2359	0.08 ± 0.00
All			**4.91 ± 0.18**
Aldehydes			
(*E*)-2-Hexenal	856	1222	0.02 ± 0.00
Octanal	1005	1284	0.08 ± 0.00

Table 1. *Cont.*

Compounds	LRI [1] on DB-5	LRI [1] on VF-WAX	Amount [2] (% ± SD)
Nonanal	1105	1384	0.40 ± 0.02
Decanal	1208	1486	0.34 ± 0.02
2-Phenyl-2-butenal	1266	1922	0.18 ± 0.01
4-Methyl-2-phenyl-2-pentenal	1364	1932	0.11 ± 0.00
5-Methyl-2-phenyl-2-hexenal	1482	2083	0.26 ± 0.01
All			**1.40 ± 0.02**
Esters			
2-Propenyl hexanoate	1088	1370	0.10 ± 0.00
Methyl 2,6-cresoate	1311	1970	0.70 ± 0.03
Massoia lactone	1471	2241	0.11 ± 0.00
(*Z*)-3-Hexenyl benzoate	1567	2093	0.50 ± 0.02
Isopropyl tetradecanoate	1824	2041	0.11 ± 0.01
2-Phenylethyl benzoate	1845	2658	0.04 ± 0.00
All			**1.56 ± 0.03**
Ketones			
3-Heptanone	888	1156	0.01 ± 0.00
2-Heptanone	893	1187	0.04 ± 0.00
3-Methyl-2-cyclohexen-1-one	1059	1592	0.06 ± 0.00
1-Octen-3-one	979	1307	0.27 ± 0.02
6-Methyl-5-hepten-2-one	988	1342	0.11 ± 0.01
2-Octanone	993	1289	0.01 ± 0.00
Acetophenone	1066	1656	0.02 ± 0.00
2-Nonanone	1093	1389	0.03 ± 0.00
(*E*,*E*)-3,5-Octadien-2-one,	1097	1521	0.04 ± 0.00
6-Methyl-3,5-heptadien-2-one	1108	1582	0.04 ± 0.00
2-Nonen-4-one	1127	1470	0.43 ± 0.02
3-Nonen-2-one	1142	1518	0.10 ± 0.00
Benzophenone	1621	2427	0.17 ± 0.01
All			**1.34 ± 0.03**
Furanoic compounds			
Furfural	831	1467	0.01 ± 0.00
2,5-Diethyl-tetrahydrofuran isomer I	889	-	0.13 ± 0.01
2,5-Diethyl-tetrahydrofuran isomer II	898	-	0.16 ± 0.01
6-Methyl-6-(5-methylfuran-2-yl) heptan-2-one	1480	-	0.24 ± 0.01
All			**0.54± 0.01**

[1] Linear retention indexes calculated according to the Van Den Dool and Kratz equation; [2] percentage of peak area in TIC chromatogram (mean ± standard deviation of 3 replicates).

The class of terpenoids was represented almost exclusively by oxygenated monoterpenes, with terpinen-4-ol, linalool, eucalyptol, thymol, α-terpineol, and menthone as the main compounds. Caryophyllene oxide and α-bisabolol oxide B were the only sesquiterpenes here identified, although they occurred at low percentages.

Among the sulphur compounds, isothiocyanates, methyl polysulfides, and cyclic octatonic sulphur were found, the latter being already reported as a constituent of pickled capers [25]. Isothiocyanates arise from the enzymatic hydrolysis of glucosinolates by the action of myrosinase. Glucosinolates are anionic secondary metabolites principally found in the plant order Brassicales, which includes the large families of Brassicaceae and Capparaceae [26]. Bianco et al. [27] reported glucocapparin and glucobrassicin as the main glucosinolates in *C. spinosa* from Southern Italy; from glucocapparin hydrolysis, methyl isothiocyanate arises, whereas from glucobrassicin hydrolysis, a highly unstable isothiocyanate occurs which rapidly degrades into indole-3-methanol and thiocyanate ions [28]. Methyl isothiocyanate and indole-3-methanol were present in the caper leaf EO emulsion at the highest percentages among the glucosinolate derivatives; indeed, ethyl isothiocyanate, isopropyl isothiocyanate, butyl isothiocyanate, and isobutyl isothiocyanate were present at levels below 0.1%.

Dimethyl disulfides, dimethyl trisulfides, dimethyl tetrasulfides, and dimethyl pentasulfides were among the main constituents of the volatile profile of our sample. Dimethyl polysulfides originate from methiin, namely (+)-S-methyl-L-cysteine sulfoxide, by the action of cysteine sulfoxidelyase. Methiin has been reported as the only S-alk(en)yl-L-cysteine sulfoxide compound in plants of the Brassicaceae family; its level can differ greatly among the genera of the family and, in the case of cruciferous vegetables, it is much higher than that of glucosinolates [29,30]. To the best of our knowledge, there are no studies focused on methiin occurrence and levels in Capparaceae plants even if dimethyl sulfide and dimethyl polysulfides have been reported among the volatile compounds of caper buds [11,25,31]. Very recent studies reported the formation of dimethyl disulfide and dimethyl trisulfide through thermal degradation of glucoerucin in rapeseed oil at 150 °C [32]; however, concerning our sample, this route for polysulfide formation seems unlikely since the leaves were not subjected to such high temperatures.

Dihydroedulan I and (*E*)-β-ionone quantitatively prevailed among the C_{13}-norisoprenoids identified in the caper leaf EO emulsion. C_{13}-norisoprenoids are 13-carbon skeleton compounds arising from carotenoid breakdown through enzymatic (by the action of carotenoid cleavage dioxygenases) and/or non-enzymatic (photo-oxygenation, thermal degradation in aqueous medium, and acid hydrolysis of intermediate megastigma precursors) pathways. C_{13}-norisoprenoids are common plant volatile compounds, present in leaves, flowers, and fruits. They have pleasant floral and fruity notes and, thanks to their very low odour thresholds (at ppt level, in water), strongly contribute to the aroma of flowers and fruits. In particular, C_{13}-norisoprenoids are the most important aroma compounds in some grape varieties and their wines, such as Cabernet Sauvignon, Merlot, Syrah, and Chardonnay [33]. Dihydroedulan I, probably biosynthesised from α-ionone, has a camphoraceous aroma and it is responsible for the characteristic odour of elderberry juice where it occurs at relatively high concentrations; moreover, it has also been detected in the pulp and juice of purple passion fruit [34,35]. β-ionone is formed directly from the enzymatic and/or non-enzymatic degradation of β-carotene [36], and it is responsible for violet and woody notes. Moreover, recent studies have highlighted its anti-inflammatory and cancer-preventing activities and the related benefits for human health [37].

Data on the chemical composition of caper leaf EO are currently scarce and only refer to the EO from *C. spinosa* grown in the Middle East [14,15]. In all cases, only a few compounds, i.e., less than fifteen, were reported; thymol, octanoic acid, methyl isothiocyanate, and 2-hexenal were the major constituents of the EO from leaves of *C. spinosa* grown in Syria [14], whereas thymol, isopropyl isothiocyanate, 2-hexenal, and butyl isothiocyanate prevailed in the leaf EO of *C. spinosa* var. *mucronifolia* from Iran [15]. In addition, the EO composition of leaves and flower buds of *C. spinosa* from Croatia was investigated by Kulisic-Bilusic et al. [16], identifying 10 compounds overall, with methyl isothiocyanate at a level over 90%. A greater number of compounds were instead reported in the EO extracted from the plant's aerial parts: 29 components were reported by Muhaidat et al. [19] in the EO of *C. spinosa* aerial parts harvested in the arid bottoms of the Dead Sea valley (Jordan), among which, isopropyl isothiocyanate, methyl isothiocyanate, butyl isothiocyanate, 3-p-menthene, 2-butenyl isothiocyanate, and 3-methylthio-1-hexanol prevailed; later Benachour et al. [18] identified 33 compounds, mainly palmitic acid, nonanal, 2,5-dimethoxy-p-cymene, and octacosane, in the aerial part EO of *C. spinosa* grown in Algeria.

This is the first report that fully characterised the EO extracted by hydrodistillation from the leaves of *C. spinosa*, identifying more than 100 compounds, most of which are reported for the first time in the EO of caper leaves.

3.2. Antibacterial Activity

The antibacterial activity of the caper leaf EO emulsion was tested against *L. monocytogenes* ATCC 7644, *S. aureus* ATCC 25923, *S.* Enteritidis ATCC 13076, *E. coli* ATCC 25922, and

P. fluorescens ATCC 13525. The MIC and MBC values were the same in all three replicates performed for each strain tested.

The emulsion showed inhibitory effects against all the strains tested (both Gram-positive and Gram-negative strains (Table 2, Figure S1), and revealed an efficacy against the foodborne pathogens *S.* Enteritidis and *L. monocytogenes* which were, respectively, the second and fifth most reported foodborne zoonotic agents in Europe in 2022 [38]. In particular, *L. monocytogenes* was the most sensitive among the tested strains with an MIC value of 0.02 mg/mL, whereas *E. coli* was the less sensitive (MIC value equal to 0.06 mg/mL). Antimicrobial activity was also detected against *P. fluorescens*, one of the most common spoilage agents in refrigerated foods [39].

Table 2. Minimum inhibitory concentrations (MICs) and minimum bactericidal concentrations (MBCs) of EO emulsion from caper (*Capparis spinosa* L.) leaves against different ATCC bacterial strains.

ATCC Strain	MIC (mg/mL)				MBC (mg/mL)			
	0.06	0.04	0.02	0.01	0.06	0.04	0.02	0.01
L. monocytogenes 7644	NG [1]	NG	NG	G	NG	G [2]	G	G
S. aureus 25923	NG	NG	G	G	NG	G	G	G
E. coli 25922	NG	G	G	G	NG	G	G	G
S. Enteritidis 13076	NG	NG	G	G	NG	G	G	G
P. fluorescens 13525	NG	NG	G	G	NG	G	G	G

[1] NG = no growth. [2] G = growth.

The antibacterial activity of the caper leaf EO emulsion was consistent with its chemical composition as many of the identified compounds have proven antibacterial activity. Ak(en)yl polysulfides display an antimicrobial activity that increases as the number of sulphur atoms increases and the number of carbon atoms in the alk(en)yl group decreases [40]. Isothiocyanates also exert antimicrobial activity against both pathogenic and food spoilage bacteria, with methyl isothiocyanate able to inhibit *Salmonella* spp., *E. coli*, and *L. monocytogenes* [41]. Thus, we can suppose that dimethyl trisulfide, dimethyl tetrasulfide, dimethyl pentasulfide, and methyl isothiocyanate, which are among the most abundant compounds of caper leaf essential oil, as well as the main oxygenated monoterpenes linalool, α-terpineol, terpinen-4-ol, thymol, and eucalyptol made a marked contribution to the observed antimicrobial activity. Linalool, α-terpineol, terpinen-4-ol, and thymol have broad-spectrum antibacterial activity and were found to be effective against, among others, *S. aureus*, *E. coli*, *S. enterica*, and *L. monocytogenes* [42,43]. Regarding eucalyptol, little information is present in the literature on the antimicrobial activity of its pure compounds; however, studies on eucalyptus EO (eucalyptol $\geq$ 60%) revealed antibacterial activity against Gram-positive and Gram-negative bacteria that are resistant to commonly used antimicrobial agents [44].

Although several studies are available in the literature on the antimicrobial activity of *C. spinosa* extracts, only a few have investigated the antibacterial effects of its essential oils, which exclusively focused on the essential oil extracted from the aerial part of the plant [18,19]. Instead, the antibacterial activity of caper leaf EO has never been explored before. Muhaidat et al. [19] tested the antibacterial activity of EO from the aerial parts of *C. spinosa* harvested in Jordan against different Gram-positive and Gram-negative ATCC strains; a greater inhibitory efficacy was observed against Gram-positive bacteria (*Staphylococcus aureus*, *Staphylococcus epidermidis*, *Streptococcus faecalis*, *Bacillus cereus*, and *Micrococcus luteus*) compared to Gram-negative bacteria (*Salmonella* Typhimurium, *Pseudomonas aeruginosa*, *Klebsiella pneumoniae*, and *Enterobacter aerogenes*), with the lowest activity against *E. coli* More recently, Benachour et al. [18] investigated the antibacterial activity of the essential oil from the aerial parts of *C. spinosa* collected in Algeria they reported antibacterial effects against the ATCC strains of *S. aureus*, *Klebsiella pneumoniae*, *B. cereus*, *P. aeruginosa*, *S. enterica*, *Proteus mirabilis*, and *Enterococcus faecalis*, whereas no activity was detected against *E. coli*.

These two studies revealed a low antibacterial efficacy of the caper aerial part EO against *E. coli* in contrast with the inhibitory effects against *E. coli* exhibited herein by the leaf EO emulsion. Indeed, the EOs from caper aerial parts tested in the above-cited studies had a chemical composition that is extremely different from that of our caper leaf EO emulsion, lacking compounds with proven antimicrobial activity, such as sulphur compounds and monoterpene alcohols.

Considering the results obtained herein, we speculate that the use of caper leaf EO emulsions in foods could improve their quality and safety. On the one hand, the effectiveness against spoilage bacteria could be exploited to extend foods' shelf life. On the other hand, they could be evaluated as a natural substitute for common additives that fight foodborne pathogens. The effectiveness against *P. fluorescens* would suggest the use of the caper leaf EO emulsion in some foods such as fishery products in which this psychrotrophic bacterium represents one of the main spoilage agents. In this regard, Shafaghi Rad and Nouri [17] added *C. spinosa* fruit essential oil in fish burgers and reported inhibitory effects against the psychrophilic flora that, after 8 days of refrigerated storage, was 6 log CFU/g lower than in untreated control burgers. In addition, the antimicrobial activity against *L. monocytogenes* suggests its use in ready-to-eat foods, such as some meat or fishery products, where the use of caper leaf essential oil can also add an appreciable pungent and sulphureous flavour to the product. Ready-to-eat foods are those most at risk for *L. monocytogenes* contamination [34] and the essential oil activity could, on the one hand, inhibit the growth of the microorganism and, on the other hand, act as a flavouring agent. Furthermore, the emulsification makes the direct use of the EO more likely in foods, which is different from other more studied extracts such as ethanolic ones whose use is affected by the alcohol content which could influence consumer acceptance and perception.

However, since the in vitro results may not reflect the antibacterial activity in a much more complex matrix such as food—due to the influence of factors such as pH, water activity, and interactions with food components—the EO emulsion applications in different food matrices are under investigation.

3.3. Antioxidant Activity

Concerning the primary antioxidant capacity, specifically the ability to scavenge free radicals, the caper leaf EO emulsion exhibited dose-dependent DPPH radical inhibition, equal to 42.98 $\pm$ 0.02% at the 0.08 mg/mL level (Figure 1) and an of 0.099 mg/mL which revealed a lower scavenging activity than the BHT standard (IC_{50} 0.049 mg/mL). However, 0.099 mg/mL could be considered a very good IC_{50} value considering that, according to extraction yield, an amount of 0.5 g of caper leaves was enough to obtain 0.105 mg of EO. In addition, the values of the percentage of DPPH inhibition and IC_{50} obtained for the caper leaf EO emulsion were higher than those reported in the literature for essential oils from both caper and other plants, suggesting a good antioxidant capacity of the caper leaf EO emulsion compared to other essential oils. In particular, Kulisic-Bilusic et al. [16] described DPPH inhibitory percentages of 7.55% and 10.45% at the concentrations of 0.2 g/L and 2 g/L, respectively, for the EOs of caper floral buds and leaves. The leaf EO of *Capparis ovata* Desf. var. *canescens* cultivated in Turkey inhibited 73.5% of DPPH radicals at the 500 μg/mL level [45]. Additionally, Anthony et al. [46] analysed 423 essential oils of 48 different botanical families and reported IC_{50} values $\leq$ 100 μg/mL for only 10 of them. The scavenging activity of the caper leaf EO suggests a possible use as an antioxidant agent for fresh and processed meat and meat products. In these foodstuffs, the oxidation process proceeds through a free radical autocatalytic mechanism involving unsaturated fatty acids, and it leads to lipid degradation, development of rancidity flavours, and changes in meat colour. The lipid oxidation process is the main factor responsible for quality loss during meats' shelf life and consumer rejection; thus, new solutions for controlling or minimising this process are always of great interest to the meat industry [47].

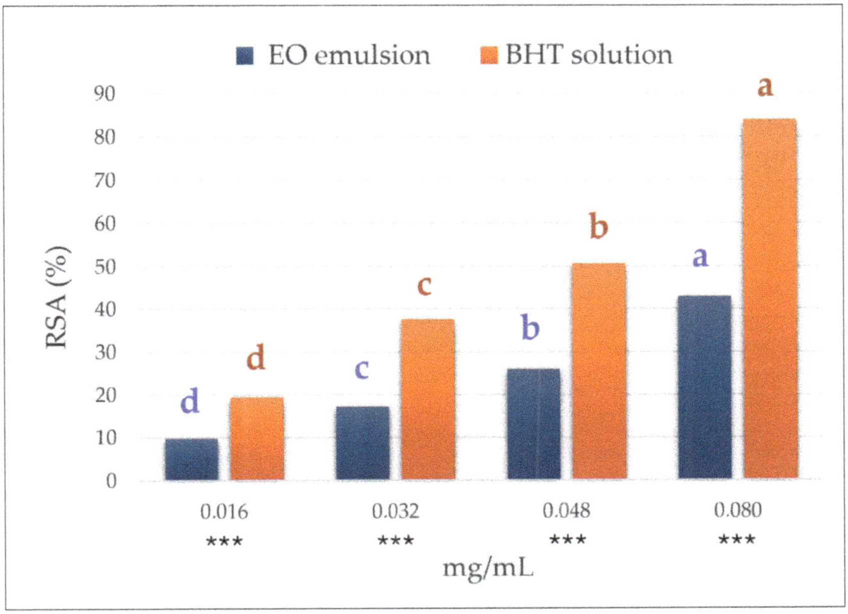

Figure 1. Antioxidant activity (DPPH assay) of caper leaf EO emulsion and BHT solution. Different letters indicate statistically significant differences among concentrations according to Duncan's multiple range test at $p < 0.001$. *** = statistically significant differences between EO emulsion and BHT solution at $p < 0.001$.

The antioxidant activity of the caper leaf EO can be related to its terpenoid profile. It is well known that, thanks to their antioxidant behaviour, terpenes provide protection against oxidative stress in ageing processes and several diseases, like cancer, diabetes, and neurodegenerative and cardiovascular diseases [48]. The antioxidant properties of terpenoids are dependent on their chemical structure and functional group, for example, (1) monoterpenes have better radical scavenging properties than sesquiterpenes; (2) among monoterpenes, hydrocarbons have the highest activity, whereas oxygenated monoterpenes' scavenging activity follows the decreasing order of alcohols > aldehydes and ketones; (3) the oxygenated sesquiterpenes exhibit good scavenging properties, similar to those of oxygenated monoterpenes; and (4) sesquiterpenes hydrocarbons have very low radical scavenging properties [48]. The terpenoid profile of the caper leaf EO emulsion was constituted almost entirely by oxygenated monoterpenes, of which, about 60% was alcohols. The most quantitatively represented terpenes were terpinen-4-ol (7.13%), linalool (5.03%), eucalyptol (3.27%), α-terpineol (1.91%), menthone (1.71%), and 1,5,9,9-tetramethyl-2-methylene-spiro[3.5]non-5-ene (1.73%); the latter was the only hydrocarbon present in the terpene fraction of the caper leaf EO emulsion. The scarce presence of hydrocarbon monoterpenes can account for the moderate antioxidant activity of the emulsion with respect to BHT. The antioxidant activity of isothiocyanates and organopolysulfides is also well known [49,50], although the compounds identified herein have never been tested for their antioxidant activity as pure compounds.

Therefore, it is possible to suppose that both the main monoterpenes and sulphur compounds were responsible for the antioxidant activity exhibited by the caper leaf EO emulsion as well as its antibacterial effects that were discussed in the previous section.

4. Conclusions

The results of this study revealed that an oil-in-water emulsion of leaf EO extracted by hydrodistillation from *C. spinosa* harvested in the Aeolian Archipelago had a very complex volatile profile, consisting of over 100 compounds and dominated by compounds with recognised antibacterial and antioxidant properties, such as methyl isothiocyanate, dimethyl polysulfides, and oxygenated monoterpenes. Indeed, the leaf EO exhibited a good antioxidant capacity (IC_{50} 0.099 mg/mL) and was effective against all the tested bacterial strains at low concentrations ($\leq$0.06 mg/mL), especially against *L. monocytogenes*.

This is the first study to deeply characterise the leaf EO of *C. spinosa*, identifying a large number of chemical compounds, most of which were described for the first time in the EO of caper leaves; the study also correlated the EO composition with its in vitro antibacterial and antioxidant capacities. The current study demonstrated that caper leaf essential oil in the form of an oil-in-water emulsion has the potential to be proposed as a natural alternative to chemical preservatives and other synthetic antioxidants in the food industry. Despite the promising results, this study is only preliminary and further investigations are ongoing to comprehensively explore the antimicrobial and antioxidant activities of the EO emulsion in food matrices.

Supplementary Materials: The following supporting information can be downloaded at: https://www.mdpi.com/article/10.3390/antiox13060718/s1.

Author Contributions: Conceptualisation, C.C., G.Z. and A.V.; methodology, M.M., C.C. and L.N.; validation, M.M., C.C., F.C. and A.V.; formal analysis, M.M., C.C., F.C. and L.N.; investigation, M.M., C.C. and L.N.; resources, C.C., F.C. and A.V.; data curation, F.C., L.N., G.Z. and A.V.; writing—original draft preparation, M.M., C.C. and L.N.; writing—review and editing, M.M., C.C. and G.Z.; visualisation, M.M. and C.C.; supervision, C.C.; project administration, C.C.; funding acquisition, C.C. and A.V. All authors have read and agreed to the published version of the manuscript.

Funding: This research received no external funding.

Institutional Review Board Statement: Not applicable.

Informed Consent Statement: Not applicable.

Data Availability Statement: Data is contained within the article and Supplementary Materials.

Acknowledgments: The authors are thankful to Fabio Mondello (Department of Chemical, Biological, Pharmaceutical and Environmental Sciences, University of Messina, Messina, Italy) for the taxonomic identification of the plant.

Conflicts of Interest: The authors declare no conflicts of interest.

References

1. Falleh, H.; Jemaa, M.B.; Saada, M.; Ksouri, R. Essential oils: A promising eco-friendly food preservative. *Food Chem.* **2020**, *330*, 127268. [CrossRef]
2. Ed-Dra, A.; Filali, F.R.; Lo Presti, V.; Zekkori, B.; Nalbone, L.; Elsharkawy, E.R.; Bentayeb, A.; Giarratana, F. Effectiveness of essential oil from the *Artemisia herba-alba* aerial parts against multidrug-resistant bacteria isolated from food and hospitalized patients. *Biodiversitas* **2021**, *22*, 2295–3005. [CrossRef]
3. Nefzi, N.; Pagliari, S.; Campone, L.; Megdiche-Ksouri, W.; Giarratana, F.; Cicero, N.; Ziino, G.; Nalbone, L. Chemical composition and comprehensive antimicrobial activity of an ethanolic extract of propolis from Tunisia. *Antibiotics* **2023**, *23*, 802. [CrossRef] [PubMed]
4. Rodilla, J.M.; Rosado, T.; Gallardo, E. Essential oils: Chemistry and food Applications. *Foods* **2024**, *13*, 1074. [CrossRef]
5. Saeed, K.; Pasha, I.; Jahangir Chughtai, M.F.; Ali, Z.; Bukhari, H.; Zuhair, M. Application of essential oils in food industry: Challenges and innovation. *J. Essent. Oil Res.* **2022**, *4*, 97–110. [CrossRef]
6. Frederico, C.; Silva dos Santos Franciscato, L.M.; Pereira Ruiz, S. Essential oils as natural food additives: Stability and safety. *Arq. Ciênc. Saúde UNIPAR* **2023**, *27*, 5739. [CrossRef]
7. Maurya, S.; Cornejo, X.; Lee, C.; Kim, S.Y.; Choudhary, R.K. Molecular phylogenetic tools reveal the phytogeographic history of the genus *Capparis* L. and suggest its reclassification. *Evol. Syst.* **2023**, *58*, 125720. [CrossRef]
8. Chedraoui, S.; Abi-Rizk, A.; El-Beyrouthy, M.; Chalak, L.; Ouaini, N.; Rajjou, L. *Capparis spinosa* L. in a systematic review: A xerophilous species of multi values and promising potentialities for agrosystems under the threat of global warming. *Front. Plant Sci.* **2017**, *8*, 1845. [CrossRef]
9. Sakcali, M.S.; Bahadir, H.; Ozturk, M. Eco-physiology of *Capparis spinosa* L.: A plant suitable for combating desertification. *Pak. J. Bot.* **2008**, *40*, 1481–1486.

10. Condurso, C.; Mazzaglia, A.; Tripodi, G.; Cincotta, F.; Dima, G.; Lanza, M.C.; Verzera, A. Sensory analysis and head-space aroma volatiles for the characterization of capers from different geographic origin. *J. Essent. Oil Res.* **2016**, *28*, 185–192. [CrossRef]
11. Cincotta, F.; Merlino, M.; Verzera, A.; Gugliandolo, E.; Condurso, C. Innovative process for dried caper (*Capparis spinosa* L.) powder production. *Foods* **2022**, *11*, 3765. [CrossRef] [PubMed]
12. Moghaddasi, M.S. Caper (*Capparis* spp.) importance and medicinal usage. *Adv. Environ. Biol.* **2011**, 872–880.
13. Nabavi, S.F.; Maggi, F.; Daglia, M.; Habtemariam, S.; Rastrelli, L.; Nabavi, S.M. Pharmacological effects of *Capparis spinosa* L. *Phytother Res.* **2016**, *30*, 1733–1744. [CrossRef] [PubMed]
14. El-Naser, Z. Analysis of essential oil of *Capparis spinosa* L. leaves and interaction between *Pieris brassicae* L. (Lepidopteran) which attack caper and natural enemy *Cotesia glomerata* (L). *Int. J. Chemtech Res.* **2016**, *9*, 477–485.
15. Afsharypuor, S.; Jeiran, K.; Jazy, A.A. First investigation of the flavour profiles of the leaf, ripe fruit and root of *Capparis spinosa* var. *mucronifolia from Iran. Pharm. Acta Helv.* **1998**, *72*, 307–309. [CrossRef]
16. Kulisic-Bilusic, T.; Blažević, I.; Dejanović, B.; Miloš, M.; Pifat, G. Evaluation of the antioxidant activity of essential oils from caper (*Capparis spinosa*) and sea fennel (*Crithmum maritimum*) by different methods. *J. Food Biochem.* **2010**, *34*, 286–302. [CrossRef]
17. Shafaghi Rad, M.; Nouri, M. Inspection of Capparis spinosa essential oils for quality assurance of fish burgers during refrigerated storage. *Food Sci. Nutr.* **2023**, *11*, 7229–7241. [CrossRef] [PubMed]
18. Benachour, H.; Ramdani, M.; Lograda, T.; Chalard, P.; Figueredo, G. Chemical composition and antibacterial activities of *Capparis spinosa* essential oils from Algeria. *Biodiversitas J. Biodiver.* **2020**, *21*, 161–169. [CrossRef]
19. Muhaidat, R.; Al-Qudah, M.A.; Al-Shayeb, A.; Jacob, J.H.; Al-Jaber, H.I.; Hussein, E.; Al-Tarawneh, I.N.; Abu Orabi, S.T. Chemical profile and antibacterial activity of crude fractions and essential oils of *Capparis ovata* Desf. and *Capparis spinosa* L. (Capparaceae). *Int. J. Integr. Biol.* **2013**, *14*, 39–47.
20. Cincotta, F.; Verzera, A.; Tripodi, G.; Condurso, C. Non-intentionally added substances in PET bottled mineral water during the shelf-life. *Eur. Food Res. Technol.* **2018**, *244*, 433–439. [CrossRef]
21. Kowalska-Krochmal, B.; Dudek-Wicher, R. The minimum inhibitory concentration of antibiotics: Methods, interpretation, clinical relevance. *Pathogens* **2021**, *10*, 165. [CrossRef]
22. Salvia-Trujillo, L.; Rojas-Graü, A.; Soliva-Fortuny, R.; Martín-Belloso, O. Physicochemical characterization and antimicrobial activity of food-grade emulsions and nanoemulsions incorporating essential oils. *Food Hydrocoll.* **2015**, *43*, 547–556. [CrossRef]
23. Clavijo-Romero, A.; Quintanilla-Carvajal, M.X.; Ruiz, Y. Stability and antimicrobial activity of eucalyptus essential oil emulsions. *Food Sci. Technol. Int.* **2019**, *25*, 24–37. [CrossRef]
24. Moghimi, R.; Ghaderi, L.; Rafati, H.; Aliahmadi, A.; McClements, D.J. Superior antibacterial activity of nanoemulsion of *Thymus daenensis* essential oil against E. coli. *Food Chem.* **2016**, *194*, 410–415. [CrossRef]
25. Brevard, H.; Brambilla, M.; Chaintreau, A.; Marion, J.P.; Diserens, H. Occurrence of elemental sulphur in capers (*Capparis spinosa* L.) and first investigation of the flavour profile. *Flavour Frag. J.* **1992**, *7*, 313–321. [CrossRef]
26. Mithen, R.; Bennett, R.; Marquez, J. Glucosinolate biochemical diversity and innovation in the Brassicales. *Phytochemistry* **2010**, *71*, 2074–2086. [CrossRef]
27. Bianco, G.; Lelario, F.; Battista, F.G.; Bufo, S.A.; Cataldi, T.R. Identification of glucosinolates in capers by LC-ESI-hybrid linear ion trap with Fourier transform ion cyclotron resonance mass spectrometry (LC-ESI-LTQ-FTICR MS) and infrared multiphoton dissociation. *J. Mass Spectrom.* **2012**, *47*, 1160–1169. [CrossRef]
28. Brindisi, L.J.; Lyu, W.; Juliani, H.R.; Wu, Q.; Tepper, B.J.; Simon, J.E. Determination of glucosinolates and breakdown products in Brassicaceae baby leafy greens using UHPLC-QTOF/MS and GC/MS. *Food Chem. Adv.* **2023**, *3*, 100389. [CrossRef]
29. Friedrich, K.; Wermter, N.S.; Andernach, L.; Witzel, K.; Hanschen, F.S. Formation of volatile sulfur compounds and S-methyl-l-cysteine sulfoxide in Brassica oleracea vegetables. *Food Chem.* **2022**, *383*, 132544. [CrossRef]
30. Marks, H.S.; Hilson, J.A.; Leichtweis, H.C.; Stoewsand, G.S. S-Methylcysteine sulfoxide in Brassica vegetables and formation of methyl methanethiosulfinate from Brussels sprouts. *J. Agric. Food Chem.* **1992**, *40*, 2098–2101. [CrossRef]
31. Romeo, V.; Ziino, M.; Giuffrida, D.; Condurso, C.; Verzera, A. Flavour profile of capers (*Capparis spinosa* L.) from the Eolian Archipelago by HS-SPME/GC–MS. *Food Chem.* **2007**, *101*, 1272–1278. [CrossRef]
32. Zhou, Q.; Zheng, C.; Wei, F.; Yang, Y. Flavor precursors identification and thermal degradation mechanisms of glucoerucin in fragrant rapeseed oil. *Food Chem.* **2024**, *435*, 137484. [CrossRef]
33. Tufariello, M.; Fragasso, M.; Pico, J.; Panighel, A.; Castellarin, S.D.; Flamini, R.; Grieco, F. Influence of non-Saccharomyces on wine chemistry: A focus on aroma-related compounds. *Molecules* **2021**, *26*, 644. [CrossRef]
34. Jensen, K.; Christensen, L.P.; Hansen, M.; Jørgensen, U.; Kaack, K. Olfactory and quantitative analysis of volatiles in elderberry (*Sambucus nigra* L) juice processed from seven cultivars. *J. Sci. Food Agric.* **2001**, *81*, 237–244. [CrossRef]
35. Casimir, D.J.; Kefford, J.F.; Whitfield, F.B. Technology and flavor chemistry of passion fruit juices and concentrates. In *Advances in Food Research*; Chichester, C.O., Mrak, E.M., Stewart, G.F., Eds.; Academic Press: New York, NY, USA, 1981; Volume 27, pp. 243–295. [CrossRef]
36. Mendes-Pinto, M.M. Carotenoid breakdown products the—Norisoprenoids—In wine aroma. *Arch. Biochem. Biophys.* **2009**, *483*, 236–245. [CrossRef]
37. Paparella, A.; Shaltiel-Harpaza, L.; Ibdah, M. β-Ionone: Its occurrence and biological function and metabolic engineering. *Plants* **2021**, *10*, 754. [CrossRef]

38. European Food Safety Authority (EFSA); European Centre for Disease Prevention and Control (ECDC). The European Union One Health 2022 Zoonoses Report. *EFSA J.* **2023**, *21*, e8442. [CrossRef]
39. Kumar, H.; Franzetti, L.; Kaushal, A.; Kumar, D. *Pseudomonas fluorescens*: A potential food spoiler and challenges and advances in its detection. *Ann. Microbiol.* **2019**, *69*, 873–883. [CrossRef]
40. Kim, J.W.; Huh, J.E.; Kyung, S.H.; Kyung, K.H. Antimicrobial activity of alk(en)yl sulfides found in essential oils of garlic and onion. *Food Sci. Biotechnol.* **2004**, *13*, 235–239.
41. Wilson, A.E.; Bergaentzlé, M.; Bindler, F.; Marchioni, E.; Lintz, A.; Ennahar, S. In vitro efficacies of various isothiocyanates from cruciferous vegetables as antimicrobial agents against foodborne pathogens and spoilage bacteria. *Food Control.* **2013**, *30*, 318–324. [CrossRef]
42. Kotan, R.; Kordali, S.; Cakir, A. Screening of antibacterial activities of twenty-one oxygenated monoterpenes. *Z. Für Naturforschung C* **2007**, *62*, 507–513. [CrossRef] [PubMed]
43. Marchese, A.; Orhan, I.E.; Daglia, M.; Barbieri, R.; Di Lorenzo, A.; Nabavi, S.F.; Gortzi, O.; Izadi, M.; Nabavi, S.M. Antibacterial and antifungal activities of thymol: A brief review of the literature. *Food Chem.* **2016**, *210*, 402–414. [CrossRef]
44. Trivedi, N.A.; Hotchandani, S.C. A study of the antimicrobial activity of oil of eucalyptus. *Indian J Pharmacol.* **2004**, *36*, 93.
45. El-Ghorab, A.; Shibamoto, T.; Ozcan, M. Chemical composition and antioxidant activities of buds and leaves of capers (*Capparis ovata* Desf. var. *canescens*) cultivated in Turkey. *J. Essent. Oil. Res.* **2007**, *19*, 72–77. [CrossRef]
46. Anthony, K.P.; Deolu-Sobogun, S.A.; Saleh, M.A. Comprehensive assessment of antioxidant activity of essential oils. *J. Food Sci* **2012**, *77*, C839–C843. [CrossRef]
47. Amaral, A.B.; Silva, M.V.D.; Lannes, S.C.D.S. Lipid oxidation in meat: Mechanisms and protective factors–a review. *Food Sci. Technol.* **2018**, *38*, 1–15. [CrossRef]
48. Gonzalez-Burgos, E.; Gomez-Serranillos, M.P. Terpene compounds in nature: A review of their potential antioxidant activity. *Curr. Med. Chem.* **2012**, *19*, 5319–5341. [CrossRef]
49. Cedrowski, J.; Grebowski, J.; Litwinienko, G. Antioxidant activity of edible isothiocyanates. In *Lipid Oxidation in Food and Biological Systems: A Physical Chemistry Perspective*; Bravo-Diaz, C., Ed.; Springer: Berlin/Heidelberg, Germany, 2022; pp. 277–303. [CrossRef]
50. Wu, Z.Q.; Li, K.; Ma, J.K.; Huang, Q.; Tian, X.; Li, Z.J. Antioxidant activity of organic sulfides from fresh *Allium macrostemon* Bunge and their protective effects against oxidative stress in Caenorhabditis elegans. *J. Food Biochem.* **2020**, *44*, e13447. [CrossRef]

antioxidants

MDPI

Article

Inclusion of a Catechol-Derived Hydrazinyl-Thiazole (CHT) in β-Cyclodextrin Nanocavity and Its Effect on Antioxidant Activity: A Calorimetric, Spectroscopic and Molecular Docking Approach

Mihaela Mic [1], Adrian Pîrnău [1,*], Călin G. Floare [1], Mariana Doina Palage [2], Ovidiu Oniga [3] and Gabriel Marc [3]

1 National Institute for Research and Development of Isotopic and Molecular Technologies, 67-103 Donat, 400293 Cluj-Napoca, Romania; mihaela.mic@itim-cj.ro (M.M.); calin.floare@itim-cj.ro (C.G.F.)

2 Department of Therapeutical Chemistry, "Iuliu Hațieganu" University of Medicine and Pharmacy, 12 Ion Creangă Street, 400347 Cluj-Napoca, Romania; mpalage@umfcluj.ro

3 Department of Pharmaceutical Chemistry, "Iuliu Hațieganu" University of Medicine and Pharmacy, 41 Victor Babeș Street, 400012 Cluj-Napoca, Romania; ooniga@umfcluj.ro (O.O.); marc.gabriel@umfcluj.ro (G.M.)

* Correspondence: apirnau@itim-cj.ro; Tel.: +40-264-58-40-37

Abstract: The aim of the present research was to obtain a supramolecular complex between a strong antioxidant compound previously reported by our group, in order to extend its antioxidant activity. The formation of the inclusion complex of a catechol hydrazinyl-thiazole derivative (CHT) and β-cyclodextrin in aqueous solution has been investigated using isothermal titration calorimetry (ITC), spectroscopic and theoretical methods. The stoichiometry of this inclusion complex was established to be equimolar (1:1) and its equilibrium constant was determined. An estimation of the thermodynamic parameters of the inclusion complex showed that it is an enthalpy and entropy-driven process. Our observations also show that hydrophobic interactions are the key interactions that prevail in the complex. ^{1}H NMR spectroscopic method was employed to study the inclusion process in an aqueous solution. Job plots derived from the ^{1}H NMR spectral data demonstrated 1:1 stoichiometry of the inclusion complex in a liquid state. A 2D NMR spectrum suggests the orientation of the aromatic ring of CHT inside the β-CD cavity. The antiradical activity of the complex was evaluated and compared with free CHT, indicating a delayed activity compared with free CHT. To obtain additional qualitative and visual insight into the particularity of CHT and β-CD interaction, molecular docking calculations have been performed.

Keywords: catechol; antioxidant; thiazole; cyclodextrin; ITC; NMR; molecular docking

1. Introduction

Nowadays, the inclusion of compounds in cyclodextrins (CDs) is a common approach to optimize the properties of the incorporated compounds such as solubility, stability, reduction in tissue irritation, odor, or taste masking in the field of food, cosmetics, pharmaceutics, agriculture, or wastewater treatment [1–4]. Due to their low toxicity, this field attracted the interest of researchers for years before and we can assume that the interest of obtention of various supramolecular complexes with CD will be of interest in the next years [5].

Cyclodextrins are macrocyclic oligosaccharides comprised of 6, 7, or 8 units of 1,4-linked glycosidic α-D-glucopyranose units, named α, β, and γ-cyclodextrins, respectively [6]. The chemical structures of α-CD, β-CD, and γ-CD are presented in Figure 1. The shape of the CDs is described in the literature in many ways, from a doughnut to a truncated cone [3,7], while the inner cavity has a cone-like shape [4]. The diameter of the inner cavity increases with the number of glucose units from 0.78 nm for α-CD, 0.78 nm for β-CD, and 0.95 nm for γ-CD, respectively [1].

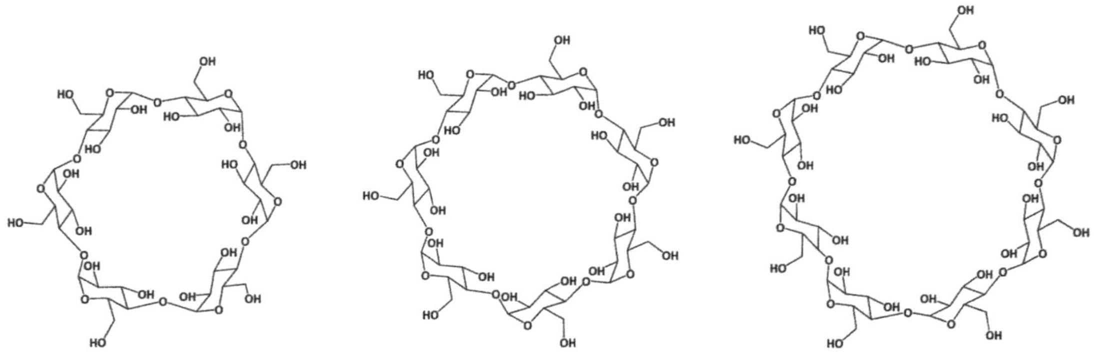

Figure 1. Chemical structure of α-CD (**left**), β-CD (**center**), and γ-CD (**right**).

CDs are amphiphilic—the internal cavity of the cyclodextrins is hydrophobic, where various hydrophobic small guest compounds can accommodate, while the outside of CDs is hydrophilic favoring the interaction of the complex with water. Based on these properties, CDs are used for an increase in solubility in water of organic compounds used as drugs with low solubility [1,8,9].

The inclusion in the CD of interesting compounds resulting in the supramolecular complexes would act as a potential reservoir by trapping the compounds inside and releasing controlled kinetics [10]. This is linked with the way the guest compound interacts with CD, the main driving interactions between them are hydrophobic interactions: van der Waals forces, electrostatic interactions, and hydrogen bonds, all of them being weak, non-covalent bonds. CDs on the other hand, due to its property of encapsulating molecules that are hydrophobic in nature, are widely applied in the field of drug delivery [11].

Applying this principle in the field of antioxidants would allow CDs to be considered "secondary antioxidants" [12]. The inclusion in CD of strong polyphenolic antioxidants such as quercetin, rutin, catechin, epicatechin, brazilin, resveratrol, kaempferol, myricetin, naringenin, baicalein, gallates, and caffeic acid which are susceptible to degradation via oxidation or photodegradation was reported in the literature, with consistent research regarding the polyphenol–CD interactions [2,3,7,13–19]. Also, the encapsulation of CD of vegetal extracts with antioxidant properties rich in polyphenols was reported [3,20,21].

In our previous publication, we reported the obtention, characterization, and evaluation of a new phenolic compound linked via a hydrazine bridge to a thiazole ring (IUPAC: 4-((2-(4-methylthiazol-2-yl)hydrazineylidene)methyl)benzene-1,2-diol) with strong antioxidant and antiradical activity [22]. The chemical structure of CHT is presented in Figure 2. The present paper focuses on the inclusion of CHT in cyclodextrin to prolong the antioxidant activity in time, compared to the free state of the compound. This working hypothesis is based on our observation during the in vitro evaluation of the activity of CHT, that the scavenging of radicals used as reagents in the antiradical assays such as DPPH (2,2-diphenyl-1-picrylhydrazyl) and ABTS (2,2′-azino-bis(3-ethylbenzothiazoline-6-sulfonic acid) occurred at a very high speed. The endpoint of the reactions was reached much sooner than the time needed to complete the assays, according to the reports about protocols of performing the respective assays [22,23].

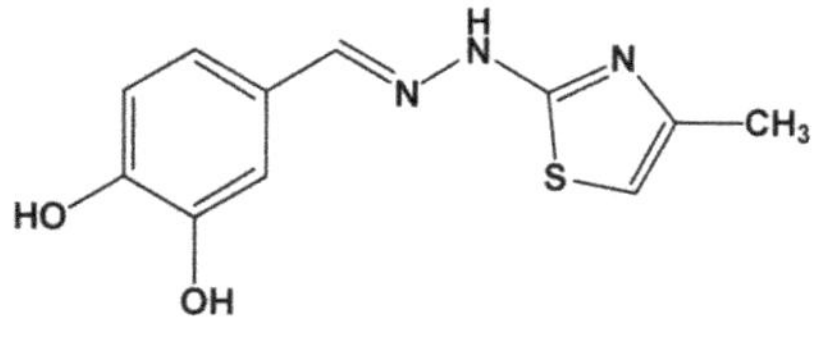

Figure 2. Chemical structure of CHT [22].

2. Materials and Methods

The reagents and solvents used for all activities presented in the research were purchased from local suppliers. For ITC and NMR measurements, we used β-CD from Sigma-Aldrich Chemie GmbH (St. Louis, MO, USA) and aqueous solutions were prepared in double distilled water and deuterium oxide, respectively.

2.1. Antioxidant Activity

The antioxidant activity of the β-CD complex of CHT was assessed using a modified protocol of DPPH scavenging assay, based on the initial report of Brand-Williams et al. [24–26]. The principle of the assay consists of the transfer of a hydrogen atom from the antioxidant compound to DPPH˙ (2,2-diphenyl-1-picrylhydrazyl). Following the binding of the hydrogen atom to the radicalized nitrogen atom, the resulting compound loses the intense purple color characteristic to the radical. The loss of the absorbance of the reagent at λ = 517 nm is proportional to the amount of DPPH˙ neutralized. Three measurements were made against a blank sample and the results are presented as averages, based on Equation (1). The absorbance of the samples was made using a UV-VIS spectrophotometer Jasco V-530 (Jasco International Co., Tokyo, Japan) in 10 mm single-use plastic cuvettes.

The stock solution of the stock CHT-β-CD complex was diluted using water at the final 250 μM concentration. Additionally, a solution of CHT 250 μM was obtained by dissolving solid CHT in water.

The working solution of the DPPH reagent was made by dissolving solid DPPH in methanol and later was diluted in water, resulting in a DPPH solution in 30% methanol and 70% water. The choice of this mixture of solvents was made to avoid the use of ethanol to avoid disruption of the CHT-β-CD complex. Similar solvent mixtures were reported in the literature for DPPH scavenging evaluation of β-CD complexes with other phenolic compounds [27,28]. Because DPPH is not soluble in water, methanol was chosen and the dilution of the initial methanolic solution with water was made to minimize the potential effect of methanol on the stability of the CHT-β-CD complex. The reagent absorbance at λ = 517 nm was made by dilution of the reagent solution with methanol:water 30:70 (*v*/*v*), until it reached approximatively A = 1.

In total, 50 μL of 62.5 μM, 125 μM, and 250 μM solutions of CHT-β-CD complex and CHT were mixed with 1 mL of DPPH reagent. The absorbance of the solutions at λ = 517 nm was measured every 30 s for 15 times, resulting in a total 450 s between the mixture of the samples with the working reagent and the last absorbance read.

$$\text{DPPH}^{\cdot}\ \text{scavenging}(\%) = \frac{\text{control absorbance} - \text{sample absorbance}}{\text{control absorbance}} \times 100 \quad (1)$$

2.2. Isothermal Titration Calorimetry (ITC)

The binding between CHT and β-CD was investigated through ITC experiments carried out with a Nano ITC2G calorimeter (TA Instruments, New Castle, DE, USA) at a constant temperature of 25 °C. Prior to the titration experiments, both solutions were degassed under a vacuum for 30 min. The experiment consisted typically of injecting 10 μL of CHT ligand (4 mM) into the calorimetric cell that contains the β-CD solution (1 mM) by using a computer-controlled Hamilton micro-syringe. The reference cell was loaded with water. The interval of 300 s between injections was set to allow the heat signal to return to the baseline. During the titration, the reaction mixture was continuously stirred at 250 rpm to ensure proper mixing after each injection. The control experiment was performed by titrating CHT into a cell containing only water and the reference signal was subtracted from corresponding experimental data. The integrated heat data, after the correction for control, were analyzed using an independent binding model supplied by Nano Analyze 3.1.2 (TA Instruments, New Castle, DE, USA). The binding constant (K), the binding stoichiometry (*n*), the change in enthalpy (ΔH), and the change in entropy (ΔS) were thus obtained.

2.3. ^1H NMR Analysis

NMR measurements were performed with Bruker AVANCE III spectrometer in a liquid state, operating at 500.13 MHz for protons which is equipped with a BBO (broadband observe) probe head with Z-gradient. All samples were prepared in D_2O solutions at 298 $\pm$ 0.1 K and all chemical shifts were measured relative to TMSP (3-(trimethyl-silyl)-propionic acid sodium salt), referencing the chemical shift to 0 ppm. The determination of the association constant and the stoichiometric ratio between CHT and β-CD by NMR, were made with two stock solutions in D_2O, both having a concentration of 10 mM. Thus, 11 separate samples were prepared based on these two equimolar solutions which contain CHT and β-CD in a different ratio, by mixing the two stock solutions to constant volume in a complete interval ($0 < r < 1$) of the ratio $r = [X]/([H] + [G])$, where X = H or G and [H] and [G] are the total concentrations of the host (β-CD) and guest (CHT), respectively, with total concentration kept constant for each sample $[H]_t + [G]_t = 10$ mM.

^1H NMR experiments were recorded using 64 scans collected into 65 K data points over a 5000 Hz spectral window, using a 4 s relaxation delay and an excitation pulse of 10.1 μs. The ROESY NMR spectrum was obtained using the pulse program *roesyphpr* with residual water suppression and the following acquisition parameters: 16 scans, 8192/2048 data points, 5000 Hz spectral width (10 ppm), spin-lock mixing time 350 ms.

2.4. Molecular Docking

The molecular docking of the CHT molecule to β-cyclodextrin has been performed following a protocol similar to that used in a preceding study where we analyzed the complexation of CHT and human serum albumin (HSA) [22]. The same molecular structure of CHT was used which was optimized using Gaussian 09 software [29] with density functional theory (DFT) based M06-2X Minnesota exchange-correlation energy functional which proved to perform better than B3LYP functional for model systems with dispersion and hydrogen-bonding interactions [30] and 6-311 ++ G(d,p) basis set. No imaginary frequencies were obtained. In Figure 3, the optimized structure of CHT is presented.

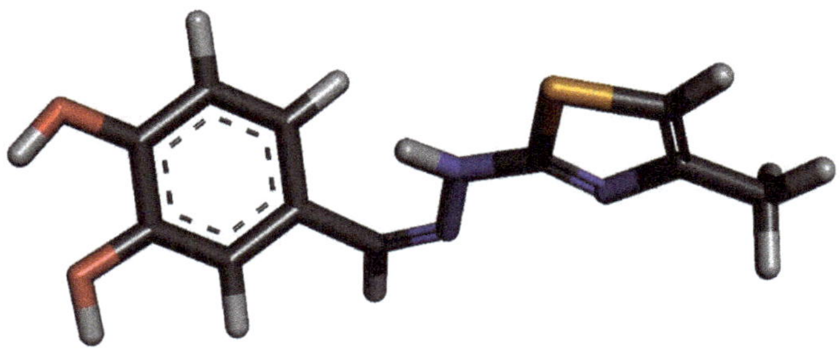

Figure 3. Gaussian 09 optimized molecular structure of CHT at the level M06-2X/6-311 ++ G(d,p).

The β-cyclodextrin molecular structure was also the same as the ones used in some of our previous publications [31,32] and was extracted from a crystallographic inclusion compound with Mg^{2+} and Ca^{2+} salts of meclofenamic acid, determined by Caira et al. [33] (CSD entry: WERGUW).

A molecular docking calculation aim is to identify the preferred orientation of a smaller molecule (ligand or guest) to a large molecule or protein (receptor or host) and uses an empirical Monte Carlo simulated annealing search, estimating, in this way, the ligand affinity and activity. AutoDockTools v1.5.6 user interface of AutoDock v4.2 [34] was used to prepare the initial pdbqt files for the ligand and host molecule and the input files required by AutoDock [35]. While the atoms of the β-CD molecule were kept rigid, the flexibility of the ligand had been considered by setting up 5 torsion angles around rotatable bonds. AutoGrid was used in the precalculation of the grid maps of interaction energies of various atom types.

The search box with a cubic form and centered on the β-CD molecule, had a grid size of 60 × 60 × 60 points with a grid point spacing of 0.375 Å. All hydrogens atoms were added to the host molecule and the ligand structure was optimized including the hydrogens (as can be seen in Figure 2) but, during the preparation stage, the non-polar ones were automatically merged. The atomic charges were computed using the Gasteiger-Marsili method [36] implemented in AutoDockTools. The iterative generation and optimization of ligand conformations have been performed using the Lamarckian genetic algorithm [37,38]. We performed 2000 runs, each with a random generation seed, with the aim to obtain good statistics and optimized conformations clustering.

The analysis and visualization of the results, apart from AutoDockTools use, have been also realized with Chimera v1.14 [39] and Biovia Discovery Studio Visualizer v20.1.3.

3. Results

3.1. Antioxidant Activity

The antiradical activity of the CHT-β-CD complex was determined using the DPPH neutralization assay in comparison to free CHT, to evaluate how the complexation with β-CD influences its antiradical activity in time. The antiradical activity at the three concentrations of free CHT and CHT-β-CD complexes was evaluated, and their results are presented in Figure 4.

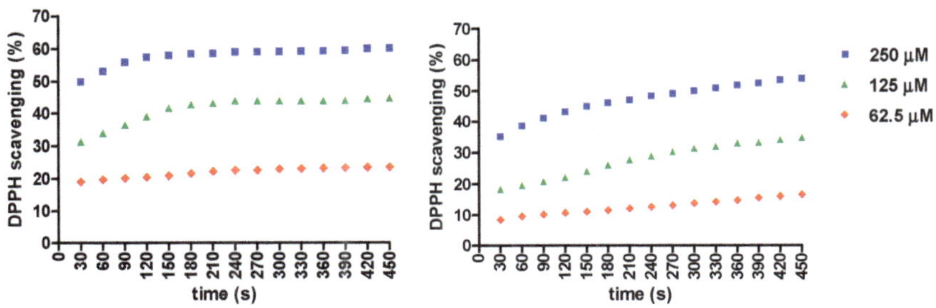

Figure 4. DPPH scavenging (%) of the CHT free (**left**) and CHT-β-CD complex (**right**) evaluated at three concentrations.

Free CHT reaches the endpoint of the DPPH scavenging reaction in approximately 150 s, after that the neutralization of the radical stays in a constant plateau. The final plateau value is dependent on the concentration of the sample involved in each assay.

On the other hand, the evaluation of the antiradical activity of the CHT-β-CD complex exhibits a quasi-linear inhibition of the DPPH reagent dependent on time. Compared with free CHT it can be observed that DPPH neutralization occurs in time, with sustained release of CHT from its β-CD complex. At the end of the assay (t_{15} = 450 s) the percentage of DPPH scavenged by the CHT-β-CD complex was lower than the CHT free, indicating that some of CHT was still sequestered in the β-CD complex at the end of assay, at all three concentrations evaluated.

3.2. Binding Affinity and Thermodynamics of CHT—β-CD Binding

Isothermal titration calorimetry (ITC) is the most sensitive and accurate analytical technique for the determination of binding constant and various thermodynamic parameters in macromolecule–bioligand complexation with precise accuracy [40–43].

To evaluate the binding affinity and the thermodynamic profile of the inclusion complex of the CHT inside the β-CD in solution, isothermal calorimetry titration was carried out. To this effect, the representative calorimetric titration profile of the β-CD with CHT is depicted in Figure 5.

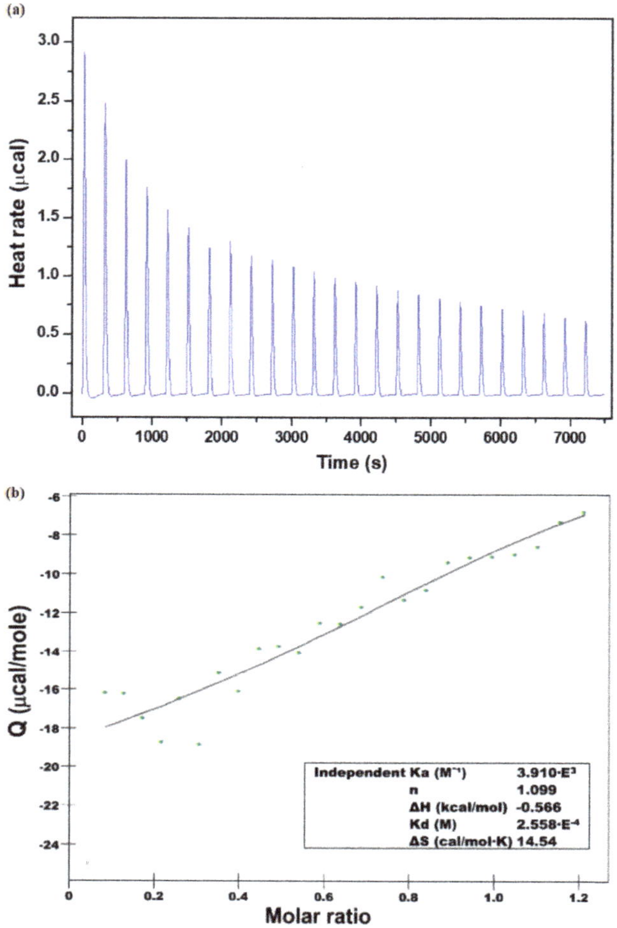

Figure 5. Isothermal titration calorimetry data obtained for the binding interaction in water at 25 °C of CHT (4 mM) to β-CD (1 mM). (**a**) Shows exothermic heat release upon injection of 10 μL aliquots of CHT into β-CD solution. (**b**) Shows integrated heat data. The solid line corresponds to the best-fit curve.

The binding isotherm (panel a) was obtained by monitoring the heat associated with injections. The magnitude of the released heat decreases progressively with which injection. The second quadrant of the figure shows the plot of the amount of heat release per injection as a function of the molar ratio of the drug to the β-CD (β-CD /CHT). An analysis of the binding isotherm using an independent binding model allowed for a description of the values of the binding and thermodynamic parameters (Table 1).

Table 1. The thermodynamic parameters of CHT/β-CD.

N	K_a (M^{-1})	ΔH (kJ mol^{-1})	TΔS (kJ mol^{-1})	ΔG (kJ mol^{-1})
1.098	3.92×10^3	−2.36	18.13	−20.5

From the analysis of the thermodynamic data (Table 1), it can be seen that the stoichiometric ratio (n) which determines the number of β-CD molecules which includes the CHT molecules is $n = 1.098$. This may suggest that one mole of CHT molecule interacts with one mole of β-CD molecule. The equilibrium constant obtained from ITC measurement was $K = 3.92 \times 10^3\ M^{-1}$, which corresponds to a Gibbs free energy $\Delta G = -20.5$ kJ/mol, indicating spontaneous inclusion of the drug into β-cyclodextrin.

The complexation process in the real system is exothermic, as confirmed by the negative change of the molar enthalpy value ($\Delta H = -2.36$ kJ/mol) of the interaction between β-CD and CHT. The entropic effect of the binding process is $T\Delta S = 18.13$ kJ/mol and brings an additional contribution to the negative Gibbs free energy. The positive value of the entropy change is due to disturbing the ordered aqueous microenvironment surrounding the hydrophobic parts of the guest molecule after the binding to cyclodextrin. Furthermore, the enthalpic effect of binding for the discussed molecules has been dominated by the entropic effect of complex formation ($|\Delta H| < |T\Delta S|$) and the releasing the water molecules from the cavity.

ITC data show that the encapsulation is both enthalpically and entropically driven, confirming once again that the hydrophobic interaction prevailing for this complex formation, which accounts for the negative sign of the enthalpy and the positive sign of the entropy changes.

3.3. Determination of the Stoichiometry by ^1H NMR

^1H NMR spectroscopy was used to highlight the formation in solution of the molecular complex CHT:β-CD, by including the CHT molecule in the hydrophobic cavity of β-CD. This fact is highlighted by changing the values of the chemical shifts of some protons of CHT and β-CD—those involved in this interaction, compared to the chemical shifts of the same protons in the free state. The determination of the molecular structure of the CHT was reported by our group in the paper of Mic et al. [22], where we find the total ^1H NMR spectrum of the CHT molecule in D_2O, with the assignment of the signals. Therefore, we will not resume this discussion and based on this assignment and the numbering of the protons in the CHT molecule presented in Figure 5, we summarize this information in Table 2, where it could be found the chemical shifts of the protons of the CHT molecule in the free state (in the absence of the β-CD).

Table 2. Chemical shifts of CHT protons in the free state (ppm).

Proton	Ha	Hb	Hc	Hd	He	Hf
δ	7.029	6.801	7.137	7.843	6.387	2.145

Figures 6 and 7 show ^1H NMR spectra plotted in different molar ratios of CHT and β-CD, considering the numbering of the atoms of CHT and β-CD from Figure 8. Thus, the shifts of the signals specific to both CHT and β-CD molecules can be observed, compared to other signals that do not show changes in chemical shifts, including the residual proton signal from D_2O.

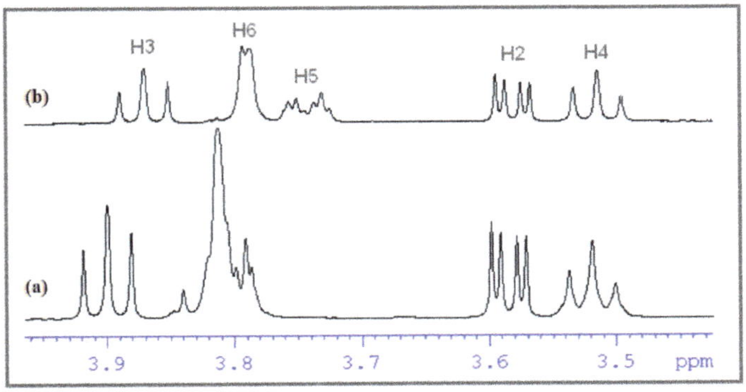

Figure 6. Partial ^1H NMR spectrum of: (**a**) 10 mM β-CD; (**b**) 5 mM CHT and 5 mM β-CD.

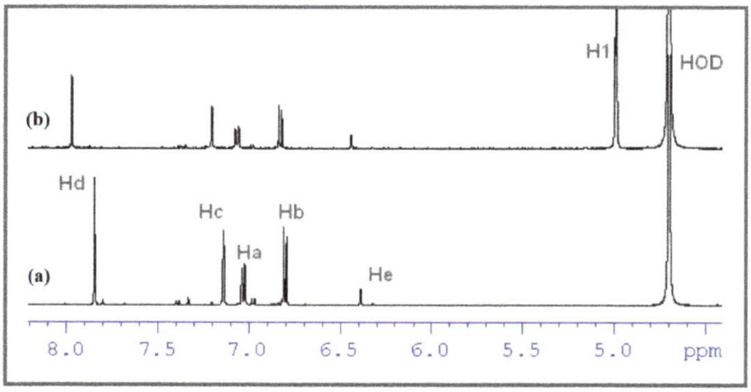

Figure 7. Partial ^{1}H NMR spectrum of: (**a**) 10 mM CHT; (**b**) 5 mM CHT and 5 mM β-CD.

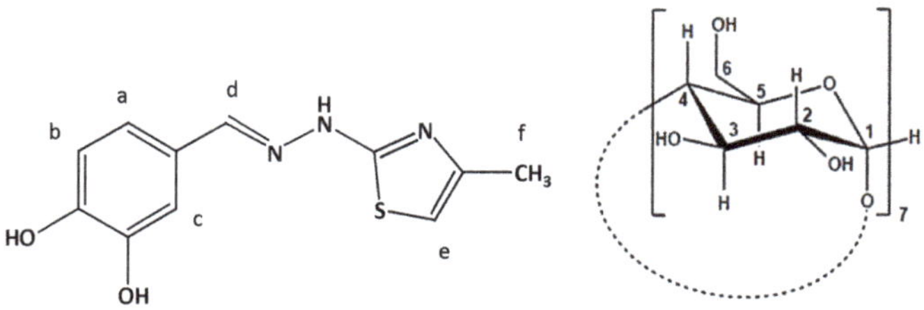

Figure 8. Chemical structure and atomic numbering of CHT (**left**) and β-CD (**right**).

As can be seen, in the spectra we have variations in the chemical shifts of some protons from the CHT molecule in the presence of β-CD. Also, the protons on the interior of the β-CD (H3 and H5) show a variation of the chemical shifts in the presence of CHT, which certifies an interaction between the CHT and the interior of the β-CD cavity, particularly the formation of the inclusion complex CHT:β-CD. The stoichiometry of the CHT:β-CD complex was determined using the continuous variation method, which involves the induced chemical shift, Δδ, which is dependent on the concentration of the complex [44]. Since the continuous variation graphs (Figure 9) have a symmetrical shape, with r having a maximum value of 0.5, we can assume that the inclusion complex has a stoichiometry of 1:1. This was highlighted both for protons belonging to the CHT molecule, and for H3 and H5 belonging to the interior of the β-CD cavity.

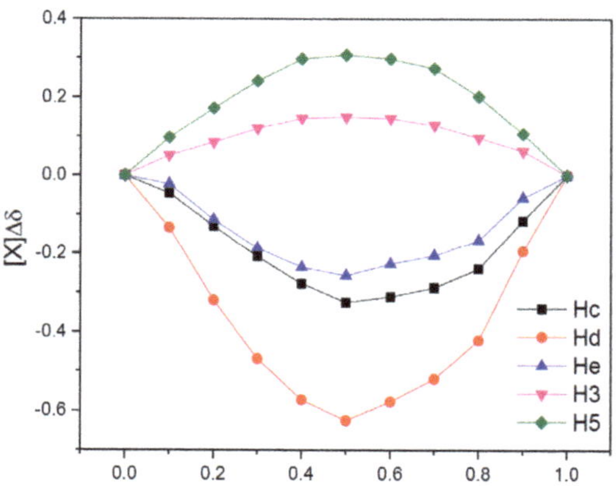

Figure 9. Concentrations variation plot for: H3 and H5 protons of β-CD; Hc, Hd, and He protons of CHT, where [X] = [CHT] or [β-CD].

3.4. ROESY Experiments

A ROESY NMR spectrum being a two-dimensional (^{1}H-^{1}H), can prove the intermolecular interactions through space such as the non-covalent bonds between two different molecular species. Thus, being able to observe which specific protons from the two types of molecules present such interactions. As can be seen in Figure 10, the H3 and H5 protons inside the β-CD cavity interact with the Ha, Hb, Hc, and Hd protons belonging to the CHT molecule, which indicates that the CHT molecule is included inside the β-CD cavity with the aromatic ring, in agreement with our previous results on similar structures [8,31,43,44] and also in agreement with the theoretical results.

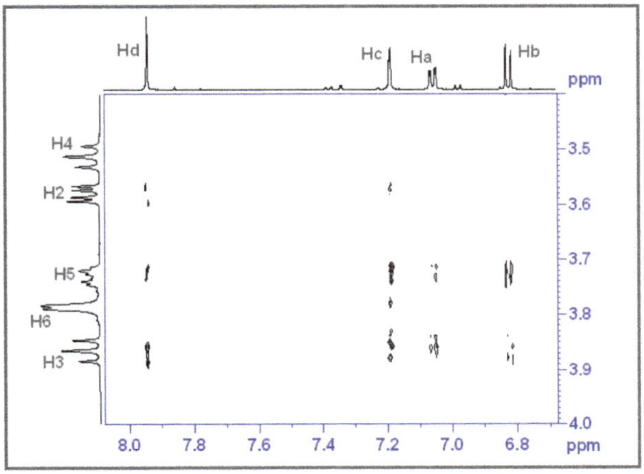

Figure 10. Expanded region of the ROESY spectrum of CHT:β-CD complex. [CHT] = 4 mM; [β-CD] = 6 mM.

3.5. Evaluation of the Association Constant

The association constant was determined based on the same set of samples used to determine the stoichiometry of the CHT:β-CD complex which was found to be 1:1. With the help of the following equation, which is based on the variation of the chemical shift $\Delta\delta^{(i,j)}$ depending on the concentration, the association constant can be determined [45]:

$$\Delta\delta^{(i,j)} = \frac{\Delta\delta_c^{(j)}}{2[X]}\left\{[C] + \frac{1}{K} - \left[\left([C] + \frac{1}{K}\right)^2 - 4[H]^{(i)}[G]^{(i)}\right]^{1/2}\right\} \tag{2}$$

with i being the sample number and j the investigated proton. The variation of chemical shifts:

$$\Delta\delta^{(i,j)} = \delta_{free}^{(i,j)} - \delta_{obs}^{(i,j)}$$

for a proton j, in a complexed $_{(obs)}$ or free $_{(free)}$ state.

$$C = H + G \tag{3}$$

where G = guest and H = host

In order to calculate the association constant, a script developed by our group was used [46], based on an iteration procedure following specific algorithms, which takes into account the experimental values of the variations chemical shifts depending on the concentration, of the protons that show the most pronounced variations, both from the CHT molecule and H3 and H5 from β-CD.

Each iteration sets up a quadratic program which determines the direction of search and the loss function, until convergence.

$$E = \sum_{i,j} (\Delta\delta^{(i,j)} - \Delta\delta^{(i,j)}_{calc})^2 \quad (4)$$

Using the procedure described above, an association constant was determined with the value K = 6.181 × 10^3 M^{-1}, with a correlation factor r = 0.991 and E = 3.656 × 10^{-3}.

3.6. Molecular Docking

Cyclodextrins have found numerous applications in food, pharmaceutical, drug delivery, and chemical industries, as well as in agriculture and environmental engineering and are consequently produced in large quantities. Supramolecular inclusion compounds of cyclodextrins are consequently intensively studied and very diverse. Practically, cyclodextrins are already ingredients in 30 different approved medicines [47] and α-, β-, and γ-cyclodextrin are recognized to be generally safe by the U.S. FDA [48]. Having a hydrophobic interior and hydrophilic exterior, cyclodextrins form complexes with hydrophobic compounds, providing increased solubility and stability to the included drugs.

β-cyclodextrin, a cyclic heptasaccharide produced from starch by enzymatic conversion is the most used natural cyclodextrin. Because the internal cavity of β-CD has an appropriate dimension to include phenyl and/or thiazole moieties, in the actual study, we concentrated our attention on a close experimental and theoretical investigation of its complex with CHT which, as mentioned, we synthesized and previously studied in interaction with human serum albumin [22]. In this subchapter, we will present the theoretical molecular docking simulations performed with the goal of better understanding the interactions at a molecular level.

Out of the mentioned 2000 docking runs, we obtained 51 distinct conformational clusters, using a root mean square deviation of 2.0 Å. In Figure 11 we have presented the histogram of the binding energy distribution of the identified clusters, the intensity representing the number of conformations belonging to each cluster. If we compare this relatively reduced number of clusters with more than 1000 obtained in our previous study, were we analyzed the inclusion compound of deferoxamine B chelator (DFOB) with β-CD [32], we can infer, from the beginning, that the conformational degrees of freedom of CHT vs. DFOB are drastically reduced. This fact can surely help and constrain our search to identify the most probable binding conformation but, as we will see in the following, even when we analyze small ligand molecules some strained conformations can be favored too, and this is most probably due to the approximations considered in our approach.

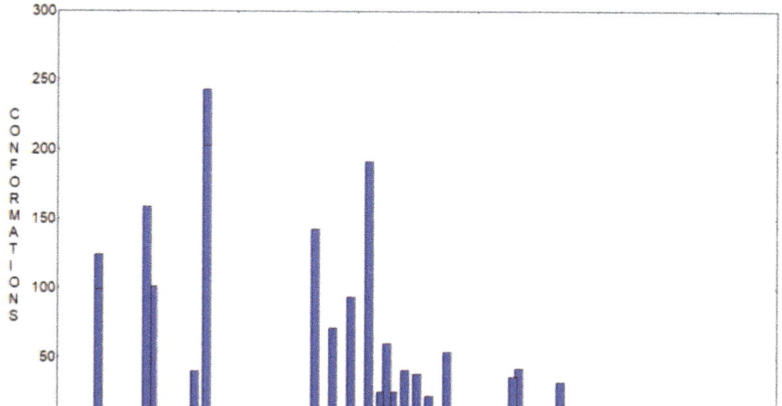

Figure 11. Histogram of the binding energy distribution to β-cyclodextrin of the cluster conformations of CHT.

If we analyze the lowest binding energy conformations represented in Figure 11, appearing at −5.63 kcal/mol, which correspond to the strongest binding conformers, to our surprise we observed that we don't have a unique most favorable conformation, but we have two clusters superposed, one with 99 members and the other with 124 members, both having the leading conformation with the same lowest binding energy. The mean binding energy value corresponding to these two clusters is also very similar, −5.10 kcal/mol vs. −5.21 kcal/mol, which means that, even if the conformations are practically different, they appear with a close probability. We will continue our analysis by taking a closer look at these two conformations. We represent them comparatively in Figure 12 presenting two views, a secondary rim view and a side view, for better visibility. Two important structural details can be remarked if we analyze these docking conformations. Firstly, we observe that the dihedral angle across the hydrazone group linking the two moieties of the CHT molecule was twisted in such a way that catechol and thiazole are nearly facing each other and, in these constrained conformations, the CHT molecule fits inside the β-CD cavity. Secondly, we observe that between the two conformations we don't have a rotation symmetry only but mainly a reflection and rotation, because the sulfur atom of the thiazole group points outwards of the β-CD cavity, in one conformation, and in the other conformation, it is pointing inwards.

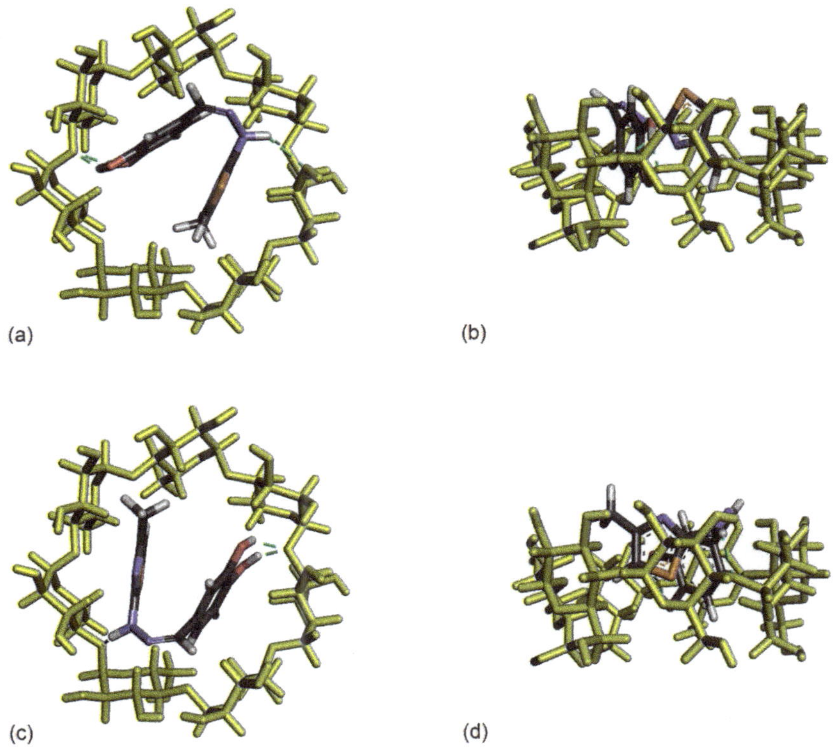

Figure 12. Inclusion compounds of CHT into the cavity of β-cyclodextrin as resulted from molecular docking calculations. (**a**,**c**) Secondary rim view and (**b**,**d**) side view of the leading clusters conformations having both an estimated binding energy of −5.63 kcal/mol. The cyclodextrin molecule was maintained fixed for a better comparison of these two similar conformations.

How can we rationalize these results? Firstly, we can understand that these packed conformations were favored by the docking algorithm because, in this way, not only the intermolecular interactions but also, the intramolecular π–π and π–sulfur CHT interactions can exist and contribute, consequently, to minimize the total energy of the complex. Here, we are aware that, in practice, such a constrained conformation probably has a short lifetime, mainly due to the dynamic interaction with the solvent molecules, which were not taken into account in our docking analysis. It is possible, therefore, as water molecules play a natural and definite role in the formation of the complex, other more relaxed conformations to be statistically favored. And, secondly, why not a simple rotation between these two conformations having practically the same binding energy, and we observe also the reflection? In our opinion this is due to the fact that the structure of β-CD used in our analysis and maintained rigid during our simulations, is not perfectly symmetric, being extracted, as mentioned, from an X-rays determined crystallographic inclusion compound with Mg^{2+} and Ca^{2+} salts of meclofenamic acid. It is known that atoms slightly move and adapt as a result of the interaction, and that the β-CD cavity has a small flexibility, which again was artificially ignored in our simulations. The crystallographic structure of the pure β-cyclodextrin cannot be perfectly symmetric too, because it depends on the strain in the macrocyclic ring of glucose subunits joined by α-1,4 glycosidic bonds, apart from the fact that, it presents hydrogen bonds with adjacent cyclodextrins and with some of the water molecules of crystallization. It is consequently surprising that, for our particular system, these two packed conformations, similar but different at the same time, turned out to have the same theoretical binding energy.

To explore further and continue the analysis, in Figure 13, we present these two conformations where we additionally represented all types of close intra- and inter-molecular interactions. Here, we can observe, even visually, that the torsion angles across the hydrazone group slightly differ between the two conformations, practically with about 3.5°, and, also, that the catechol and thiazole rings are not perfectly parallel. If we were to attempt a comparison between the NMR experimental results and our actual theoretical calculations, we can observe that the most sensitive proton, presenting the largest chemical shift variation was the proton Hd belonging to hydrazone group and close to the catechol ring. Also, the next protons presenting large chemical shift variations were the proton Hc from catechol group and the proton He of the thiazole ring. All these experimental results converge to the conclusion that both rings of the CHT molecule are included in the β-CD cavity with the hydrazone proton, Hd, most involved in the interaction. Both of our theoretical structures, also considering the uncertainties due to the missing water, presented previously, can effectively confirm the experimental results.

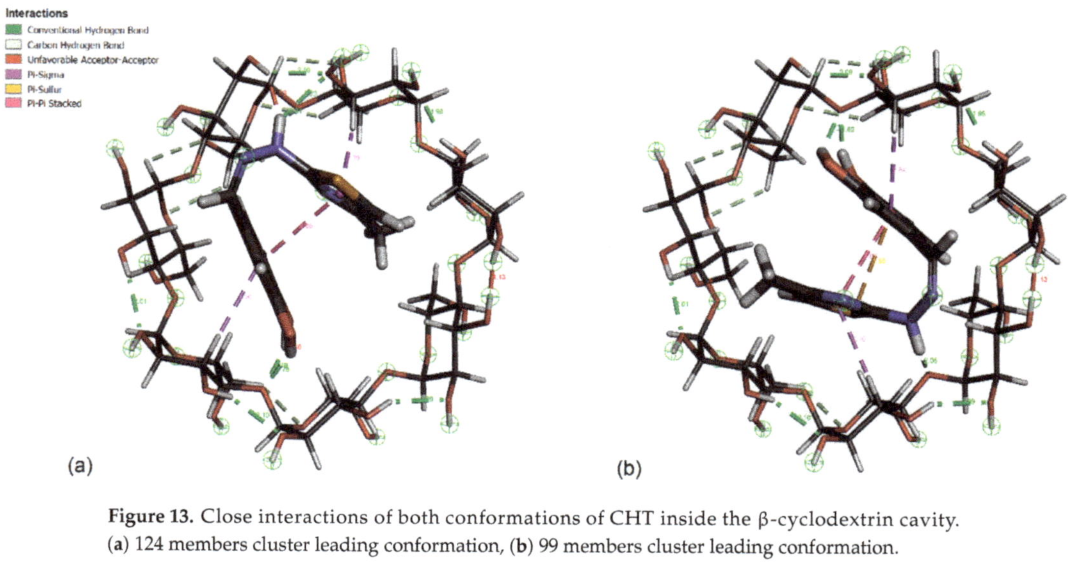

Figure 13. Close interactions of both conformations of CHT inside the β-cyclodextrin cavity. (**a**) 124 members cluster leading conformation, (**b**) 99 members cluster leading conformation.

4. Discussion

The scavenging of DPPH by the CHT-β-CD complex compared to free CHT is obviously delayed due to the complexation. It could be observed that there is a kinetics of release of CHT from the complex over time, compared to its free form. At the lowest concentration in the assay, the effect is almost linear in time, but at the average concentration and the highest concentration, it can be observed that the antiradical effect is more intense in the first part of the experiment (higher slope), after which the effect has a greater increase slow (smaller slope). It could be said that the graph would have an asymptotic trend.

The value of the CHT-β-CD association constant determined by both methods (ITC and NMR) is high if we compare it with the values obtained for other β-CD inclusion complexes investigated by us (procaine, benzocaine, tolmetin, flurbiprofen, 3-carboxy-1-[(2-phenyl-1,3-thiazol-4-yl) methyl]pyridin-1-ium iodide), having values between 10^2 and 10^3 M^{-1}.

From a theoretical point of view, surely, a molecular dynamics investigation where the solvent molecules are explicitly considered and where all atoms can move and relax as a result of the interactions, can refine and improve these minimal conformations but this analysis exceeds the purpose of this study being too computationally intensive.

5. Conclusions

The encapsulation of CHT into β-CD was studied by ITC in combination with the spectroscopic method NMR and molecular docking simulation. Based on the presented studies, it can be concluded that the CHT inclusion complex with β-CD was formed in a molar ratio of 1:1. ITC measurements of CHT–β-CD complex indicate spontaneity ($\Delta G < 0$) of the ligand–receptor process. The formation of the complex is controlled by enthalpy ($\Delta H < 0$) and entropic ($\Delta S > 0$) effects. Thus, the inclusion complexation of CHT: β-CD is predominantly governed by hydrophobic interactions driven by entropy energy. The association constants (K) obtained from both the ^{1}H NMR and ITC experiments for CHT complexation with β-CD are in the same order of magnitude.

The CHT: β-CD complex in aqueous solution has been studied by ^{1}H NMR spectroscopy. An analysis of our data by the continuous variation method indicates that the inclusion occurs and the complex has 1:1 stoichiometry. The ROESY NMR experiment highlights intermolecular interactions through space between the H3 and H5 protons inside the β-CD and the Ha, Hb, Hc, and Hd protons belonging to the CHT molecule, which indicate that the CHT molecule is included inside the β-CD cavity with the aromatic ring. The association constant K obtained by ^{1}H NMR and ITC are in agreement and both methods sustain a 1:1 stoichiometry.

The antiradical activity of CHT when included in a complex with β-CD is extended in time, compared to free CHT.

As a result of the docking analysis, we identified two possible conformations of the inclusion compound which agree with the largest chemical shift variations of the protons, corresponding to the CHT molecule, obtained from NMR experiments. We observed that these rough but fast simulations are full of insights and contain rather precise details Molecular dynamics simulations have the ability to improve these estimations because, in this case, all atoms of the host and guest molecules can freely move and the solvent molecules are considered explicitly.

Author Contributions: Conceptualization, A.P.; methodology; software, C.G.F.; validation, O.O. and M.D.P.; formal analysis, A.P., M.M., C.G.F. and G.M.; investigation, A.P., M.M., C.G.F. and G.M.; resources, A.P. and G.M.; data curation, A.P., M.M. and G.M.; writing—original draft preparation, A.P., M.M., C.G.F. and G.M.; writing—review and editing, A.P., M.M., C.G.F. and G.M.; supervision, O.O.; project administration, A.P. and G.M.; funding acquisition, A.P. and G.M. All authors have read and agreed to the published version of the manuscript.

Funding: This research was funded from to MCID through the "Nucleu" Program within the National Plan for Research, Development and Innovation 2022–2027, project PN 23 24 01 05 and project PN-III-

P2-2.1-PTE-2021-0303 and by "Iuliu Hațieganu" University of Medicine and Pharmacy, Cluj-Napoca, Romania through project 35182/17.12.2021.

Institutional Review Board Statement: Not applicable.

Informed Consent Statement: Not applicable.

Data Availability Statement: The data presented in this study is available in the article.

Conflicts of Interest: The authors declare no conflict of interest.

References

1. Antony Muthu Prabhu, A.; Suresh Kumar, G.S.; Fatiha, M.; Sorimuthu, S.; Sundar Raj, M. Encapsulation of phenylalanine and 3,4-dihydroxyphenylalanine into β-cyclodextrin: Spectral and molecular modeling studies. *J. Mol. Struct.* **2015**, *1079*, 370–382. [CrossRef]
2. Alper Öztürk, A.; Başaran, E.; Şenel, B.; Demirel, M.; Sarıca, Ş. Synthesis, characterization, antioxidant activity of Quercetin, Rutin and Quercetin-Rutin incorporated β-cyclodextrin inclusion complexes and determination of their activity in NIH-3T3, MDA-MB-231 and A549 cell lines. *J. Mol. Struct.* **2023**, *1282*, 135169. [CrossRef]
3. Aree, T.; Jongrungruangchok, S. Crystallographic evidence for β-cyclodextrin inclusion complexation facilitating the improvement of antioxidant activity of tea (+)-catechin and (-)-epicatechin. *Carbohydr. Polym.* **2016**, *140*, 362–373. [CrossRef] [PubMed]
4. Hong, J.H.; Kim, M.J.; Oh, W.Y.; Lee, J.H. Evaluation of deodorization techniques using cyclodextrins on the headspace volatiles and antioxidant properties of onion. *Food Chem.* **2023**, *410*, 135416. [CrossRef] [PubMed]
5. Aguado, R.; Murtinho, D.; Valente, A.J.M. Association of antioxidant monophenolic compounds with β-cyclodextrin-functionalized cellulose and starch substrates. *Carbohydr. Polym.* **2021**, *267*, 118189. [CrossRef]
6. Suresh Kumar, G.S.; Antony Muthu Prabhu, A.; Bhuvanesh, N. Studies on the self-catalyzed Knoevenagel condensation, characterization, DPPH radical scavenging activity, cytotoxicity, and molecular properties of 5-arylidene-2,2-dimethyl-1,3-dioxane-4,6-diones using single crystal XRD and DFT techniques. *J. Mol. Struct.* **2014**, *1075*, 166–177. [CrossRef]
7. Liang, Y.; Hou, D.; Ni, Z.; Cao, M.; Cai, L. Preparation, characterization of naringenin, β-cyclodextrin and carbon quantum dot antioxidant nanocomposites. *Food Chem.* **2022**, *375*, 131646. [CrossRef]
8. Mic, M.; Pîrnău, A.; Floare, C.G.; Miclăuș, M.; Kacso, I.; Palage, M.; Bogdan, M. Interaction of 1-methyl-1-({2-[4-(trifluoromethyl)phenyl]-1,3-thiazol-4-yl}methyl) piperidinium chloride with β-CD: Spectroscopic, calorimetric and molecular modeling approaches. *J. Incl. Phenom. Macrocycl. Chem.* **2018**, *92*, 195–204. [CrossRef]
9. Maragos, S.; Archontaki, H.; Macheras, P.; Valsami, G. Effect of Cyclodextrin Complexation on the Aqueous Solubility and Solubility/Dose Ratio of Praziquantel. *AAPS PharmSciTech* **2009**, *10*, 1444. [CrossRef]
10. Das, S.; Gazdag, Z.; Szente, L.; Meggyes, M.; Horváth, G.; Lemli, B.; Kunsági-Máté, S.; Kuzma, M.; Kőszegi, T. Antioxidant and antimicrobial properties of randomly methylated β cyclodextrin–captured essential oils. *Food Chem.* **2019**, *278*, 305–313. [CrossRef]
11. Siva, S.; Li, C.; Cui, H.; Lin, L. Encompassment of isoeugenol in 2-hydroxypropyl-β-cyclodextrin using ultrasonication: Characterization, antioxidant and antibacterial activities. *J. Mol. Liq.* **2019**, *296*, 111777. [CrossRef]
12. Navarro-Orcajada, S.; Conesa, I.; Matencio, A.; Rodríguez-Bonilla, P.; García-Carmona, F.; López-Nicolás, J.M. The use of cyclodextrins as solubility enhancers in the ORAC method may cause interference in the measurement of antioxidant activity. *Talanta* **2022**, *243*, e123336. [CrossRef] [PubMed]
13. Junior, O.V.; Dantas, J.H.; Barão, C.E.; Zanoelo, E.F.; Cardozo-Filho, L.; de Moraes, F.F. Formation of inclusion compounds of (+)catechin with β-cyclodextrin in different complexation media: Spectral, thermal and antioxidant properties. *J. Supercrit. Fluids* **2017**, *121*, 10–18. [CrossRef]
14. Yang, L.J.; Chang, Q.; Zhou, S.Y.; Yang, Y.H.; Xia, F.T.; Chen, W.; Li, M.; Yang, X.D. Host–guest interaction between brazilin and hydroxypropyl-β-cyclodextrin: Preparation, inclusion mode, molecular modelling and characterization. *Dye. Pigment.* **2018**, *150*, 193–201. [CrossRef]
15. Lu, Z.; Cheng, B.; Hu, Y.; Zhang, Y.; Zou, G. Complexation of resveratrol with cyclodextrins: Solubility and antioxidant activity. *Food Chem.* **2009**, *113*, 17–20. [CrossRef]
16. Lucas-Abellán, C.; Fortea, I.; López-Nicolás, J.M.; Núñez-Delicado, E. Cyclodextrins as resveratrol carrier system. *Food Chem.* **2007**, *104*, 39–44. [CrossRef]
17. Mercader-Ros, M.T.; Lucas-Abellán, C.; Fortea, M.I.; Gabaldón, J.A.; Núñez-Delicado, E. Effect of HP-β-cyclodextrins complexation on the antioxidant activity of flavonols. *Food Chem.* **2010**, *118*, 769–773. [CrossRef]
18. Hu, X.; Wang, X.; Han, L.; Li, S.; Zhou, W. Antioxidant and antimicrobial polyvinyl alcohol electrospun nanofibers containing baicalein-hydroxypropyl-β-cyclodextrin inclusion complex. *Colloids Surfaces A Physicochem. Eng. Asp.* **2021**, *614*, 126135. [CrossRef]
19. Shiozawa, R.; Inoue, Y.; Murata, I.; Kanamoto, I. Effect of antioxidant activity of caffeic acid with cyclodextrins using ground mixture method. *Asian J. Pharm. Sci.* **2018**, *13*, 24–33. [CrossRef] [PubMed]

20. Kalogeropoulos, N.; Yannakopoulou, K.; Gioxari, A.; Chiou, A.; Makris, D.P. Polyphenol characterization and encapsulation in β-cyclodextrin of a flavonoid-rich Hypericum perforatum (St John's wort) extract. *LWT Food Sci. Technol.* **2010**, *43*, 882–889. [CrossRef]
21. de Lima Paula, P.; de Lemos, A.S.O.; Queiroz, L.S.; Rocha, V.N.; Coimbra, E.S.; Fabri, R.L.; Denadai, Â.M.L. Supramolecular complexes between Plinia cauliflora (DC.) Kausel extracts and β-cyclodextrin: Physicochemical characterization and antioxidant and anti-inflammatory properties. *J. Drug Deliv. Sci. Technol.* **2023**, *84*, 104533. [CrossRef]
22. Mic, M.; Pîrnău, A.; Floare, C.G.; Borlan, R.; Focsan, M.; Oniga, O.; Bogdan, M.; Vlase, L.; Oniga, I.; Marc, G. Antioxidant Activity Evaluation and Assessment of the Binding Affinity to HSA of a New Catechol Hydrazinyl-Thiazole Derivative. *Antioxidants* **2022**, *11*, 1245. [CrossRef] [PubMed]
23. Alam, M.N.; Bristi, N.J.; Rafiquzzaman, M. Review on in vivo and in vitro methods evaluation of antioxidant activity. *Saudi Pharm. J.* **2013**, *21*, 143–152. [CrossRef] [PubMed]
24. Brand-Williams, W.; Cuvelier, M.E.; Berset, C. Use of a free radical method to evaluate antioxidant activity. *LWT Food Sci. Technol.* **1995**, *28*, 25–30. [CrossRef]
25. Marc, G.; Stana, A.; Oniga, S.D.; Pîrnău, A.; Vlase, L.; Oniga, O. New Phenolic Derivatives of Thiazolidine-2,4-dione with Antioxidant and Antiradical Properties: Synthesis, Characterization, In Vitro Evaluation, and Quantum Studies. *Molecules* **2019**, *24*, 2060. [CrossRef]
26. Marc, G.; Stana, A.; Franchini, A.H.; Vodnar, D.C.; Barta, G.; Tertiş, M.; Şanta, I.; Cristea, C.; Pîrnău, A.; Ciorîţă, A.; et al. Phenolic Thiazoles with Antioxidant and Antiradical Activity. Synthesis, In Vitro Evaluation, Toxicity, Electrochemical Behavior, Quantum Studies and Antimicrobial Screening. *Antioxidants* **2021**, *10*, 1707. [CrossRef]
27. Da Ferreira, F.R.; Valentim, I.B.; Ramones, E.L.C.; Trevisan, M.T.S.; Olea-Azar, C.; Perez-Cruz, F.; de Abreu, F.C.; Goulart, M.O.F. Antioxidant activity of the mangiferin inclusion complex with β-cyclodextrin. *LWT Food Sci. Technol.* **2013**, *51*, 129–134. [CrossRef]
28. Jullian, C.; Moyano, L.; Yañez, C.; Olea-Azar, C. Complexation of quercetin with three kinds of cyclodextrins: An antioxidant study. *Spectrochim. Acta Part A Mol. Biomol. Spectrosc.* **2007**, *67*, 230–234. [CrossRef]
29. Frisch, M.J.; Schlegel, H.B.; Scuseria, G.E.; Robb, M.A.; Cheeseman, J.R.; Scalmani, G.; Barone, V.; Mennucci, B.; Petersson, G.A.; Nakatsuji, H.; et al. *Gaussian 09 Revision D.01*; Gaussian Inc.: Wallingford, CT, USA, 2009.
30. Walker, M.; Harvey, A.J.A.; Sen, A.; Dessent, C.E.H. Performance of M06, M06-2X, and M06-HF Density Functionals for Conformationally Flexible Anionic Clusters: M06 Functionals Perform Better than B3LYP for a Model System with Dispersion and Ionic Hydrogen-Bonding Interactions. *J. Phys. Chem. A* **2013**, *117*, 12590–12600. [CrossRef]
31. Bogdan, M.; Floare, C.G.; Buimaga-Iarinca, L.; Morari, C.; Pirnau, A. NMR study and computational assays of meclofenamic Na salt and β-cyclodextrin inclusion complex. *J. Incl. Phenom. Macrocycl. Chem.* **2016**, *85*, 111–120. [CrossRef]
32. Floare, C.G.; Bogdan, M.; Tomoaia-Cotişel, M.; Mocanu, A. 1H NMR spectroscopic characterization of inclusion complex of desferrioxamine B chelator and β-cyclodextrin. *J. Mol. Struct.* **2022**, *1248*, 131477. [CrossRef]
33. Caira, M.R.; Miller, J.L.; Nassimbeni, L.R. β-Cyclodextrin Inclusion Complexes of Mg^{2+} and Ca^{2+} Salts of Meclofenamic Acid: Preparation and Structural Characterisation. *Supramol. Chem.* **2006**, *18*, 553–559. [CrossRef]
34. Morris, G.M.; Huey, R.; Lindstrom, W.; Sanner, M.F.; Belew, R.K.; Goodsell, D.S.; Olson, A.J. AutoDock4 and AutoDockTools4: Automated Docking with Selective Receptor Flexibility. *J Comput Chem.* **2010**, *30*, 2785–2791. [CrossRef] [PubMed]
35. Stoica, C.I.; Marc, G.; Pîrnău, A.; Vlase, L.; Araniciu, C.; Oniga, S.; Palage, M.; Oniga, O. Thiazolyl-oxadiazole derivatives targeting lanosterol 14α-demethylase as potential antifungal agents: Design, synthesis and molecular docking studies. *Farmacia* **2016**, *64*, 390–397.
36. Gasteiger, J.; Marsili, M. Iterative partial equalization of orbital electronegativity—A rapid access to atomic charges. *Tetrahedron* **1980**, *36*, 3219–3228. [CrossRef]
37. Morris, G.M.; Goodsell, D.S.; Halliday, R.S.; Huey, R.; Hart, W.E.; Belew, R.K.; Olson, A.J. Automated docking using a Lamarckian genetic algorithm and an empirical binding free energy function. *J. Comput. Chem.* **1998**, *19*, 1639–1662. [CrossRef]
38. Fuhrmann, J.; Rurainski, A.; Lenhof, H.-P.; Neumann, D. A new Lamarckian genetic algorithm for flexible ligand-receptor docking. *J. Comput. Chem.* **2010**, *31*, 1911–1918. [CrossRef]
39. Pettersen, E.F.; Goddard, T.D.; Huang, C.C.; Couch, G.S.; Greenblatt, D.M.; Meng, E.C.; Ferrin, T.E. UCSF Chimera—A visualization system for exploratory research and analysis. *J. Comput. Chem.* **2004**, *25*, 1605–1612. [CrossRef]
40. Hazra, S.; Kumar, G.S. Physicochemical properties of inclusion complexes of sanguinarine with natural cyclodextrins: Spectroscopy, calorimetry and NMR studies. *RSC Adv.* **2015**, *5*, 1873–1882. [CrossRef]
41. Mic, M.; Pîrnău, A.; Floare, C.G.; Marc, G.; Franchini, A.H.; Oniga, O.; Vlase, L.; Bogdan, M. Synthesis and molecular interaction study of a diphenolic hidrazinyl-thiazole compound with strong antioxidant and antiradical activity with HSA. *J. Mol. Struct.* **2021**, *1244*, 131278. [CrossRef]
42. Mic, M.; Pîrnău, A.; Floare, C.G.; Bogdan, M. Study of the binding affinity between imatinib and α-1 glycoprotein using nuclear spin relaxation and isothermal titration calorimetry. *Int. J. Biol. Macromol.* **2020**, *147*, 326–332. [CrossRef] [PubMed]
43. Mic, M.; Pîrnău, A.; Neamţu, S.; Palage, M.; Bogdan, M. Molecular interaction of β-CD with 3-carboxy-1-[(2-phenyl-1,3-thiazol-4-yl) methyl]pyridin-1-ium iodide analyzed by isothermal titration calorimetry and NMR spectroscopy. *J. Incl. Phenom.* Macrocycl. Chem. **2015**, *83*, 257–265. [CrossRef]
44. Pirnau, A.; Floare, C.G.; Bogdan, M. The complexation of flurbiprofen with β-cyclodextrin: A NMR study in aqueous solution. *J. Incl. Phenom. Macrocycl. Chem.* **2014**, *78*, 113–120. [CrossRef]

45. Bogdan, M.; Caira, M.R.; Farcas, S.I. Inclusion of the Niflumic Acid Anion in β-cyclodextrin: A Solution NMR and X-ray Structural Investigation. *Supramol. Chem.* **2002**, *14*, 427–436. [CrossRef]
46. Floare, C.G.; Balibanu, F.; Bogdan, M. CONSTEQ-A program for the calculation of the equilibrium constants using spectroscopic data. *Stud. Univ. Babes-Bolyai Phys.* **2005**, *4*, 451.
47. Gu, A.; Wheate, N.J. Macrocycles as drug-enhancing excipients in pharmaceutical formulations. *J. Incl. Phenom. Macrocycl. Chem.* **2021**, *100*, 55–69. [CrossRef]
48. Uekama, K.; Hirayama, F.; Irie, T. Cyclodextrin Drug Carrier Systems. *Chem. Rev.* **1998**, *98*, 2045–2076. [CrossRef]

antioxidants

MDPI

Review

Lipid Nanovesicles for Antioxidant Delivery in Skin: Liposomes, Ufasomes, Ethosomes, and Niosomes

Agnese Ricci [1], Luca Stefanuto [1], Tecla Gasperi [1], Fabio Bruni [2] and Daniela Tofani [1,*]

1 Department of Science, Section of Nanoscience and Nanotechnologies, "Roma Tre" University, Via della Vasca Navale 79, 00146 Rome, Italy; agnese.ricci@uniroma3.it (A.R.); luca.stefanuto@uniroma3.it (L.S.); tecla.gasperi@uniroma3.it (T.G.)

2 Department of Science, Section of Nanoscience and Nanotechnologies, "Roma Tre" University, Via della Vasca Navale 84, 00146 Rome, Italy; fabio.bruni@uniroma3.it

* Correspondence: daniela.tofani@uniroma3.it; Tel.: +39-0657333371

Abstract: The skin, being the largest organ of the human body, serves as the primary barrier against external insults, including UV radiation, pollutants, and microbial pathogens. However, prolonged exposure to these environmental stressors can lead to the generation of reactive oxygen species (ROS), causing oxidative stress, inflammation, and ultimately, skin aging and diseases. Antioxidants play a crucial role in neutralizing ROS and preserving skin health by preventing oxidative damage. In recent years, nanotechnology has emerged as a powerful tool for enhancing the delivery of antioxidants onto the skin. In particular, liposomal formulations have offered unique advantages such as improved stability, controlled release, and enhanced penetration through the skin barrier. This has led to a surge in research focused on developing liposomal-based antioxidant delivery systems tailored for skin health applications. Through a comprehensive analysis of the literature from the 2019–2024 period, this review provides an overview of emerging trends in the use of liposomal delivery systems developed for antioxidants aimed at improving skin health. It explores the latest advancements in liposomal formulation strategies, vesicle characterization, and their applications in delivering antioxidants to combat oxidative stress-induced skin damage and other associated skin pathologies. A comparison of various delivery systems is conducted for the most common antioxidants. Finally, a brief analysis of lipid nanovesicles used in the cosmeceutical industry is provided.

Keywords: antioxidants; liposomes; ethosomes; niosomes; delivery systems

1. Introduction

Skin cells play a crucial role as the body's first line of defense against external stressors. These stressors, including UV rays, pollution, and various infections, can increase the production of reactive oxygen species (ROS) such as $O_2^{\bullet}$, $HOO^{\bullet}$, H_2O_2, and $HO^{\bullet}$. These species are naturally produced as byproducts in cell metabolism and are involved in redox signaling [1]. Cells are able to regulate ROS concentration using antioxidants (vitamin E, and C, glutathione, coenzyme Q10, carotenoids) and enzymes (superoxide dismutase, catalases, glutathione peroxidase, etc.) that help maintain redox homeostasis [2–4]. However, this regulatory capacity declines with age [5], leading to an imbalance in redox equilibrium, which results in oxidative stress and inflammation. In the skin, chronic oxidative stress can contribute to the development of wrinkles, loss of elasticity, and hyperpigmentation as well as to various pathological conditions such as psoriasis, dermatitis, and even cancer [6,7].

It is well known that the use of exogenous antioxidants supplied by the diet enhances the body's response against ROS [2,8]. However, many natural antioxidants are not as effective in vivo due to their short lifetime, low absorption through cell membranes, rapid metabolic degradation, and excretion [9]. Topical application to the skin suffers from similar disadvantages, because many natural antioxidants are unable to penetrate the stratum corneum (SC). The structure and composition of the skin layer are shown in Figure 1.

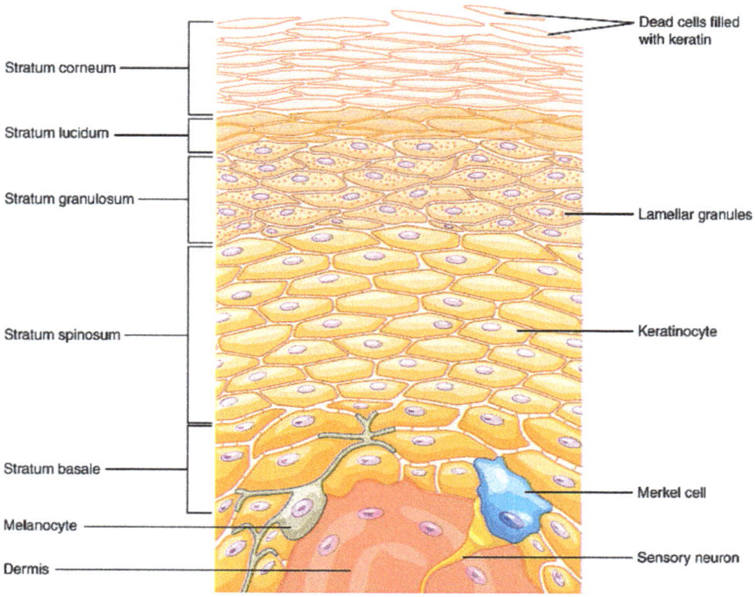

Figure 1. Structure of layers of epidermis (https://commons.wikimedia.org/wiki/File:502_Layers_of_epidermis.jpg (accessed on 16 October 2024)).

In recent decades, several nanomaterials have been developed, gaining significant attention for their applications across various fields. Notably, extensive research has also been conducted for delivering molecules with poor solubility [10]. Some examples of nano-delivery systems include vesicles [11,12], liposomes [13], polymer nanoparticles [14], solid lipid nanoparticles [15], and liquid crystals [16]. These systems are designed to encapsulate and deliver drugs or other active compounds to specific target sites in the body, offering improved efficacy, reduced side effects, and enhanced therapeutic outcomes. On the other hand, lipid vesicles experience instability over time, limited active ingredient loading capacity, challenges in retaining the active ingredient, and high costs. Each system has its own unique properties, advantages, and drawbacks, making it suitable for different applications. In the case of skincare, specialized liposomal carriers have been tailored to target the unique characteristics of the skin's external surface and to pass through the specific structure of the stratum corneum.

This review aims to provide a synopsis of various liposomal systems developed for delivering antioxidants into skin cells. It offers a general description of the critical quality attributes (vesicle size, size distribution, stability), the potential benefits and limitations of each type of vesicle, along with recent applications for delivering specific antioxidants to the skin and the use of lipid nanovesicles in the cosmeceutical industry.

2. Natural and Synthetic Vesicles

A natural vesicle is a small, membrane-bound cavity. Liposome vesicles are formed from the phospholipids of cell membranes and can vary in size from nanometers to micrometers. These sacs play crucial roles in transporting, storing, and processing various substances within the cells. Since Bangham's discovery of the possibility of producing synthetic liposomes in the laboratory [14], their use as drug delivery systems has expanded significantly. Initially, liposomes were formed only from natural phospholipids (PL), such as dipalmitoyl phosphatidylcholine (lecithin or DPPC), phosphatidylethanolamine (cephalin), and phosphatidylserine derived from soybeans, eggs, and milk. Later, liposomes have also been produced using synthetic or semisynthetic molecules.

More recent formulations have envisaged the use of organic solvents (ethosomes) or surfactants (transethosomes) to enhance the vesicle deformability. On the other hand, the use of polyethylene glycol bound to the membrane to produce stealth liposomes allows for evasion of immune system recognition. Additionally, specific proteins can be added to the outer layer to enhance cell specificity and facilitate phagocytosis (actively targeted liposomes) [17].

The administration of liposomal formulations can occur via intravenous or lumbar injection or through topical administration [18]. Their biocompatibility and biodegradability have made them a popular choice in nanomedicine [19,20]. This has proven extremely useful in anticancer therapies with common drugs [21] and is paving the way for the pharmacological application of natural bioactive molecules that were previously overlooked due to their low solubility and poor pharmacokinetics [22]. Many natural antioxidants face similar issues and are seldom used, despite being highly effective against inflammation and diseases associated with ROS imbalance. Liposomes and the other lipid-based vesicles specifically designed for antioxidant delivery to skin cells allow for direct application on the epidermis, enhancing efficacy and absorption. This review traces the development of various lipid-based nanodelivery systems studied over the past five years, focusing specifically on skin treatments. The antioxidants examined are listed in Table 1.

Table 1. List of antioxidants analyzed with the liposomal delivery system used.

Antioxidant	Formulation	Delivery System [1]	Purpose	Ref.
Moringa oleifera	extract, cholesterol	LIP	hydration, anti-aging	[13]
quercetin	pure	LIP	skin pressure ulcer	[23]
Myrtus communis	extract	LIP	skin antioxidant	[24]
propolis	extract	LIP	antioxidant, photoprotection	[25]
propolis	extract, cyclodextrins	LIP	antioxidant, anti-aging	[26]
onhadi pistachio	extract	LIP	antioxidant, anti-inflammatory	[27]
resveratrol	pure	LIP	antioxidant	[28]
naringenin	pure, Tween 20	TLIP	antioxidant, anti-inflammatory	[29]
coenzime Q10	pure	ETH	antioxidant	[30]
rutin	pure	ETH	antioxidant	[31]
rutin	pure	ETH	antioxidant	[32]
euphorbia characias	extract	ETH	photoprotective, enzyme inhibition	[33]
curcumin, piperine	pure	ETH	skin antioxidant	[34]
mangiferin	pure	ETH, TETH	skin antioxidant	[35]
quercetin	pure	ETH, TETH	anti-melanoma activity	[36]
hydroxytirosol	pure	NIS	antioxidant	[37]
epigallocatechin gallate	pure	NIS	antioxidant	[38]
glicirrate	pure	UFS	antioxidant	[39]

[1] LIP: liposomes; ETH: ethosomes; TETH: transethosomes; NIS: niosomes; UFS: unsaturated fatty acids liposomes.

2.1. Liposomes

Synthetic liposomes (LIP) can be produced with various methodologies [40–42]. The most commonly used methods include (i) thin film hydration; (ii) reverse phase evaporation; (iii) solvent injection; and (iv) micro hydrodynamic focusing [17]. The first method, used by Bargham, involves blending phospholipids and other lipophilic substances in an organic solvent. After gently evaporating the solvent, a thin phospholipid layer adheres to the walls of the flask. By slowly adding a buffered aqueous solution, vesicles are formed

(Figure 2a). The reverse phase method initially produces inverted micelles by adding the organic phospholipid solution to the aqueous phase containing the antioxidant compound. The evaporation of both solvents facilitates the formation of unilamellar and multilamellar vesicles (Figure 2b). In the solvent injection method, the lipid fraction, diluted in an organic solvent, is injected into a boiling aqueous solution containing the active ingredient. The evaporation of the organic phase leads to the aggregation of the lipid bilayer (Figure 2c). A more recent system employs two thin channels with a diameter of less than 500 μm to allow the organic and aqueous solutions to flow. When the channels are brought into contact through a Y-shaped connector, the mixing of the phases produces small vesicles of similar diameter, with a low polydispersity index (Figure 2d).

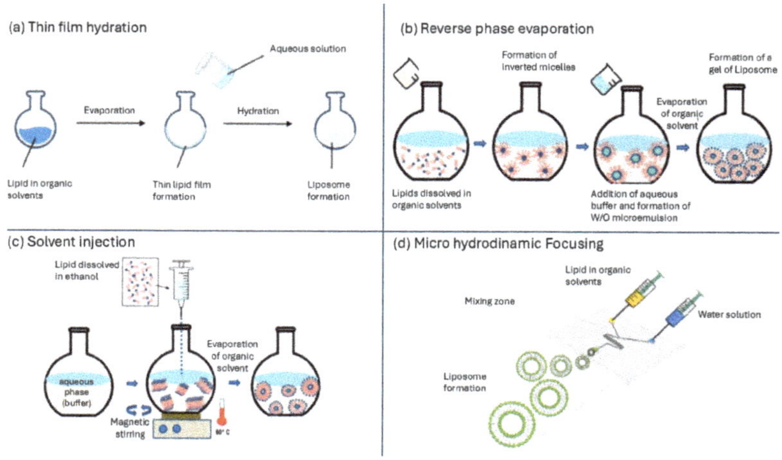

Figure 2. Principal techniques for producing liposomes [42].

LIP are composed of one or multiple bilayer membranes with variable diameters, made up of natural or synthetic phospholipids [43]. The choice of phospholipids used to prepare these vesicles significantly influences their physicochemical properties. Liposomes formed from saturated phospholipids exhibit a higher gel-to-liquid crystalline phase transition temperature, increased rigidity, and a greater ability to retain drugs within their bilayer. One potential limitation of these vesicles is their excessive stability, which can result in reduced release of the drug trapped inside.

On the contrary, LIP made from unsaturated phospholipids have properties that are the opposite. Therefore, the proper design of a vesicle nanocarrier involves a balanced mixture of phospholipids to achieve stability and effective drug release [44]. LIP can carry hydrophilic or hydrophobic molecules in their core or bilayer, making them versatile carriers that enhance drug efficacy while reducing side effects [45]. The efficiency of drug absorption can be quantified by calculating the encapsulation efficiency (EE), which represents the percentage of the drug that remains in the vesicles compared to the amount initially present in the solution. A schematic overview of various liposome types is shown in Figure 3.

Since the initial use of LIP as a drug delivery system, several drawbacks have been identified, and many efforts have been pursued to address them. To enhance bilayer stability against mechanical and chemical stress, additives such as cholesterol and vitamin E have been added. Cholesterol modulates membrane rigidity by reducing interactions between saturated chains and enhancing those between unsaturated ones [46]. More recently surfactants have also been used to balance membrane stability and deformability [29]. Additionally, because vesicles can aggregate, charged phospholipids have been preferred to maintain separation between vesicles through electrostatic repulsion [47]. The surface potential of vesicles is influenced by the choice of phospholipid head groups in the bilayer and the presence of a drug with a specific charge. The ζ potential (ZP), typically determined by microelectrophoresis, provides insight into this parameter. In particular, anionic LIP are easily phagocytized by macrophages and can penetrate cells via endocytosis, making these two aspects particularly useful in drug delivery through the skin [45]. The size of the vesicle is another important parameter to control during LIP preparation. Vesicles with a mean diameter (MD) in the range of 20–40 nm, referred to as small unilamellar vesicles (SUV, Figure 4), can evade immune system attacks, resulting in longer clearance times. In contrast, only vesicles smaller than 100 nm, known as medium unilamellar vesicles (MUV), can prolong their circulation in the blood and are more easily absorbed by cells [19]. Multilamellar vesicles (MLV) can also be produced. In the field of dermal drug delivery, even larger vesicles (LUV) can be used, provided they can penetrate the SC. It is important to note that the physicochemical properties of the entrapped organic compounds can affect the dimensions of the vesicles, especially when these molecules interact with the bilayer. In some cases, vesicle dimensions were increased due to the loosening of interactions between phospholipids from the intercalation of the drug molecule [24,29], while other drugs exhibited the opposite effect [13,31].

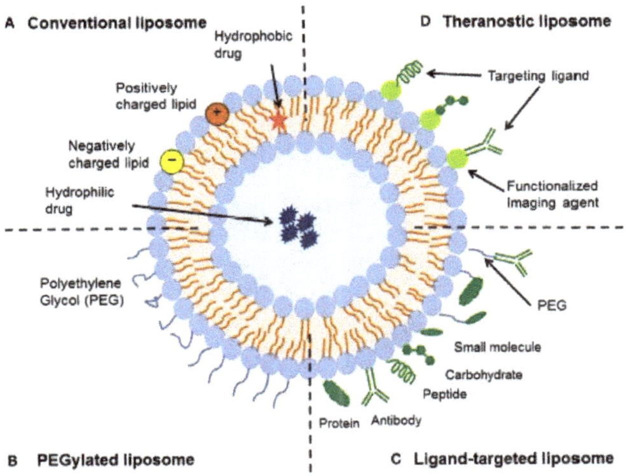

Figure 3. Schematic representation of the different types of liposomal drug delivery systems (https://commons.wikimedia.org/wiki/File:Liposomas_articulo_liberaci%C3%B3n_activos.jpg (accessed on 16 October 2024)).

One important parameter for assessing vesicle size distribution is the polydispersity index (PDI). It characterizes the distribution of molecular mass among the vesicles in a sample, providing insight into the heterogeneity of their volume. Measured using Dynamic Light Scattering (DLS), a PDI value lower than 0.1–0.2 indicates a narrow size distribution and a more uniform population of liposomes. The uniformity is often desirable for consistent drug delivery and predictable behavior in biological systems. Techniques generally used to enhance liposomal size homogeneity include sonication during mixing operations [24] and filtration through filters smaller than the average diameter of the liposomes [26].

Many liposomal formulations have been used to deliver antioxidants, either pure or derived from complex natural extracts. Bavarsad et al. [23] developed liposomal formulations of pure quercetin for treating skin pressure ulcers. They employed the fusion method using soy DPPC (20% *w*/*w*), cholesterol (2% *w*/*w*), propylene glycol (6% *w*/*w*), and vitamin E (0.3% *w*/*w*), in the presence of quercetin buffer solutions at concentrations ranging from 0.01 to 0.04 g/mL, without employing organic solvents. The small unilamellar vesicles (SUV) produced exhibited a homogeneous particle size with a MD of 10–16 nm, demonstrated good encapsulation efficiency (EE 65–78%), and were sufficiently viscous for direct application to the affected areas. This delivery system demonstrated a high efficiency of retention (RE) into the skin (RE 46–51%) and a good penetration ratio (retention/penetration ratio of 7.1 to 9.0). This indicates that the majority of quercetin-loaded LIP were effectively retained in the skin. The active ingredient in the liposomes helped regulate vascular circulation and necrosis, reducing inflammation and shortening recovery time for pressure ulcers. Additionally, with minimal systemic absorption, the liposomal formulation exhibited fewer side effects compared to phenytoin cream, which was used as the reference drug. An example of research on plant extract delivery was explored through the study of *Moringa oleifera* leaves [13]. Rich in carotenoids, glucosinolates, vitamins, amino acids, and various phenolic antioxidants, *Moringa oleifera* is well known for its antimicrobial, anticholesterolemic, and anti-inflammatory properties. In the skin, this plant extract helps maintain collagen and hydration levels, thereby enhancing defense against aging and photodamage. The hydroethanolic extracts of the *Moringa oleifera* leaves were encapsulated in LIP formed with a 4:1 ratio of DPPC to cholesterol. The presence of the extracts reduced the diameter of the LIP from that of a giant unilamellar vesicle (GUV, 736 $\pm$ 100 nm MD) to that of a large unilamellar vesicle (LUV, 154 $\pm$ 1 nm MD). The loaded vesicles possessed a low polydispersity index (PDI 0.17 $\pm$ 0.01), a weak positive charge (ZP +6.6 $\pm$ 0.1), and a good retention rate (RR > 80% at 35 °C for 6 h). The antioxidant activity (AA), as determined by the DPPH assay, of the pure extract (AA 85.8 $\pm$ 1.5%) was similar to that of the reference, gallic acid, while the liposomal formulation showed lower activity (AA 70.0 $\pm$ 1.0%). Additionally, the antioxidant effect of liposomes, determined by ORAC experiments, diminished more rapidly (60 min instead of 120 min for the pure extract). A gel formulation containing propylene glycol, hydroxyethyl cellulose, and glycerin was used to analyze effects on skin. The results demonstrated that although the free hydroethanolic extracts exhibited greater antioxidant properties and enhanced hydration more effectively, the liposomal formulation was superior in preventing transepidermal water loss.

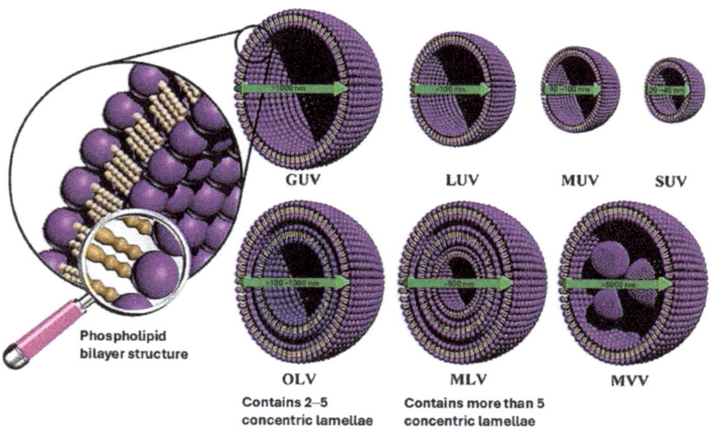

Figure 4. Different sizes and lamellarity of liposomes. V: vesicles, U: unilamellar; G: giant, L: large; M: medium; S: small; OLV: oligolamellar vesicles; MLV: multilamellar vesicles; MVV: multi vesicular vesicles (https://commons.wikimedia.org/wiki/File:1-s2.0-S0168365921005034-gr6_lrg.jpg#filelinks (accessed on 16 October 2024)).

Another interesting application of liposomal delivery was the preparation of vesicles loaded with Myrtus Communis berry extract [24], renowned for its rich antioxidant profile, confirmed by HPLC-DAD analysis (Figure 5) revealing high concentrations of anthocyanins, galloyl derivatives, and flavonoids in the macerated ethanolic extract (1:1 *w*/*v*). Employing a straightforward sonication method with fat-free soybean phospholipids (70% phosphatidylcholine) in PBS, liposomes with a mean diameter (MD) of 102 $\pm$ 5.6 nm were produced. These liposomes, while slightly larger than empty vesicles, exhibited comparable polydispersity index. The research group determined the entrapment efficiency (EE) of each main antioxidant of the myrtle extract: the EE ranges from 71.4% of Myricetin-3-O-galactoside to over 95% of most of the anthocyanins. The antioxidant activity of the loaded liposomes, measured by DPPH analysis, was only slightly lower (326 ± 17 µg Trolox Equivalent/mL) than that of the free myrtle extract. The reducing power remained unchanged, as confirmed by FRAP analysis. Furthermore, the 3-(4,5-Dimethylthiazol-2-yl)-2,5-diphenyltetrazolium bromide (MTT) assay on cell viability demonstrated the low cell cytotoxicity of myrtle extracts and the even higher vitality of cells treated with loaded liposomes vs. control cells. Moreover, in fibroblasts exposed to the AAPH radical generator (i.e., anti-ROS activity experiments), liposomes loaded with increasing quantities of myrtle extracts (from 0.1 to 10 µg) significatively mitigated the oxidative stress, even better than the pure myrtle extract after 24 h. Unfortunately, no direct tests on skin were performed by authors even with these encouraging results.

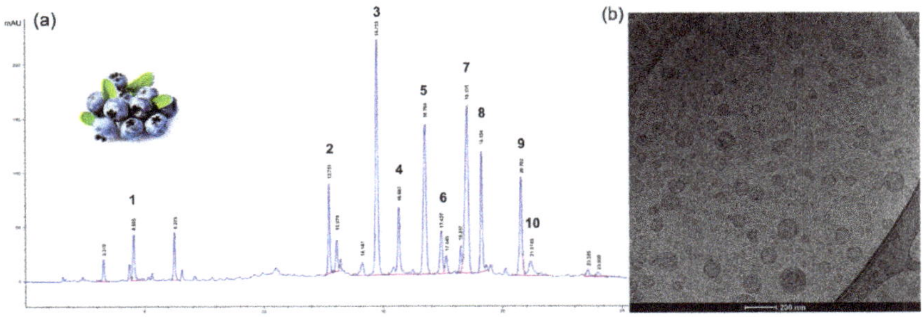

Figure 5. (**a**) HPLC–DAD chromatogram of myrtle berry extract at λ = 280 nm. Chromatographic conditions are described in the reference article. 1: gallic acid; 2: gallic acid derivative; 3: Delphinidin-3-O-glucoside; 4: Cyanidin-3-O-glucoside; 5: Petunidin-3-O-glucoside; 6: Peonidin-3-O-glucoside; 7: Malvidin-3-O-glucoside; 8: Myricetin-3-O-galactoside; 9: Myricetin-3-O-rhamnoside; and 10: ellagic acid. (**b**) Myrtle liposomes through cryo-TEM observation [24].

Permana, Rahman et al. [25] investigated the use of liposomes to enhance the antioxidant activity, permeability, and skin retention of propolis extracts (PE). The more efficient extraction (75% *v*/*v* of ethanol/water, PE75) in terms of total polyphenolic content (TPC 179.3 ± 9.3 mg/g PE) and total flavonoid concentration (TFC 277.2 ± 3.4 mg/g PE) turned out to be the more antioxidant, either as radical scavenger capacity (DPPH IC_{50} 43.2 ± 3.0) or as lipid peroxidation inhibition (LP, 65.3 ± 3.4 µg/mL). These PE are rich in caffeic acid (maximum 4.15 ± 0.01 mg/g in PE75), quercetin (0.72 ± 0.01 mg/g in PE75), and kaempferol (2.62 ± 0.01 in PE75), as confirmed by HPLC analyses. 1 g of PE was encapsulated into liposomal vesicles using different amounts (1–3 g) of phosphatidylcholine and 0.5 g of cholesterol. The preparation applied the thin-film hydration technique with subsequent solvent evaporation. Higher concentrations of phospholipids, increasing the viscosity of the solution, generally produced larger vesicles due to the higher surface tension. On the other hand, the use of a lower weight ratio of PE/phosphatidylcholine combined with a longer duration of homogenization or bath sonication allowed for a reduction in liposomal size from 620 ± 54 to 248 ± 22 nm in diameter, while PDI and ZP remained constant. Antioxidant entrapment efficiency (EE) correlated strongly with the ability of each molecule to form hydrogen bonds with the liposomal bilayer, acting either as a hydrogen bond donor or acceptor. Quercetin demonstrated superior encapsulation (77 ± 6%), exceeding that of kaempferol (45 ± 4%) and caffeic acid (55 ± 5%). Liposome size and processing parameters (homogenization and sonication time) also influenced EE. The obtained resulting vesicles enhanced the water solubility of caffeic acid, kaempferol, and quercetin by 5-fold, 11-fold, and 12-fold, respectively. LIP were dissolved in a hydrogel

(Carbomer 940 and glycerol in water) in order to analyze their potentiality as skin formulations. Antioxidant analysis confirmed that their antioxidant activity, either as a radical scavenger (DPPH IC_{50} 48 ± 5) or as a lipid peroxidation inhibitor (LP, 68 ± 5 μg/mL), remained similar to that of propolis extract hydrogel. However, studies in vitro on mouse skin revealed that the use of LIP hydrogel enhanced the skin permeation and retention of all antioxidants by 4- to 8-fold compared to the hydrogel containing PEs. This was ascribed to the poor absorption of the free antioxidants into the skin and the opposite effect of the phospholipidic bilayer that facilitates the penetration into the stratum corneum. Further, this hydrogel formulation containing propolis extract also acted as a photoprotector against UVA and UVB aggression, as stated by the measured sun protection factor of 17 ± 2.

The combined use of liposomes and other carriers, such as cyclodextrin (CD), has been designed in pharmaceutical applications to enhance the stability of loaded drugs and modulate their release (Modulatory Liposomal Controlled Release, MCLR) [48].

Gardikis et al. [26] analyzed the use of liposomes and CD to enhance the adsorption of propolis antioxidants and study their biochemical activity in skin. In their research they used micronized propolis extracted (PE) in the presence of CD. The commercial LIP (Pro-Lipo^{TM} Neo) formed by propanediol (75.0% *w*/*w*), lecithin (25.0% *w*/*w*), vitamin E (0.25% *w*/*w*), and sunflower oil (0.15% *w*/*w*), mixed with the CD/propolis extract, produced large micelles, the size of which was reduced through membrane filtration. Using 10% *w*/*w* propolis concentration, the obtained liposomes showed an initial MD of 255 ± 36 nm, decent PDI (0.366 ± 0.051), and good ZP (−21 ± 5 mV). This preparation gave higher EE (84 ± 4%) with respect to previous works, thanks to the presence of CD. The TPC and the antioxidant activity determined by Folin–Ciocalteu's method and DPPH assay, respectively, evidenced a high polyphenol concentration of 100.0 ± 0.7 mg gallic acid eq/mL and a significant antioxidant activity of 33.7 ± 0.2 mM Trolox equivalent. However, both parameters decreased over time or when vesicles were stored at temperatures higher than 6 °C. The liposomal formulation was also studied in vitro on human skin cells (HaCaT) by biochemical and genetic tests (Figure 6).

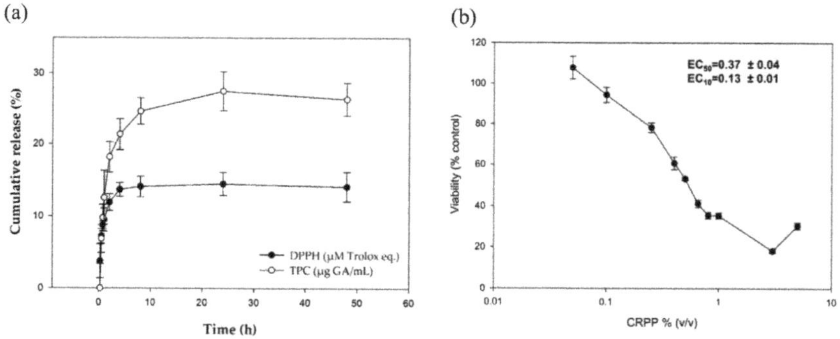

Figure 6. (**a**) Release of antioxidant activity (DPPH) and total phenolic compounds (TPC) of propolis extract (PE) loaded in LIP during 48 h. (**b**) Cytotoxicity profile of PE loaded in human immortalized keratinocyte (HaCaT) cells. HaCaT cells were incubated for 72 h with increasing concentrations (0–5%) of PE loaded in LIP, and their cell viability was estimated via sulforhodamine B (SRB) assay. The EC50 and EC10 values (efficient concentration that causes 50% and 10% decrease in cell viability, respectively) of PE loaded in LIP were determined from the dose–response curves. The results are shown as the mean ± SD of three independent experiments [26].

LIP enhance skin cell metabolism, increases ATP levels, and protects cells against UVB, thereby reducing photoaging. The upregulation of various genes connected with cell proliferation (AQP3), antiaging (TNFα), cell defense against pathogens (IL4), and immune protection (ITGB2) confirmed the beneficial effect of the propolis and the efficacy of the liposomal carrier in enhancing the absorbance of the active principles into the skin. Finally, a cosmetic formulation containing 1% LIP proved that the loaded liposomes were non-irritating for human skin cells, as tested by MTT on a reconstructed skin model, indicating potential for market application.

Oskoueiana, Karimi, and coworkers explored the potential of nanoliposomes loaded with Owhadi pistachio hull extract (PHE) with the aim of enhancing the stability and applicability of a complex polyphenolic antioxidant mixture (TPC 52 ± g GAE/100 g dry extract) including gallic acid, rutin, quercetin, myricetin, catechin, and pyrogallol [27]. The LIP were formed by mixing a hot water solution of lecithin (2% *w/w*) and ethanolic pistachio hull extract (0.2% final concentration) and then sonicating. While the resulting vesicles exhibited desirable characteristics (MD: 96.3 ± 7.2 nm; ZP: −48.0 ± 0.8 mV), entrapment efficiency was modest (EE of 38.4 ± 2.7%). As indicated by in vitro studies (MTT assay), the liposomal formulations maintained high cell viability in murine hepatocytes at extract concentrations up to 50 μg/mL. The antioxidant activity was confirmed by DPPH assay, even if LIP radical scavenging capacity resulted to be lower (IC_{50} 12.6 ± 2.9 μg/L) in comparison with gallic acid reference (IC_{50} 8.8 ± 0.6 μg/L). The LIP also up-regulate the expression of endogenous antioxidant enzyme genes in murine hepatocytes: SOD (+3.2-fold change), CAT (+2.1), GPX (+1.9), and Nrf2 (+2.7), although not to the same extent as gallic acid. Furthermore, the PHE liposomal formulation demonstrated activity against melanogenesis: it inhibited tyrosinase activity in mushroom cells (IC_{50} 21.8 ± 4.3 μg/mL) and reduced melanin production. Additionally, it partially suppressed the expression of the tyrosinase, MTIF, TRP-1, and TRP-2 genes in melanoma cells, suggesting potential applications in skin whitening processes.

Addressing the crucial issue of scalability and the challenging endeavor of industrial production, Caddeo, Pucci, and coworkers [28] analyzed the preparation of vesicles loaded with resveratrol from a commercial liposomal preparation to check the possibility of an easy-to-handle formulation for skin application. The Pronanosome LIPO-N (PLIPO-N) by Nanovex Biotechnologies SL (Llanera, Spain) is a commercial powder of dry LIP. The easy preparation of the liposomal carrier, through an overnight mixing of resveratrol (2 mg/mL), Tween 80 surfactant (10 mg/mL), and PLIPO-N (100 mg) in water, followed by sonication, yielded highly homogeneous nanovesicles (PDI 0.19) having MD (80 nm) and ZP (−25 mV) statistically consistent with the same regenerated empty liposomes. The loading of the resveratrol was high (EE 83%), and long-term tests confirmed the stability of the physicochemical properties of the vesicles for more than three months. The formulation maintained resveratrol's antioxidant activity (DPPH assay: 321.0 ± 7.0 μg of Trolox equivalents/mL; FRAP assay: 8.1 ± 0.7 mg of Fe^{2+} equivalents/mL) and significantly reduced its cytotoxicity compared to the ethanolic solution, as demonstrated by 3T3-L1 cell viability assays. Indeed, surprisingly, while resveratrol ethanolic solution lowered 3T3-L1 cell viability up to 20% at 10 μg maximum concentration, the loaded LIP proved to be less toxic with a 55% viability in the same condition. The ability of resveratrol to protect fibroblasts from ROS injuries was analyzed by dichlorofluorescein (DCF) assay. In the DCF experiments, the fluorescence lowered as resveratrol concentration increased much faster than with the liposomal formulation. On the other hand, microscope images confirmed that, using free resveratrol, the cell mortality increased so much as to invalidate recorded data. Instead, the ROS scavenging capacity of liposomal resveratrol did not produce cell death at the higher concentration, thus supporting the usefulness of the liposomal formulation.

Surfactants, which are amphiphilic, generally are used to stabilize the vesicles, reduce their diameter, and increase their deformability by lowering the energy required to modify membrane curvature. For this reason, adding surfactants to liposomal formulation could be useful for a dermal formulation.

In the case of naringenin, another antioxidant that is poorly soluble in water but shows high potential as an anti-inflammatory agent and regulator of fibroblast and melanoma cell growth, Lowry et al. [29] investigate its incorporation into liposomal formulations containing different percentages of Tween 20 surfactant. The positioning of naringenin in the bilayer increased the diameter of the vesicles. The so-formed loaded liposomes had a MD of 392 ± 32 nm that decreased to 235 ± 9 nm in the presence of 10% Tween 20. The vesicles resulted neutral or slightly negative (ZP from 4.12 to −0.22 mV) but with a PDI of 0.3%. The EE of the active principle was also decreased by Tween 20, from 90.8 ± 4.6% to 64.3 ± 5.6% at 10% Tween 20 concentration. It was supposed that the surfactant competed with lipophilic naringenin for a position within the membrane. As the authors expected, the elasticity of the different liposomal formulations increased in the presence of a higher percentage of surfactant, as evidenced by a decrease in the deformability index (DI) from 80.7 ± 2.0% (without Tween 20) to 59.2 ± 4.4% (with 10% Tween 20). Gel derived from hydroxyethyl cellulose (HEC) and hydroxypropyl methylcellulose (HPMC) was added to naringenin and to the loaded liposomal formulations to compare the release of the active principle using side-by-side diffusion cell experiments (50 nm pore size membrane). The results showed that the liposomes released naringenin faster than the solution, and the release rate increased as the concentration of Tween 20 reached 10%. On the other hand, the comparison of the formulations containing either loaded liposomes with 2% Tween 20 or without, and one cellulose gel, evidenced that HMPC gel slowed the release of the naringenin better than HEC gel even in the presence of the surfactant that enhanced the process in both formulations. The study of naringenin uptake by HaCat and HDFa cells using the sodium 3′-(1-phenylamino)-carbonyl-3,4-tetrazolium (XTT) assay as a fluorescent dye confirmed the ability of gel liposomes to penetrate the cells, likely due to ionic interactions with the cell membrane. These data illustrate the importance of both the vesicle structure and the formulation employed in the delivery system for achieving a synergistic effect on the transport and adsorption of the active ingredient.

2.2. *Unsaturated Fatty Acid Liposomes*

Unsaturated fatty acid liposomes (UFS, ufasomes), composed of a mixture of salified and non-ionized unsaturated fatty acids, represent an alternative vesicular delivery system. The use of ufasomes has followed a parallel but slower development in recent decades [49,50]. UFS formation is pH-dependent, typically requiring a pH range of 7–9, depending on the acids used, to ensure the presence of both forms and achieve stable vesicles rather than micelles (at higher pH) or oily emulsions (at lower pH). As UFS are vesicles formed by a single lipophilic chain, they are more deformable. Furthermore, there is a dynamic equilibrium between the free acid molecules in the solution and those present in the membrane bilayer. Oleic acid (C18, Δ9) and linoleic acid (C18, Δ9–12), commonly chosen for their biocompatibility and intrinsic biological activity as functional food, also exhibit skin-whitening properties via melanin production inhibition [51]. Paolino et al. [39] investigated UFS containing equal molar quantities of oleic and linoleic acids, with (WL) or without lecithin as an aggregating molecule for the topical delivery of ammonium glycyrrhizinate (AG, 3 mg/mL), a potent antioxidant and anti-inflammatory agent. Both formulations (with and without lecithin) exhibited desirable characteristics: narrow polydispersity indices and negative ZP (from −42 to −50 mV). The MD, higher for empty vesicles or without the presence of lecithin, ranges from 146 ± 1 nm to 153 ± 3 nm in the loaded samples, due to an aggregating effect of the active principle. Both compositions showed a high EE (81–85%). As expected, the deformability index of UFS was good in all formulations, and the presence of the antioxidant enhanced this property. However, in the Franz diffusion cell test, an unexpectedly low release capacity was observed in vesicles containing only unsaturated acids (6.0 ± 0.8%) and in those supplemented with lecithin (15.5 ± 1.0%). The comparison of percutaneous permeation on SC cells between the two UFS formulations and an ethanolic solution of AG demonstrated more clearly the superior permeation capacity of the vesicles. In fact, a permeation rate of 15.0 ± 4.9 μg/cm^2 h for UFS WL and 16.6 ± 5.2 μg/cm^2 h for UFS was obtained. In comparison, the permeation rate for the ethanolic solution was only half that of UFS, as AG had limited ability to penetrate the SC due to its large amphiphilic molecular structure. Experiments on NCTC2544 cells with UFS loaded with 0.03 mM AG confirmed the complete safety of both the vesicle formulations. The UFS demonstrated superior protection for cells stressed by H_2O_2 compared to the AG solution at the same concentration, resulting in cell viability exceeding 80% for all formulations. Furthermore, these vesicles succeeded in decreasing twofold the release of lactic dehydrogenase in comparison with AG solution. As the unloaded UFS did not exert any antioxidant activity, the enhancement of glycyrrhizinate's effect should be attributed to the higher internalization of the active ingredient into the cells, facilitated by the delivery system. Finally, skin tolerability testing indicated that UFS did not disrupt skin lipid structure, thus facilitating penetration without causing tissue damage.

2.3. Ethosomes and Transethosomes

In recent years the use of vesicles in therapeutic applications has significantly increased, particularly due to the development of liposomal derivatives characterized by enhanced mobility and deformability. These vesicles are particularly effective in dermal treatments. Ethosomes (ETH) are soft, malleable structures composed of phospholipids, ethanol (20–45%), and water, classified as ultra-deformable vesicles [52]. Their unique composition allows them to penetrate deeper skin layers and facilitate greater drug permeation compared to traditional liposomes (LIP). Their flexible structure and ability to enhance skin penetration is largely attributable to ethanol's ability to perturb the organization of SC lipids. The deformability index (DI) can parametrize the deformability of these vesicles, with a lower DI indicating a higher degree of deformability. This index can be determined by measuring ethosomes' diameter and weight before and after extrusion through membrane filters with pores smaller than the average vesicle size. The exceptional deformability of ETH make them a valuable tool for transdermal drug delivery in pharmaceutical and cosmetic formulations. Furthermore, ethosomes are particularly advantageous for delivering drugs with poor water solubility or high molecular weight.

Sguizzato and coworkers [30] prepared ETH containing coenzyme Q10 (CoQ10) to enhance the bioavailability of this important antioxidant in the skin. CoQ10 is insoluble in the most common solvents and poorly soluble in ethanol (0.3 mg/mL). The preparation of ETH from a solution of soybeans PC (0.6–1.5% *w*/*w*), CoQ10 (0.1% *w*/*w*), and ethanolic solution (28.5–29.5% *w*/*w*) in water (70% *w*/*w*) was performed at room temperature and enhanced the solubility of the antioxidant to 1 mg/mL, allowing for 98% adsorption into the vesicles. This procedure resulted in better performance if compared to that obtained with solid lipid nanoparticles or simple LIP, which permitted an adsorption of only 73% and 89%, respectively [53]. Regarding the EE parameter, the most stable vesicles obtained were derived from a 0.9% *w*/*w* concentration of PC in ethanol, with a 254–271 nm MD and a mixed unilamellar and multilamellar structure.

Over a three-month period, the stability of CoQ10 inside the ETH was higher than that of the free molecule or other supramolecular preparations. ETH alone or with CoQ10 was found to be safe for cells in MTT and LDH experiments. The research group demonstrated the ability of loaded vesicles to penetrate fibroblast membranes, probably by endocytosis, and protect human fibroblast cells from H_2O_2 insults in ex vivo experiments, as demonstrated by analysis of the production of 4-hydroxynonenale (4-HNE) from lipids. The same fluorescence experiments on reconstructed human epidermis and TEM images (Figure 7) confirmed ETH adsorption within the stratum corneum (SC) and even into the stratum basale, as well as their protective effect against hydrogen peroxide-induced lipid degradation, resulting in a 35% reduction in the production of 4-HNE.

To fully harness the powerful antioxidant properties of rutin, given its difficulty to go beyond the SC, the set-up of an effective delivery system is almost essential. Rolim Baby et al. [31] addressed this challenge by formulating ethosomes with PC from egg yolk (2% *w*/*v*), rutin (0.03% *w*/*v*), and a hydroalcoholic solution (20% ethanol). The vesicles produced were larger (369.5 ± 5.2 nm) than the empty one, with a lower ZP (−19.6 ± 0.4 mV) and a higher PDI (0.43 ± 0.01). In particular, the elevated PDI exceeding 0.3 could be due either to the formation of vesicle agglomerates during the addition of the rutin ethanolic solution or to a more disperse geometry of the vesicles. In vivo tape-stripping tests on human skin demonstrated that rutin-loaded ETH were not irritating and that they penetrated the lower layers of the stratum corneum (SC) three times more than the rutin solution. Even if the antioxidant activity in vitro was confirmed for the ETH formulation, the ex vivo DPPH assay, either on the upper or deeper layer of the skin, did not show relevant differences between the loaded ethosomal formulation, the free rutin solution, and the control. The authors ascribed this negative result to the low concentration of rutin in the formulation.

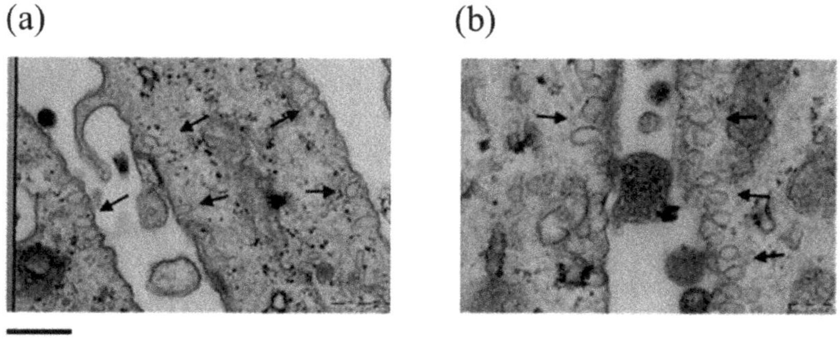

Figure 7. TEM images of fibroblasts in cells treated with ETH (0.9 concentration of PC in ethanol) (**a**,**b**). Panel (**a**), 31.5K magnification; panel (**b**), 50K magnification. ETH are indicated by arrows. Lower bar corresponds to 500 nm [30].

Two years later, Paolino et al. [32] succeeded in enhancing the rutin antioxidant efficacy with respect to the simple rutin solution by studying a different ethosomal composition. The ETH were prepared with Phospholipon 90G©, water, and rutin ethanolic solutions, in the dark to avoid rutin photodegradation. The concentration of rutin solutions, from 0.5 to 4 mg/mL, into the ETH was inversely correlated with the size (MD from 110 ± 8 nm to 575 ± 6 nm) and also with the elasticity of the vesicles. However, DI values (from 3.5 to 10%) permitted a reasonable degree of membrane elasticity. Also, the ZP of the different ETH varied as a function of the rutin adsorbed into the bilayer. In all cases, ranging from −22.4 to −29.5 mV, ZP values were consistently negative and low enough to ensure effective repulsion of the vesicles. Interestingly, in comparison with the other formulations, those ETH prepared with a higher concentration of ethanol (40% w/v instead of 30% w/v) produced smaller ZP (from −5.8 to −12.8 mV) and were discarded. The EE was dependent on different parameters: a higher concentration of rutin solution significantly enhanced the EE, while a higher ethanol content reduced it. On the other hand, rutin diminished the release of the active principle by enhancing bilayer rigidity. Considering the higher EE (67.5 ± 5.2%) and DI (3.5 ± 0.5%) obtained, the best ethosome composition was found to be 4 mg of rutin dissolved in 30 mL of EtOH, 69 mL of water and 1% w/v PL90G©. The experiments indicated that this delivery system not only stabilized the antioxidant but also significantly reduced photodegradation under UV light exposure. The MTT assay on cell lines subjected to oxidative stress via H_2O_2, demonstrated that the ETH not only maintained cell viability but also exerted antioxidant protection on the cells, unlike the pure compound. This protective effect was concentration-dependent and increased with rutin levels up to 100 µM. Finally, the anti-inflammatory effect of rutin ETH, conducted on conveniently irritated epidermis of human volunteers, confirmed the higher effect of the loaded vesicles in comparison with a simple solution of the active principle. It should be noted that rutin properties counterbalanced the initial irritating effect of ethanol on the skin due to the hydroalcoholic preparation, while the erythema recovery, in the presence of rutin ETHs, was more rapid and with better results after 4–5 h of treatment.

In the case of *Euphorbia characias* latex extract, ETH were conveniently used as a skin penetration enhancer of the quercetin glycosides [33]. This extract inhibited collagenase, hyaluronidase, and elastase enzymes responsible for skin aging. Suitable ETH were prepared with soy PC (120 mg) and *E. characias* extracts (2 mg) in ethanol (0.1 mL) and water (0.9 mL). The obtained vesicles were nanostructures (MD 101.1 ± 9.7 nm, PDI 0.29 ± 0.07%) with high ZP (−70.6 ± 7.3 mV) and a good EE of 85–86%, depending on the glycoside bound to quercetin. MTT experiments pointed out the safety of the vesicles, with cell viability > 82% for empty and even higher for loaded ETH, while the DPPH assay confirmed the maintenance of high antioxidant activity of 94.8 ± 0.4% in the liposomal formulation. The anti-melanogenic effect on tyrosinase, assessed by zymography on cells treated with the L-DOPA enzyme activator, confirmed a high level of inhibition by the plant extract, which was further enhanced when the extract was loaded in the ETH. All of these properties, along with the long-term stability of the prepared ETH, confirm their potential for delivering plant extract bioactive compounds through the skin, even though no specific experiments were conducted on skin cells in this case.

Another recent application of nano delivery systems involves the use of two or more components together for a synergistic effect. Ferrara et al. [34] explored this concept by utilizing ETH-containing curcumin (Cur) and piperine (Pip) to mitigate skin damage derived from environmental pollution. Cur is widely recognized for its powerful antioxidant, antibacterial, and antiviral properties, as well as its effects against inflammation and various pathologies [54]. Meanwhile, Pip shares these properties, but it also increases the bioavailability of other molecules [55] as well, by slowing down the cell efflux systems and enhancing skin cell permeation [56]. In their research, Ferrara and colleagues compared two methods for producing ETH: the bulk production method, in which water was added dropwise to a solution of PC (0.9% *w*/*w*) and ethanol (30% *v*/*v*), and a microfluidic technique. In the microfluidic approach, the flow ratio of the two solutions was maintained at 2:1 (*v*/*v*), while varying total flow rates were applied. During flows mixing, the LIP were formed with homogeneous size as a function of the total flow rates. Nevertheless, the bulk technique turned out to be more rapid (5 mL vs. 0.36 mL volume of 200 nm diameter ETH), also considering a possible scaling up of the production. The ETH so prepared were multilamellar and showed an EE of 97 and 79% for Cur and Pip, respectively. In a second step, the researchers compared the antioxidant effect of Cur and Pip solutions and their ETH, either alone or together. The results evidence that the mixed ETH solutions enhance the drug efficacy of Cur in the FRAP assay. However, some drawbacks can be envisaged in the release profile of the two drugs. In comparison with the drug loaded in the vesicles, the percentage of drug released from ETH was 33% in the case of Pip and 15% for Cur, lower than the values obtained by the corresponding drug solutions that were 57% and 24%, respectively. In the permeability tests, Cur and Pip behaved differently. Pip was faster in diffusing through the synthetic membrane (STRAT-M®) that simulates the SC, either in solution or in loaded ETH, the second being slower. Conversely, Cur in ETH showed a partition coefficient higher than the Cur solution even if both had a longer lag time than the Pip systems. If compared to those of Pip, Cur, either in solution or in ETH, also showed a low permeability coefficient. Nevertheless, the Cur ETH system seemed to be more efficient than the corresponding solution in providing a prolonged protective effect. Additionally, the combined use of Cur ETH and Pip ETH further enhanced this behavior against oxidative stress in skin submitted to diesel engine exhausts.

Transethosomes (TETH) represent a further variation in ethosomes containing, together with ethanol, a surfactant that acts as an SC permeation enhancer, which is particularly advantageous for dermal drug delivery [57].

To investigate the antioxidant activity of mangiferin, a glucitol xanthone found in many plants and fruits (papaya, mango), Valacchi, Esposito et al., prepared ETH and TETH as nano delivery systems, considering its higher solubility in ethanol (0.20 mg/mL) with respect to water (0.11 mg/mL) [35]. They analyzed the different loaded vesicles in their physicochemical characteristics, antioxidant activity, and anti-inflammatory properties on skin keratinocytes. The ETH were formulated by mixing PC (30% *w*/*w*) and ethanol (30% *v*/*v*) with mangiferin (3.3 mg/mL), enhancing its solubility to 3.3 mg/mL thanks to PC presence. TETH were prepared by adding, to the same ingredients, different quantities of surfactants (0.1–0.6% *w*/*w*), including hydrophilic polysorbate (Twin 20, Twin 80), lipophilic sorbitane monoleate (SP20, SP 80), or ionic dimethyldidodecylammonium bromide (DDAB). The vesicle size and zeta potential varied depending on the surfactant used, while the PDI remained below 0.2, indicating good homogeneity in vesicle volumes. Generally, a higher concentration of surfactant produced a lowering of the diameter and zeta potential. Except for DDAB, which resulted in highly positively charged vesicles, the ZP were always negative. As already stated, stability tests confirmed that low ZP values were detrimental to the stability of the vesicles, leading to their potential fusion over time due to insufficient repulsive force. Consequently, the research group focused their subsequent studies with mangiferin (0.1% *w*/*w*), only on the vesicles containing either Twin 80 (MD 169.3 ± 0.46, ZP −28.29 ± 0.4) or DDAB (MD 86.41 ± 1.09, ZP 84.08 ± 0.6) as surfactant. MTT assay revealed that while ETH combined with Tween 80 exhibited low cytotoxicity that increased with mangiferin concentration, both empty or loaded DDAB ethosomes showed high toxicity and were discarded from further experiments. Such a result is likely due to the destabilizing effects of the positively charged vesicles on keratinocytes, leading to their lysis. Diffusion tests, conducted with a Franz cell using a nylon membrane, confirmed a lower diffusion capacity for both formulations compared to a simple mangiferin solution; however, TETH showed better performance compared to ETH, because the surfactant most likely facilitated the release of the active principle from the vesicles. Finally, exposure to cigarette smoke was employed to assess the antioxidant and anti-inflammatory effects of the vesicles on HaCaT cells. The results demonstrated that both ETH and TETH permitted mangiferin to penetrate cells (TEM images) and to lower the inflammation of exposed keratinocytes as further demonstrated by the reduction in OH-1 and IL-6 inflammatory enzyme expression.

Subsequently, the same research group investigated the anti-melanoma activity of different concentrations of pure quercetin loaded into either ETH or TETH [36]. Building on previous findings, Tween 80 was employed as the only surfactant. Both vesicle types were multilamellar with similar diameters dependent on the PC and surfactant concentration but with generally uniform dispersity index. Quercetin, incorporated in the bilayer, produced a loosening of the membrane and an enlargement of vesicle diameter while Twin 80 enhances vesicle stability (ETH MD 258 ± 21 nm, TETH MD 240 ± 21 nm).

The entrapment capacity of TETH was slightly higher than that of ETH, with a percentage of 59.2% and 56.4%, respectively, while antioxidant capacity on lipid proved to be almost equal for the two delivery systems (3.16 µmol TE/g vs. 3.26 µmol TE/g) but lower than free quercetin solution. In vitro release tests (IVRT) conducted on lipophilic polytetrafluoroethylene (PTFE) membranes, which closely resembled human skin, exhibited a constant release rate over time (zero-order kinetics), confirming the ability of the vesicles to control and prolong the release of the active principle. In vitro permeation tests (IVPT) confirmed that transethosomes possessed a better capacity in letting quercetin penetrate into the SC with respect to ETH (4-fold lower) and quercetin solution (10-fold lower) (Figure 8).

The MTT experiments, conducted on both HT144 melanoma and HaCaT cells, demonstrated low cytotoxicity at quercetin concentrations below 5 µg/mL. Surprisingly, the subsequent wound healing assay, performed at 2.5 µg/mL quercetin concentration, evidenced a lower capacity of TETH and ETH to favor wound resolution in both the cell lines. This has been ascribed to the capacity of the active principle to inhibit ROS production, thus lowering the natural response of cells to repair the tissue damage, partially activated by ROS as a second messenger.

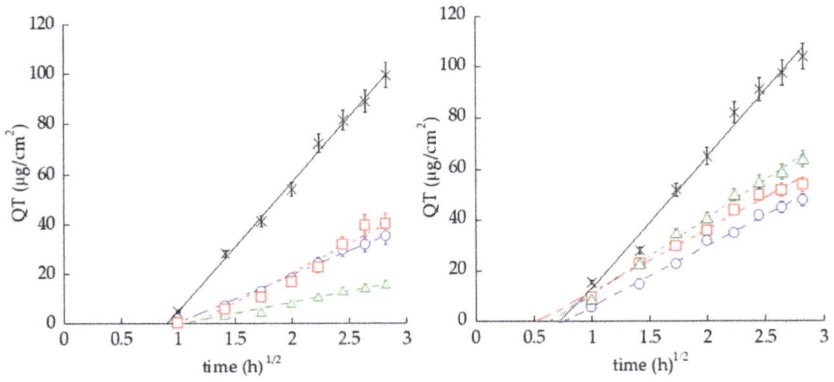

Figure 8. Quercetin (QT) release kinetics from $ETH_{0.9}$-QT (blue circles), $TETH_{0.9}$-QT (red squares), $TETH_{2.7}$-QT (green triangles), and SOL-QT (black crosses), as determined by Franz cell associated with NY (**a**) and PTFE (**b**). Data are the mean of 6 independent experiments ± s.d.

2.4. Niosomes

A different improvement in the use of synthetic vesicles is the development of non-ionic surfactant bilayers, including alkyl ethers, alkylated sugars, and polyethylene fatty acid esters, known as niosomes (NIS). Niosomes exhibit higher chemical stability compared to LIP and allow for improved control over shape and size, prolonged drug release, and enhanced bioavailability, making them a preferred option in many delivery systems. Additionally, the cost-effectiveness and scalability of NIS production contribute to their advantages over conventional LIP [58,59].

Al Saqr and coworkers [37] prepared NIS to study the skin adsorption of hydroxytyrosol, an antioxidant found in olive oil that offers various benefits, including anti-inflammatory, anti-platelet aggregation, and anti-UV-B DNA damage properties, among others [60]. Different types of vesicles were prepared by varying the concentrations of ingredients, including Span 60 surfactant (27.6–45%), lecithin (0–15%), cholesterol (5% *w*/*w*), and hydroxytyrosol (1% *w*/*w*). The vesicles were formed by adding buffered water (0–9%) to a hot solution of these components diluted in ethanol (40–45%). Span 60 was chosen as a surfactant for its higher transition temperature (53–57 °C) and its high EE. The use of different percentages of lecithin and surfactant resulted in vesicles with similar small diameters (MD around 1 nm), high negative charge (ZP between −41.2 and −56.4 mV), and high entrapment efficiency (all EE > 92%). In the TEM images, the niosomes utilized in the gel formulation appeared as non-aggregated, spherical-shaped vesicles. The release capacity of niosomes was assessed and compared with ethanolic and PBS aqueous solutions of hydroxytyrosol. In vitro results confirmed the slower release properties of the niosomes, which were associated with the Span 60/water concentration ratio and the surfactant's ability to incorporate hydroxytyrosol into the bilayer. In vivo permeation and retention experiments were conducted using human cadaver skin. The results confirmed that niosomes had a greater ability to transport hydroxytyrosol through the SC and facilitate its penetration into skin cells compared to antioxidant solutions. The authors attributed this effect to the capacity of Span 60 to deconstruct the SC. The concentration of water used in NIS preparation also played a significant role, as evidenced by the lower transdermal permeation and retention of hydroxytyrosol from niosomes prepared without water. Analogously, a higher concentration of lecithin lowered the delivery of the active principle.

The group led by Jingyuan Wen investigated the optimal niosome formulation for delivering epigallocatechin gallate (EGCG) into skin cells using mathematical methodologies together with statistical optimization techniques [38]. They examined the effects of the concentration of EGCG, surfactant (either Twin 40 or Span 60), and dihexadecyl phosphate (DCP), the cholesterol/surfactant ratio, the buffer amount (PBS solution with 10% EtOH, 7.4 pH), and hydration time on encapsulation efficiency. Different batches of niosomes were prepared, and their encapsulation efficiency (EE) was measured. A multivariable equation was developed in which EE is the dependent variable. ANOVA statistics confirmed the significance of both the amount of antioxidant and the cholesterol/surfactant ratio in achieving higher EE. The subsequent experimental data confirmed that niosomes containing 1.4 mg of EGCG and with a cholesterol/surfactant ratio of 0.9 exhibited the highest EE (53.05 $\pm$ 4.46%). These niosomes were spherical, as shown in the TEM images, and had a MD of 235.4 $\pm$ 15.6 nm, a negative ZP of -45.20 ± 0.03 mV, and a good PDI of 0.267 $\pm$ 0.053. Studies on the release of EGCG from loaded NIS, using a Franz diffusion cell, evidenced a longer release time (35% after 3 h, 73% after 21 h) compared to the EGCG solution, which released 100% in 2 h. Additionally, the deposition of EGCG from NIS onto the skin was significantly greater at both 12 (69.0 $\pm$ 13.87 μg/cm^2) and 24 h (54.38 $\pm$ 8.86 μg/cm^2), proving to be twice as effective compared to the EGCG solution. Experiments with fluorescein 5(6)-isothiocyanate (FITC), either free or loaded in the same NIS, demonstrated that the molecule penetrated deeper into the human SC when released from the vesicles rather than from the solution.

The antioxidant activity of EGCG-loaded NIS was evaluated by measuring malondialdehyde (MDA) production in UV-stressed fetal bovine cells. Cells treated with EGCG-loaded NIS exhibited MDA levels that were 38.4% lower compared to those treated with EGCG solution. This effect was attributed to the activation of glutathione peroxidase (GPX) production due to the release of EGCG in the treated cells, confirming the enhanced protective effect of the loaded antioxidant. Notably, the cellular uptake of NIS requires energy, suggesting a time- and concentration-dependent mechanism mediated by endocytosis.

3. Antioxidant Delivery System Comparison

First-generation liposomes suffer from physical instability over time, as they tend to aggregate into larger vesicles. New formulations with higher zeta potential (ZP) and surfactants have been developed to address this issue. Furthermore, many phospholipid nanoparticles exhibit low viscosity, necessitating their use in gel formulations for topical applications, in contrast to solid lipid nanoparticles. However, LIP, ETH, and their modifications exhibit a high soothing effect, facilitate easy fusion with keratinocyte membranes, and allow for rapid permeation and retention. Evaluating the optimal drug delivery system for a specific antioxidant remains challenging due to the need to compare diverse systems. Furthermore, areas such as cosmetics, skincare, and skin medicine have distinct needs and require different analytical techniques, which complicate direct comparisons. Therefore, this chapter focuses on the two most studied antioxidants: curcumin and quercetin. It examines several relevant parameters for assessing the performance and effectiveness of each delivery system, including entrapment efficiency, release profiles, efficacy, and compatibility with the skin.

3.1. Curcumin

Curcumin possesses numerous biological properties [54] and has low water solubility, making it an ideal candidate for studying new techniques to enhance its bioavailability and adsorption. Consequently, among the antioxidant delivery systems, those incorporating curcumin have been the most extensively studied [61–63].

Focusing on lipid nanovesicles, in addition to the previously presented ethosomes [34], solid lipid nanoparticles (SLN, Figure 9) containing curcumin and turmerone have recently been developed by Aydin et al. to alleviate skin soreness [64] (Table 2).

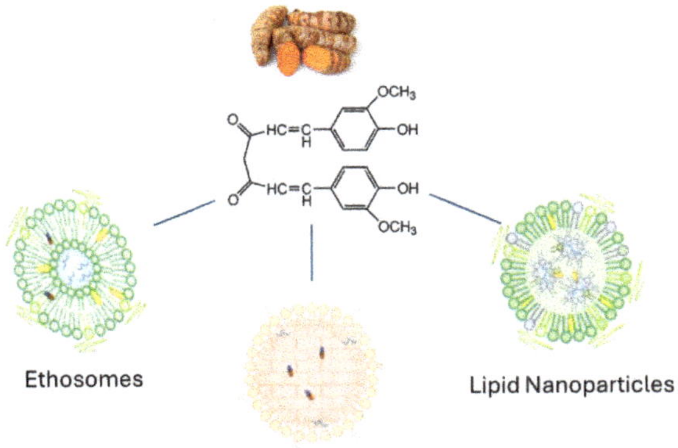

Figure 9. Curcumin lipid nanoparticle delivery systems analyzed.

Table 2. Curcumin delivery systems by lipid nanoparticles.

Compound (%)	Delivery System	EE (%)	Permeation (μg/cm^2/24 h)	Skin Model	Retention (%)	Skin Model	Application	Ref.
curcumin (0.025%)	ETH	97	24.7 ± 4.1	Strat-M	15.9 ± 2.5	PTFE	skin damage antioxidant prevention	[34]
piperine (0.025%)		79	38.4 ± 5.7		27.9 ± 6.0			
curcumin (0.02% *w/v*)	SLN	77	71.3 ± 3.7%	dialysis on cellulose membrane	/	/	skin irritation	[64]
turmerone (0.1% *w/v*)		75	67.2 ± 1.6%		/	/		
curcumin (2.4% *w/w*)	NLC	83–84 (55)	/	Strat-M	26.9 ± 1.9 22% (24 h)	Strat-M	Antioxidant activity	[65]
	NLC + gel	/	/		6.3 ± 0.7 5% (24 h)			

SLN are vesicles generally formed from a mixture of glycerides. Their particle size ranges from 40 to 1000 nm, and their higher capacity for lipophilic drug loading makes them a viable alternative to liposomal and ethosomal formulations. As they penetrate the stratum corneum and accumulate in the epidermis and dermis, SLN are often used to deliver drugs that are effective in treating psoriasis and other epidermal imbalances. In the case of curcumin, SLN were produced using curcumin (0.02% *w/v*) and turmerone (0.1% *w/v*), both dissolved in behenic acid glycerides and mixed with a nonionic surfactant. The computer-aided study, validated by experimental data, achieved encapsulation efficiency of 77% for curcumin and 75% for turmerone. Surprisingly, these data revealed a lower EE compared to that obtained with ethosomes. The release of curcumin and turmerone from SLN was initially rapid: 48.2 ± 2.3% of curcumin and 54.2 ± 4.5% of turmerone were released within the first 2 h. The remaining antioxidants were released more slowly: after 24 h, a total of 71.3 ± 3.7% of curcumin and 67.2 ± 1.6% of turmerone had passed through the dialysis membrane used for measurement. The biphasic release of lipophilic compounds is a characteristic feature of SLN. The initial rapid release can be attributed to the molecules associated with the surface of the particles, while the subsequent slowdown was due to the diffusion of molecules more deeply embedded within the nanoparticles. Unfortunately, the dialysis using a cellulose membrane did not provide information on retention within the epidermis and dermis, which would have enabled a complete comparison with the data obtained from ethosomes.

Nanostructured lipid carriers (NLC, Figure 8) are another type of delivery system used for curcumin. They are composed of solid and liquid lipids, which create a blended phase. The group of Rodriguez-Ruiz et al. [65]. produced NLC by mixing the lipid phase, composed of Precirol® ATO 5, Labrafac, Tween 80, and curcumin (2.4% *w/w*), with a water phase containing a surfactant at 70 °C. The resulting lipid nanoparticles exhibited a bimodal distribution. The encapsulation efficiency (EE) was 83–84% and 55% after size exclusion chromatography, which was employed to completely purify the lipid carriers from any free curcumin outside the particles. Like solid lipid nanoparticles, the lipophilicity of NLC facilitates the absorption of substances both within and outside the nanoparticles, making it challenging to achieve complete control over antioxidant release. The lipid nanoparticles, both alone and incorporated into a gel, were compared regarding their permeability and retention. The resulting data demonstrated an exponential penetration of curcumin into the skin over a 24 h period. In parallel, no curcumin was detected leaving the Strat-M membrane, confirming its complete retention within the synthetic membrane that mimics the skin.

The three systems analyzed demonstrate significant differences in the delivery of curcumin. SLN appear to favor skin permeation, while NLC exhibit the opposite behavior, showing good retention within the skin. Ethosomes display an intermediate behavior, as they partially penetrate the membrane while also maintaining some retention. The varying partitioning of these delivery systems through the skin can be advantageous for targeted applications. For drugs that need to remain entrapped in the epidermis and dermis the NLC or ethosomes should be preferred. Conversely, for drugs intended for transdermal absorption, the SLN are the most suitable choice.

3.2. Quercetin

Many delivery systems have been employed to enhance quercetin absorption [66]. In the field of lipid nanoparticles, which this review seeks to investigate, quercetin has been delivered not only using liposomes [23], ethosomes, and transethosomes [36], but also through solid lipid nanoparticles (SLN) to reduce skin permeation and water loss [67] as well as natural lipid nanoparticles (NLN) [68] (Table 3).

Table 3. Quercetin delivery systems by lipid nanoparticles.

Compound	Delivery System	EE (%)	Permeation	Skin Model	Retention (%)	Skin Model	Application	Ref.
quercetin (0.01–0.04 g/mL)	LIP	65–78	5.2–6.8%	mice skin	46–53	mice skin	wound healing	[23]
quercetin (0.5 mg/mL)	ETH	56	1 μg/cm^2 3 μg/cm^2	Strat-M HSCE	35.4 48	NY PTFE	melanoma cell migration	[36]
	TETH	59–64	2.2–1.8 μg/cm^2 6–8 μg/cm^2	Strat-M HSCE	16.1–40.4 54–64	NY PTFE		
quercetin (0.5 *w*/*v*%)	SLN	15–46	33.5 μg/cm^2 (21.9%)	ICR mice skin	2.20	ICR mice skin	occlusion, water retention	[67]
quercetin (12 mg)	NLN	97–98	not detected	pig ear skin	1.3 μg/cm^2 (86%)	pig ear skin	antioxidant, antiallergic	[68]

In the SLN formulation, palmitic acid (5 *w*/*v*%) was combined with a varying percentage of Tween 80 (0.5–4 *w*/*v*%) to load quercetin (0.5 *w*/*v*). The SLN demonstrated EE that depended on the percentage of Tween 80 used, with the highest value of 46.2 ± 2.2% observed in the formulation containing 2% surfactant. These efficiencies, however, are lower than those of liposomes (65–68%), ethosomes (56.4%), or transethosomes (59.4%) (Table 3). In in vitro permeation tests (IVPT), SLN demonstrated a higher permeability (33.5 μg/cm^2 or 21.9%) compared to all the other vesicles, although the different membranes used added variability to the experiment comparison. Conversely, in vitro retention tests (IVRT) yielded different results, with transethosomes exhibiting the highest retention values (54–64%). The retention of SLNs in the epidermis and dermis was only 2.2% of the total amount, while the highest quantity that passed through the skin was 18.7%.

In the NLN delivery system developed by Marcato et al., illipe butter and calendula oil were utilized to solubilize quercetin [68]. The lipid nanoparticles were produced by adding water, mixed with detergent, to the lipid/quercetin mixture at 50°C under sonication. The lipophilic quercetin was effectively encapsulated within the NLN, achieving an EE of

97–98%, which is higher than that of all previously reported systems. Additionally, the low crystallinity of the lipids employed contributed to the high stability of the delivery system, which maintained its integrity for 90 days without a significant loss of quercetin. The experiments confirmed that the antioxidant activity of the NLN was comparable to that of quercetin in solution. In in vitro permeability and retention tests, these lipid nanoparticles allowed for complete penetration of the formulation into the SC with an 86% retention in the epidermis and dermis while preventing permeation beyond the skin. These properties help avoid systemic drug release, which is particularly useful when the active principle needs to function within the skin, such as in the development of photoprotective systems that must not penetrate beyond the skin to prevent potentially dangerous systemic effects.

The last two examples illustrate how the appropriate choice of delivery system can influence the final destination of antioxidants. Systems designed for transdermal delivery require higher penetration and lower retention capabilities. Conversely, a delivery system that is less penetrating can increase the concentration of the active ingredient in the deeper layers of the skin, as happens with ethosomes and transethosomes, and even more with natural lipid nanoparticles, which exhibit good retention.

4. Lipid Nanoparticles in Cosmetic Formulations

The cosmetics industry is experiencing rapid growth, with nanotechnology playing a significant role in product development. Among the most notable innovations are nanoscale materials that exploit unique physicochemical properties to enhance the stability, efficacy, and delivery of active ingredients across the skin barrier. Lipid nanoparticles, particularly liposomes, serve a dual function in cosmeceutical formulations: they act both as vehicles for active ingredients and as active ingredients themselves [69]. When used as a delivery system in the cosmetics industry, liposomes offer a multifunctional role: they enhance the penetration, solubility, and stability of active ingredients; prolong the effects of the loaded compounds; protect them from environmental degradation; target active ingredients to specific sites of action; reduce toxicity; and improve control over pharmacokinetics and pharmacodynamics. In addition, liposomes can help make products more cost-effective [69]. Rahimpour et al. [70] reported the successful application of liposomes in a range of cosmetic products, including sunscreens, beauty creams, anti-aging formulations, and hair care products. Nanoliposomes have also been employed in fragrance delivery systems, such as antiperspirants, deodorants, and lipsticks [71]. Studies have demonstrated that the encapsulation of phenolic compounds, such as epigallocatechin-3-gallate (EGCG), in liposomal carriers can enhance their antioxidant activity and provide controlled release over time [72]. In vivo research, such as that conducted by Liu et al., has shown that loading liposomes with compounds like syringic acid can improve their bioavailability [73]. Several commercially available products utilize the liposomal encapsulation of phenolic compounds, particularly in skin care (Table 4).

For example, Capilene® is a liposome-based cream that encapsulates concentrated herbal plant extracts in a jelly-like liposomal vehicle, integrating omega-3, omega-6, omega-7, and omega-9 fatty acids, as well as ceramide precursors [74]. Sesderma®, a well-known brand, uses apple polyphenol extracts or quercetin in a liposomal formulation to treat photoaging, dehydration, wrinkles, and skin spots. Other brands, such as Decorté®, Mythos®, and Apivita®, also incorporate anthocyanins in their liposomal skin treatments [75]. These products have been shown to improve the skin's lipid layer, with liposomal formulations actively delivering key ingredients to targeted sites for enhanced skin repair. In addition to these benefits, liposomal formulations, such as niosomes, have demonstrated the potential to reversibly reduce the resistance of the skin's outer layer, facilitating the controlled delivery of active ingredients with reduced toxicity [74]. This makes liposomal formulations particularly useful for products focused on skin soothing and tanning. Despite their promising advantages, liposomal formulations still face challenges in the cosmetics industry. Issues such as low drug loading capacity, inconsistent reproducibility, and physicochemical instability have limited their broader commercial application. However, ongoing research into these challenges continues to drive innovation in the field.

Table 4. Lipid nanoparticles in cosmetic formulation.

Commercial Brand	Encapsulated Compounds	Nanoparticles Size	Application
Sesderma	Ferulic Acid (apple extract)	80–120 nm	Photoaging, dehydration, wrinkles, and skin spots.
Sesderma	Resveratrol, Quercetin, and EGCG (red grape extract)	80–120 nm	Aging, wrinkles, filmogenic effect, and dehydration.
Decortè	Anthocyanins (purple rice extract)	1–100 nm	Dehydration.
Mythos	Anthocyanins (pomegranate extract)	50–500 nm	Skin repair and protection.
Apivita	Anthocyanins (royal jelly extract)	70–100 nm	Skin regeneration, wrinkles, and increases skin elasticity and antioxidant.
Capilene	Fatty acids (sea-buckthorn–seaberry oil and rice bran oil)	50–500 nm	Skin repair.

5. Conclusions

Poorly soluble or unstable antioxidants represent an unexploited resource that could be extremely useful if an appropriate delivery system is developed to facilitate their broader application. In recent decades, research has led to the development of various nanodelivery systems. Liposomes, along with their modifications (ufasomes, ethosomes, transethosomes, and niosomes), have partially responded to the demand for a reliable, stable, and effective delivery system. It is evident that, at this stage of research, it is insufficient to solely focus on the properties of the materials used to prepare suitable liposomes. Each lipophilic antioxidant can interact with and alter the characteristics of the bilayer. However, research conducted in recent years showed several correlations between the composition of the lipid nanoparticles and the delivery system properties, which can facilitate future development. This overview has highlighted the main questions that a researcher must address when designing a liposomal delivery system, as well as the parameters (MD, ZP, PI, and EE) that should be considered before and during the research. In particular, if the antioxidant drug is lipophilic, it dissolves in the bilayer and can modify the stability of the membrane, inducing changes in tension or curvature that can either increase or decrease the size of the liposome, thereby varying the MD values. On the other hand, drug loading generally has less effect on the polydispersity index (PDI) and Z potential of the liposomal structure. The PDI is more closely related to the components used (phospholipids, surfactant, cholesterol, ethanol), the preparation procedure, and any subsequent manipulation (e.g., extrusion, sonication) that leads to a more defined distribution of vesicle diameters, resulting in a smaller PDI. The membrane charge, measured by ZP, depends on the charge of the phospholipids used and any other charged molecules present in the preparation. In the context of skin, negative zeta potentials have been shown to be more advantageous.

Focusing on transdermal delivery, liposomes have been surpassed by more malleable vesicles such as ethosomes, transethosomes, niosomes, and ufasomes. The use of ethanol allows for a higher drug loading capacity and a more deformable structure, while surfactants enhance membrane stability and provide a high deformability, which is suitable for passing through the SC in transdermal applications. The comparison of different delivery systems was conducted specifically on curcumin and quercetin. A final presentation on the lipid nanoparticles used in the cosmetic industry highlighted the significant impact of this research on the cosmetic market.

Author Contributions: Conceptualization, D.T.; investigation, D.T., A.R., L.S. and T.G.; data curation, D.T., L.S., A.R. and T.G.; writing—original draft preparation, D.T.; writing—review and editing, T.G., L.S., A.R., F.B. and D.T. All authors have read and agreed to the published version of the manuscript.

Funding: This research was also funded by the Grant Excellence Departments 2023–2027 (Department of Science, Roma Tre University, MIUR (articolo 1, commi 314-337 legge 232/2016), Rome Technopole Project (Flagship 7, research Lines 3 and 6, CUP: F83B22000040006).

Data Availability Statement: No new data were created.

Conflicts of Interest: The authors declare no conflicts of interest.

References

1. Sies, H.; Jones, D.P. Reactive Oxygen Species (ROS) as Pleiotropic Physiological Signalling Agents. *Nat. Rev. Mol. Cell Biol.* **2020**, *21*, 363–383. [CrossRef] [PubMed]
2. Snezhkina, A.V.; Kudryavtseva, A.V.; Kardymon, O.L.; Savvateeva, M.V.; Melnikova, N.V.; Krasnov, G.S.; Dmitriev, A.A. ROS Generation and Antioxidant Defense Systems in Normal and Malignant Cells. *Oxid. Med. Cell. Longev.* **2019**, *2019*, e6175804. [CrossRef] [PubMed]
3. Juan, C.A.; Pérez de la Lastra, J.M.; Plou, F.J.; Pérez-Lebeña, E. The Chemistry of Reactive Oxygen Species (ROS) Revisited: Outlining Their Role in Biological Macromolecules (DNA, Lipids and Proteins) and Induced Pathologies. *Int. J. Mol. Sci.* **2021**, *22*, 4642. [CrossRef] [PubMed]
4. Halliwell, B. Understanding Mechanisms of Antioxidant Action in Health and Disease. *Nat. Rev. Mol. Cell Biol.* **2024**, *25*, 13–33. [CrossRef]
5. Anik, M.I.; Mahmud, N.; Masud, A.A.; Khan, M.I.; Islam, M.N.; Uddin, S.; Hossain, M.K. Role of Reactive Oxygen Species in Aging and Age-Related Diseases: A Review. *ACS Appl. Bio Mater.* **2022**, *5*, 4028–4054. [CrossRef]
6. Rinnerthaler, M.; Bischof, J.; Streubel, M.K.; Trost, A.; Richter, K. Oxidative Stress in Aging Human Skin. *Biomolecules* **2015**, *5*, 545–589. [CrossRef]
7. Liu, H.-M.; Cheng, M.-Y.; Xun, M.-H.; Zhao, Z.-W.; Zhang, Y.; Tang, W.; Cheng, J.; Ni, J.; Wang, W. Possible Mechanisms of Oxidative Stress-Induced Skin Cellular Senescence, Inflammation, and Cancer and the Therapeutic Potential of Plant Polyphenols. *Int. J. Mol. Sci.* **2023**, *24*, 3755. [CrossRef]
8. Poljšak, B.; Fink, R. The Protective Role of Antioxidants in the Defence against ROS/RNS-Mediated Environmental Pollution. *Oxid. Med. Cell. Longev.* **2014**, *2014*, 671539. [CrossRef]
9. Ndhlala, A.R.; Moyo, M.; Van Staden, J. Natural Antioxidants: Fascinating or Mythical Biomolecules? *Molecules* **2010**, *15*, 6905–6930. [CrossRef]
10. Afzal, O.; Altamimi, A.S.A.; Nadeem, M.S.; Alzarea, S.I.; Almalki, W.H.; Tariq, A.; Mubeen, B.; Murtaza, B.N.; Iftikhar, S.; Riaz, N.; et al. Nanoparticles in Drug Delivery: From History to Therapeutic Applications. *Nanomaterials* **2022**, *12*, 4494. [CrossRef]
11. Li, D.; Martini, N.; Liu, M.; Falconer, J.R.; Locke, M.; Wu, Z.; Wen, J. Non-Ionic Surfactant Vesicles as a Carrier System for Dermal Delivery of (+)-Catechin and Their Antioxidant Effects. *J. Drug Target.* **2021**, *29*, 310–322. [CrossRef] [PubMed]
12. Sguizzato, M.; Ferrara, F.; Mariani, P.; Pepe, A.; Cortesi, R.; Huang, N.; Simelière, F.; Boldrini, P.; Baldisserotto, A.; Valacchi, G.; et al. "Plurethosome" as Vesicular System for Cutaneous Administration of Mangiferin: Formulative Study and 3D Skin Tissue Evaluation. *Pharmaceutics* **2021**, *13*, 1124. [CrossRef] [PubMed]
13. Gimenis, J.M.; Figueiredo, C.C.M.; Gomes, A.d.C.; Granero, F.O.; Ximenes, V.F.; Silva, L.P.; Bronzel Junior, J.L.; Leite e Silva, V.R.; Silva, R.M.G. da Microencapsulation of *Moringa oleifera* L. Extract in Liposomes: Evaluation of Antioxidant and Antiglycation Activities, Cytotoxicity and Skin Biometric Parameters. *J. Drug Deliv. Sci. Technol.* **2023**, *88*, 104972. [CrossRef]
14. Elmowafy, M.; Shalaby, K.; Elkomy, M.H.; Alsaidan, O.A.; Gomaa, H.A.M.; Abdelgawad, M.A.; Mostafa, E.M. Polymeric Nanoparticles for Delivery of Natural Bioactive Agents: Recent Advances and Challenges. *Polymers* **2023**, *15*, 1123. [CrossRef]
15. Sguizzato, M.; Mariani, P.; Ferrara, F.; Drechsler, M.; Hallan, S.S.; Huang, N.; Simelière, F.; Khunti, N.; Cortesi, R.; Marchetti, N.; et al. Nanoparticulate Gels for Cutaneous Administration of Caffeic Acid. *Nanomaterials* **2020**, *10*, 961. [CrossRef]
16. Mancuso, A.; Tarsitano, M.; Cavaliere, R.; Fresta, M.; Cristiano, M.C.; Paolino, D. Gelled Liquid Crystal Nanocarriers for Improved Antioxidant Activity of Resveratrol. *Gels* **2023**, *9*, 872. [CrossRef]
17. Nsairat, H.; Khater, D.; Sayed, U.; Odeh, F.; Al Bawab, A.; Alshaer, W. Liposomes: Structure, Composition, Types, and Clinical Applications. *Heliyon* **2022**, *8*, e09394. [CrossRef]
18. Chen, J.; Hu, S.; Sun, M.; Shi, J.; Zhang, H.; Yu, H.; Yang, Z. Recent Advances and Clinical Translation of Liposomal Delivery Systems in Cancer Therapy. *Eur. J. Pharm. Sci.* **2024**, *193*, 106688. [CrossRef]
19. Kraft, J.C.; Freeling, J.P.; Wang, Z.; Ho, R.J.Y. Emerging Research and Clinical Development Trends of Liposome and Lipid Nanoparticle Drug Delivery Systems. *J. Pharm. Sci.* **2014**, *103*, 29–52. [CrossRef]
20. Akombaetwa, N.; Ilangala, A.B.; Thom, L.; Memvanga, P.B.; Witika, B.A.; Buya, A.B. Current Advances in Lipid Nanosystems Intended for Topical and Transdermal Drug Delivery Applications. *Pharmaceutics* **2023**, *15*, 656. [CrossRef]
21. Paolino, D.; d'Avanzo, N.; Canato, E.; Ciriolo, L.; Grigoletto, A.; Chiara Cristiano, M.; Mancuso, A.; Celia, C.; Pasut, G.; Fresta, M. Improved Anti-Breast Cancer Activity by Doxorubicin-Loaded Super Stealth Liposomes. *Biomater. Sci.* **2024**, *12*, 3933–3946. [CrossRef] [PubMed]
22. Zarrabi, A.; Zarepour, A.; Khosravi, A.; Alimohammadi, Z.; Thakur, V.K. Synthesis of Curcumin Loaded Smart pH-Responsive Stealth Liposome as a Novel Nanocarrier for Cancer Treatment. *Fibers* **2021**, *9*, 19. [CrossRef]

23. Bavarsad, N.; Karampour, N.S.; Hemmati, G.; Rezaie, A. The Effect of Topical Quercetin Loaded Liposome on Pressure Ulcer Healing in Rats. *Nanomed. J.* **2021**, *8*, 187–199. [CrossRef]
24. De Luca, M.; Lucchesi, D.; Tuberoso, C.I.G.; Fernàndez-Busquets, X.; Vassallo, A.; Martelli, G.; Fadda, A.M.; Pucci, L.; Caddeo, C. Liposomal Formulations to Improve Antioxidant Power of Myrtle Berry Extract for Potential Skin Application. *Pharmaceutics* **2022**, *14*, 910. [CrossRef]
25. Permana, A.D.; Utami, R.N.; Courtenay, A.J.; Manggau, M.A.; Donnelly, R.F.; Rahman, L. Phytosomal Nanocarriers as Platforms for Improved Delivery of Natural Antioxidant and Photoprotective Compounds in Propolis: An Approach for Enhanced Both Dissolution Behaviour in Biorelevant Media and Skin Retention Profiles. *J. Photochem. Photobiol. B* **2020**, *205*, 111846. [CrossRef]
26. Spanidi, E.; Karapetsas, A.; Voulgaridou, G.-P.; Letsiou, S.; Aligiannis, N.; Tsochantaridis, I.; Kynigopoulos, S.; Lambropoulou, M.; Mourtzinos, I.; Pappa, A.; et al. A New Controlled Release System for Propolis Polyphenols and Its Biochemical Activity for Skin Applications. *Plants* **2021**, *10*, 420. [CrossRef]
27. Oskoueian, E.; Karimi, E.; Noura, R.; Ebrahimi, M.; Shafaei, N.; Karimi, E. Nanoliposomes Encapsulation of Enriched Phenolic Fraction from Pistachio Hulls and Its Antioxidant, Anti-Inflammatory, and Anti-Melanogenic Activities. *J. Microencapsul.* **2020**, *37*, 1–13. [CrossRef]
28. Caddeo, C.; Lucchesi, D.; Fernàndez-Busquets, X.; Valenti, D.; Penno, G.; Fadda, A.M.; Pucci, L. Efficacy of a Resveratrol Nanoformulation Based on a Commercially Available Liposomal Platform. *Int. J. Pharm.* **2021**, *608*, 121086. [CrossRef]
29. Marwah, M.; Badhan, R.K.S.; Lowry, D. Development of a Novel Polymer-Based Carrier for Deformable Liposomes for the Controlled Dermal Delivery of Naringenin. *J. Liposome Res.* **2022**, *32*, 181–194. [CrossRef]
30. Sguizzato, M.; Mariani, P.; Spinozzi, F.; Benedusi, M.; Cervellati, F.; Cortesi, R.; Drechsler, M.; Prieux, R.; Valacchi, G.; Esposito, E. Ethosomes for Coenzyme Q10 Cutaneous Administration: From Design to 3D Skin Tissue Evaluation. *Antioxidants* **2020**, *9*, 485. [CrossRef]
31. Cândido, T.M.; De Oliveira, C.A.; Ariede, M.B.; Velasco, M.V.R.; Rosado, C.; Baby, A.R. Safety and Antioxidant Efficacy Profiles of Rutin-Loaded Ethosomes for Topical Application. *AAPS PharmSciTech* **2018**, *19*, 1773–1780. [CrossRef] [PubMed]
32. Cristiano, M.C.; Barone, A.; Mancuso, A.; Torella, D.; Paolino, D. Rutin-Loaded Nanovesicles for Improved Stability and Enhanced Topical Efficacy of Natural Compound. *J. Funct. Biomater.* **2021**, *12*, 74. [CrossRef] [PubMed]
33. Pintus, F.; Floris, S.; Fais, A.; Era, B.; Porcedda, C.; Tuberoso, C.I.G.; Caddeo, C. Euphorbia Characias Extract: Inhibition of Skin Aging-Related Enzymes and Nanoformulation. *Plants* **2022**, *11*, 1849. [CrossRef] [PubMed]
34. Ferrara, F.; Bondi, A.; Pula, W.; Contado, C.; Baldisserotto, A.; Manfredini, S.; Boldrini, P.; Sguizzato, M.; Montesi, L.; Benedusi, M.; et al. Ethosomes for Curcumin and Piperine Cutaneous Delivery to Prevent Environmental-Stressor-Induced Skin Damage. *Antioxidants* **2024**, *13*, 91. [CrossRef]
35. Sguizzato, M.; Ferrara, F.; Hallan, S.S.; Baldisserotto, A.; Drechsler, M.; Malatesta, M.; Costanzo, M.; Cortesi, R.; Puglia, C.; Valacchi, G.; et al. Ethosomes and Transethosomes for Mangiferin Transdermal Delivery. *Antioxidants* **2021**, *10*, 768. [CrossRef]
36. Ferrara, F.; Benedusi, M.; Sguizzato, M.; Cortesi, R.; Baldisserotto, A.; Buzzi, R.; Valacchi, G.; Esposito, E. Ethosomes and Transethosomes as Cutaneous Delivery Systems for Quercetin: A Preliminary Study on Melanoma Cells. *Pharmaceutics* **2022**, *14*, 1038. [CrossRef]
37. Al Saqr, A.; Annaji, M.; Poudel, I.; Rangari, S.; Boddu, S.H.S.; Tiwari, A.K.; Babu, R.J. Niosomal Formulation of Hydroxytyrosol, a Polyphenolic Antioxidant, for Enhancing Transdermal Delivery across Human Cadaver Skin. *Pharm. Dev. Technol.* **2022**, *27*, 155–163. [CrossRef]
38. Li, D.; Martini, N.; Wu, Z.; Chen, S.; Falconer, J.R.; Locke, M.; Zhang, Z.; Wen, J. Niosomal Nanocarriers for Enhanced Dermal Delivery of Epigallocatechin Gallate for Protection against Oxidative Stress of the Skin. *Pharmaceutics* **2022**, *14*, 726. [CrossRef]
39. Cristiano, M.C.; Mancuso, A.; Fresta, M.; Torella, D.; De Gaetano, F.; Ventura, C.A.; Paolino, D. Topical Unsaturated Fatty Acid Vesicles Improve Antioxidant Activity of Ammonium Glycyrrhizinate. *Pharmaceutics* **2021**, *13*, 548. [CrossRef]
40. Patil, Y.P.; Jadhav, S. Novel Methods for Liposome Preparation. *Chem. Phys. Lipids* **2014**, *177*, 8–18. [CrossRef]
41. Has, C.; Sunthar, P. A Comprehensive Review on Recent Preparation Techniques of Liposomes. *J. Liposome Res.* **2020**, *30*, 336–365. [CrossRef] [PubMed]
42. Jiang, Y.; Li, W.; Wang, Z.; Lu, J. Lipid-Based Nanotechnology: Liposome. *Pharmaceutics* **2024**, *16*, 34. [CrossRef] [PubMed]
43. Trucillo, P.; Campardelli, R.; Reverchon, E. Liposomes: From Bangham to Supercritical Fluids. *Processes* **2020**, *8*, 1022. [CrossRef]
44. Mouritsen, O.G. Lipids, Curvature, and Nano-Medicine. *Eur. J. Lipid Sci. Technol.* **2011**, *113*, 1174–1187. [CrossRef] [PubMed]
45. Lombardo, D.; Kiselev, M.A. Methods of Liposomes Preparation: Formation and Control Factors of Versatile Nanocarriers for Biomedical and Nanomedicine Application. *Pharmaceutics* **2022**, *14*, 543. [CrossRef]
46. Andra, V.V.S.N.L.; Pammi, S.V.N.; Bhatraju, L.V.K.P.; Ruddaraju, L.K. A Comprehensive Review on Novel Liposomal Methodologies, Commercial Formulations, Clinical Trials and Patents. *Bionanoscience* **2022**, *12*, 274–291. [CrossRef]
47. Miller, C.R.; Bondurant, B.; McLean, S.D.; McGovern, K.A.; O'Brien, D.F. Liposome−Cell Interactions in Vitro: Effect of Liposome Surface Charge on the Binding and Endocytosis of Conventional and Sterically Stabilized Liposomes. *Biochemistry* **1998**, *37*, 12875–12883. [CrossRef]
48. Papagiannaros, A.; Dimas, K.; Papaioannou, G.T.; Demetzos, C. Doxorubicin-PAMAM Dendrimer Complex Attached to Liposomes: Cytotoxic Studies against Human Cancer Cell Lines. *Int. J. Pharm.* **2005**, *302*, 29–38. [CrossRef]
49. Morigaki, K.; Walde, P. Fatty Acid Vesicles. *Curr. Opin. Colloid Interface Sci.* **2007**, *12*, 75–80. [CrossRef]

50. Kaur, M.; Kaur, L.; Singh, G.; Singh, L.; Kaur, A.; Dhawan, R.K. Recent Advancements in Biomimetic Drug Delivery System of Single- Chain Fatty Acids as Ufasomes and Ufosomes: A Comprehensive Review. *Curr. Nanosci.* **2023**, *19*, 362–371. [CrossRef]
51. Ando, H.; Ryu, A.; Hashimoto, A.; Oka, M.; Ichihashi, M. Linoleic Acid and α-Linolenic Acid Lightens Ultraviolet-Induced Hyperpigmentation of the Skin. *Arch. Dermatol. Res.* **1998**, *290*, 375–381. [CrossRef] [PubMed]
52. Ascenso, A.; Raposo, S.; Batista, C.; Cardoso, P.; Mendes, T.; Praça, F.G.; Bentley, M.V.L.B.; Simões, S. Development, Characterization, and Skin Delivery Studies of Related Ultradeformable Vesicles: Transfersomes, Ethosomes, and Transethosomes. *Int. J. Nanomed.* **2015**, *10*, 5837–5851. [CrossRef] [PubMed]
53. Gokce, E.H.; Korkmaz, E.; Tuncay-Tanrıverdi, S.; Dellera, E.; Sandri, G.; Bonferoni, M.C.; Ozer, O. A Comparative Evaluation of Coenzyme Q10-Loaded Liposomes and Solid Lipid Nanoparticles as Dermal Antioxidant Carriers. *Int. J. Nanomed.* **2012**, *7*, 5109–5117. [CrossRef]
54. Sohn, S.-I.; Priya, A.; Balasubramaniam, B.; Muthuramalingam, P.; Sivasankar, C.; Selvaraj, A.; Valliammai, A.; Jothi, R.; Pandian, S. Biomedical Applications and Bioavailability of Curcumin—An Updated Overview. *Pharmaceutics* **2021**, *13*, 2102. [CrossRef] [PubMed]
55. Shi, L.; Xun, W.; Peng, W.; Hu, H.; Cao, T.; Hou, G. Effect of the Single and Combined Use of Curcumin and Piperine on Growth Performance, Intestinal Barrier Function, and Antioxidant Capacity of Weaned Wuzhishan Piglets. *Front. Vet. Sci.* **2020**, *7*, 418. [CrossRef]
56. Jantarat, C.; Sirathanarun, P.; Boonmee, S.; Meechoosin, W.; Wangpittaya, H. Effect of Piperine on Skin Permeation of Curcumin from a Bacterially Derived Cellulose-Composite Double-Layer Membrane for Transdermal Curcumin Delivery. *Sci. Pharm.* **2018**, *86*, 39. [CrossRef]
57. Ma, M.; Wang, J.; Guo, F.; Lei, M.; Tan, F.; Li, N. Development of Nanovesicular Systems for Dermal Imiquimod Delivery: Physicochemical Characterization and in Vitro/in Vivo Evaluation. *J. Mater. Sci. Mater. Med.* **2015**, *26*, 191. [CrossRef]
58. Bhardwaj, P.; Tripathi, P.; Gupta, R.; Pandey, S. Niosomes: A Review on Niosomal Research in the Last Decade. *J. Drug Deliv. Sci. Technol.* **2020**, *56*, 101581. [CrossRef]
59. Liga, S.; Paul, C.; Moacă, E.-A.; Péter, F. Niosomes: Composition, Formulation Techniques, and Recent Progress as Delivery Systems in Cancer Therapy. *Pharmaceutics* **2024**, *16*, 223. [CrossRef]
60. Wang, E.; Jiang, Y.; Zhao, C. Hydroxytyrosol Isolation, Comparison of Synthetic Routes and Potential Biological Activities. *Food Sci. Nutr.* **2024**, *12*, 6899–6912. [CrossRef]
61. Singh, H.; Dhanka, M.; Yadav, I.; Gautam, S.; Bashir, S.M.; Mishra, N.C.; Arora, T.; Hassan, S. Technological Interventions Enhancing Curcumin Bioavailability in Wound-Healing Therapeutics. *Tissue Eng. Part B Rev.* **2024**, *30*, 230–253. [CrossRef] [PubMed]
62. Salama, A.; Elsherbiny, N.; Hetta, H.F.; Safwat, M.A.; Atif, H.M.; Fathalla, D.; Almanzalawi, W.S.; Almowallad, S.; Soliman, G.M. Curcumin-Loaded Gold Nanoparticles with Enhanced Antibacterial Efficacy and Wound Healing Properties in Diabetic Rats. *Int. J. Pharm.* **2024**, *666*, 124761. [CrossRef] [PubMed]
63. Waghule, T.; Gorantla, S.; Rapalli, V.K.; Shah, P.; Dubey, S.K.; Saha, R.N.; Singhvi, G. Emerging Trends in Topical Delivery of Curcumin Through Lipid Nanocarriers: Effectiveness in Skin Disorders. *AAPS PharmSciTech* **2020**, *21*, 284. [CrossRef] [PubMed]
64. Aydin, B.S.; Sagiroglu, A.A.; Civelek, D.O.; Gokce, M.; Bahadori, F. Development of Curcumin and Turmerone Loaded Solid Lipid Nanoparticle for Topical Delivery: Optimization, Characterization and Skin Irritation Evaluation with 3D Tissue Model. *Int. J. Nanomed.* **2024**, *19*, 1951–1966. [CrossRef]
65. Calderon-Jacinto, R.; Matricardi, P.; Gueguen, V.; Pavon-Djavid, G.; Pauthe, E.; Rodriguez-Ruiz, V. Dual Nanostructured Lipid Carriers/Hydrogel System for Delivery of Curcumin for Topical Skin Applications. *Biomolecules* **2022**, *12*, 780. [CrossRef]
66. Wadhwa, K.; Kadian, V.; Puri, V.; Bhardwaj, B.Y.; Sharma, A.; Pahwa, R.; Rao, R.; Gupta, M.; Singh, I. New Insights into Quercetin Nanoformulations for Topical Delivery. *Phytomed. Plus* **2022**, *2*, 100257. [CrossRef]
67. Han, S.B.; Kwon, S.S.; Jeong, Y.M.; Yu, E.R.; Park, S.N. Physical characterization and in vitro skin permeation of solid lipid nanoparticles for transdermal delivery of quercetin. *Int. J. Cosmet. Sci.* **2014**, *36*, 588–597. [CrossRef]
68. Pivetta, T.P.; Silva, L.B.; Kawakami, C.M.; Araújo, M.M.; Del Lama, M.P.F.M.; Naal, R.M.Z.G.; Maria-Engler, S S.; Gaspar, L.R.; Marcato, P.D. Topical Formulation of Quercetin Encapsulated in Natural Lipid Nanocarriers: Evaluation of Biological Properties and Phototoxic Effect. *J. Drug Deliv. Sci. Technol.* **2019**, *53*, 101148. [CrossRef]
69. Ashtiani, H.R.A.; Bishe, P.; Lashgari, N.-A.; Nilforoushzadeh, M.A.; Zare, S. Liposomes in Cosmetics. *J. Skin Stem Cell* **2016**, *3*, e65815. [CrossRef]
70. Rahimpour, Y.; Hamishehkar, H. Liposomes in Cosmeceutics. *Expert Opin. Drug Deliv.* **2012**, *9*, 443–455. [CrossRef]
71. Fytianos, G.; Rahdar, A.; Kyzas, G.Z. Nanomaterials in Cosmetics: Recent Updates. *Nanomaterials* **2020**, *10*, 979. [CrossRef] [PubMed]
72. Pires, F.; Geraldo, V.P.N.; Rodrigues, B.; Granada-Flor, A.d.; de Almeida, R.F.M.; Oliveira, O.N., Jr.; Victor, B.L.; Machuqueiro, M.; Raposo, M. Evaluation of EGCG Loading Capacity in DMPC Membranes. *Langmuir* **2019**, *35*, 6771–6781. [CrossRef] [PubMed]
73. Liu, Y.; Sun, C.; Li, W.; Adu-Frimpong, M.; Wang, Q.; Yu, J.; Xu, X. Preparation and Characterization of Syringic Acid–Loaded TPGS Liposome with Enhanced Oral Bioavailability and In Vivo Antioxidant Efficiency. *AAPS PharmSciTech* **2019**, *20*, 98. [CrossRef] [PubMed]

74. Bilal, M.; Iqbal, H.M.N. New Insights on Unique Features and Role of Nanostructured Materials in Cosmetics. *Cosmetics* **2020**, *7*, 24. [CrossRef]
75. Figueroa-Robles, A.; Antunes-Ricardo, M.; Guajardo-Flores, D. Encapsulation of Phenolic Compounds with Liposomal Improvement in the Cosmetic Industry. *Int. J. Pharm.* **2021**, *593*, 120125. [CrossRef]

MDPI AG
Grosspeteranlage 5
4052 Basel
Switzerland
Tel.: +41 61 683 77 34

Antioxidants Editorial Office
E-mail: antioxidants@mdpi.com
www.mdpi.com/journal/antioxidants

www.ingramcontent.com/pod-product-compliance
Lightning Source LLC
Chambersburg PA
CBHW061612210926
51850CB00009B/2302
* 9 7 8 3 7 2 5 8 7 0 3 0 1 *